Handbook of Modern Biophysics

Series Editor

Thomas Jue
Davis, CA, USA

The Modern Handbook in Biophysics series will present to students and researchers leading concepts at the interface of biology, medicine, chemistry, mathematics, and physics. Each book will focus on a distinct topic. Each chapter will establish the fundamental concepts, describe the instrumentation, and illustrate current application in solving biomedical problems. A problem set and additional references for further study augment the didactic structure. These elements will make the books in the series exceptionally useful in the classroom, for self-study, in a laboratory setting, and in research preparation.
The presentation avoids an encyclopedic style and helps the reader navigate through physical science formalism to reach an understanding of the biomedical application.

More information about this series at http://www.springer.com/series/7845

Thomas Jue • Dirk Mayer
Editors

Dynamic Hyperpolarized Nuclear Magnetic Resonance

Springer

Editors
Thomas Jue
Department of Biochemistry and Molecular Medicine
University of California, Davis
Davis, CA, USA

Dirk Mayer
Diagnostic Radiology and Nuclear Medicine
University of Maryland, Baltimore
Baltimore, MD, USA

Handbook of Modern Biophysics
ISBN 978-3-030-55041-7 ISBN 978-3-030-55043-1 (eBook)
https://doi.org/10.1007/978-3-030-55043-1

This Springer imprint is published by the registered company Springer Nature Switzerland AG
The registered company address is: Gewerbestrasse 11, 6330 Cham, Switzerland

Preface

Teaching biophysics presents a challenge for educators, who must introduce many novel scientific ideas and methods in the rapidly evolving field of biophysics. The Handbook of Modern Biophysics series has focused on filling a curriculum need and has presented special topics at the frontier of biophysics. To facilitate teaching of and self-learning by graduate students and researchers, the chapter format has balanced physical science/mathematics formalism with applications. Each chapter divides into two major sections: The first part establishes the conceptual framework and describes the instrumentation or technique. The second part illustrates the biomedical applications. With additional sections on further reading, problems, and references, the chapter serves as a guide for the interested reader to further explore chapter ideas.

In the sixth volume of the series, the authors explain concisely the biophysical principles and biomedical application of "Dynamic Hyperpolarized Nuclear Magnetic Resonance." The book opens with chapters explaining the chemical physics and engineering principles that have led to the development of the dynamic hyperpolarized NMR technique (Chaps. 1 and 2). Chapters 3 and 4 transform these principles into chemical shift imaging, acquisition, and sampling methodologies. Chapter 5 traces the kinetic modeling required to recharacterize the experiment data as enzymatic reactions. The book then turns to biomedical applications and starts with a transition chapter that provides insight on how researchers use dynamic hyperpolarized NMR to address biomedical questions from cells to humans (Chap. 6). Securing insights into these biomedical questions requires the identification and development of innovative metabolic biomarkers (Chap. 7). Chapters 8–10 show how hyperpolarized NMR has helped to clarify critical questions in metabolic regulation, heart function, and lactate metabolism.

Vol. 6 then continues the philosophy behind the "Handbook of Modern Biophysics" series and provides the reader with a fundamental grasp of a novel development in biophysics, dynamic hyperpolarized NMR.

Davis, CA Thomas Jue

Contents

Problem Solutions

Contributors

Jan Henrik Ardenkjaer-Larsen, Ph.D. Department of Electrical Engineering and Biomedical Engineering, Technical University of Denmark, Lyngby, Denmark

Department of Health Technology, Technical University of Denmark, Lyngby, Denmark

Celine Baligand, Ph.D. Molecular Imaging Research Center, CEA-Atomic Energy and Alternative Energies Commission, Fontenay-aux-Roses, France

Radiology and Biomedical Imaging, University of California San Francisco, San Francisco, CA, USA

Kevin M. Brindle, Ph.D. Molecular Imaging (MRI and MRS) Laboratory, Cancer Research UK Cambridge Institute, University of Cambridge, Li Ka Shing Centre, Cambridge, UK

Department of Biochemistry, University of Cambridge, Cambridge, UK

Albert P. Chen, Ph.D. General Electric Healthcare, Toronto, ON, Canada

Arnaud Comment, Ph.D. Centre d'Imagerie BioMedicale, Ecole Polytechnique Federale de Lausanne, EPFL SB IPSB GR-CO, Lausanne, Switzerland

Cancer Research UK Cambridge Institute, Li Ka Shing Centre, Cambridge, UK

General Electric Healthcare, Buckinghamshire, UK

Stephen J. DeVience, Ph.D. Scalar Magnetics, LLC, Cuyahoga Falls, OH, USA

Department of Biomedical Engineering, Case Western Reserve University, Cleveland, OH, USA

Richard L. Hesketh, Ph.D. Department of Radiology, University College London Hospital, London, UK

CRUK Cambridge Institute, Cambridge, UK

Ralph E. Hurd, Ph.D. Radiological Imaging Laboratory, Stanford University, Stanford, CA, USA

Thomas Jue, Ph.D. Department of Biochemistry and Molecular Medicine, University of California Davis, Davis, CA, USA

John Kurhanewicz, Ph.D. Department of Radiology & Biomedical Imaging, Pharmaceutical Chemistry, and Urology, UCSF Mission Bay Campus, San Francisco, CA, USA

Angus Lau, Ph.D. Sunnybrook Health Sciences Centre, Toronto, ON, Canada

Department of Physiology Anatomy and Genetics, University of Oxford, Oxford, UK

Oxford Centre for Clinical Magnetic Resonance Research, University of Oxford, Oxford, UK

Craig R. Malloy, M.D. Advanced Imaging Research Center Lab, Department of Radiology, University of Texas Southwestern Medical Center, Dallas, TX, USA

Department of Internal Medicine, University of Texas Southwestern Medical Center, Dallas, TX, USA

Veterans Affairs North Texas Health Care System, Dallas, TX, USA

Dirk Mayer, Ph.D. Diagnostic Radiology and Nuclear Medicine, University of Maryland, Baltimore, Baltimore, MD, USA

Fischell Department of Bioengineering, University of Maryland, College Park, MD, USA

Jae Mo Park, Ph.D. Advanced Imaging Research Center, Radiology, University of Texas Southwestern Medical Center, Dallas, TX, USA

Electrical and Computer Engineering, University of Texas Dallas, Dallas, TX, USA

A. Dean Sherry, Ph.D. Advanced Imaging Research Center, Department of Radiology, University of Texas Southwestern Medical Center, Dallas, TX, USA

Department of Chemistry and Biochemistry, University of Texas at Dallas, Richardson, TX, USA

Daniel M. Spielman, Ph.D. Radiology and Electrical Engineering, The Lucas Center for MR Spectroscopy and Imaging, Stanford University, Stanford, CA, USA

Renuka Sriram, Ph.D. Radiology and Biomedical Imaging, University of California San Francisco, San Francisco, CA, USA

Kerstin Timm, Ph.D. Department of Physiology, Anatomy and Genetics, University of Oxford, Oxford, UK

Damian Tyler, Ph.D. Department of Physiology Anatomy and Genetics, University of Oxford, Oxford, UK

Oxford Centre for Clinical Magnetic Resonance Research, University of Oxford, Oxford, UK

Alan J. Wright, Ph.D. CRUK Cambridge Institute, Cambridge, UK

About the Editors

Dirk Mayer is Professor of Diagnostic Radiology and Nuclear Medicine at the University of Maryland and is the Director of Metabolic Imaging. He is a recognized expert on dynamic nuclear polarization (DNP) MRI-based imaging techniques and has optimized acquisition and reconstruction techniques, has constructed kinetic modeling for quantitative analysis, and has developed new probes. His research focuses on using DNP to discover new approaches in tumor diagnosis and treatment monitoring and in the assessment of cardiovascular and liver pathologies, inflammatory diseases, and brain metabolism.

Thomas Jue is Professor of Biochemistry and Molecular Medicine at the University of California Davis. He is an internationally recognized expert in developing and applying magnetic resonance techniques to study animal as well as human physiology in vivo and has published extensively in the field of magnetic resonance spectroscopy and imaging, near-infrared spectroscopy, bioenergetics, cardiovascular regulation, exercise, and marine biology. He served as a Chair of the Biophysics Graduate Group Program at UC Davis, where he started to redesign a graduate curriculum that balances physical science/mathematics formalism and biomedical perspective in order to promote interest at the interface of physical science, engineering, mathematics, biology, and medicine. The *Handbook of Modern Biophysics* represents an aspect of that effort.

Chapter 1
Hyperpolarization by Dissolution Dynamic Nuclear Polarization

Jan Henrik Ardenkjaer-Larsen

Abbreviations

dDNP	Dissolution dynamic nuclear polarization
DNP	Dynamic nuclear polarization
EPA	Electron paramagnetic agent
EPR	Electron paramagnetic resonance
MAS-DNP-NMR	Magic angle spinning dynamic nuclear polarization nuclear magnetic resonance
MR	Magnetic resonance
MRI	Magnetic resonance imaging
NMR	Nuclear magnetic resonance
OP	Optical pumping
PHIP	Parahydrogen-induced polarization

1.1 Introduction

Hyperpolarized magnetic resonance (MR) grew out of a vision that MR has excellent properties for measuring biochemistry noninvasively. The MR spectrum provides both identification and quantification of metabolites and can characterize disease for early diagnosis, staging, and response monitoring. However, MR spectroscopy suffers from several limitations: poor sensitivity, leading to long scan time and poor spatial resolution, and limited spectral resolution due to a crowded spectrum. The long scan time means that only steady-state concentrations or slow dynamic changes can be measured. To overcome the problem of sensitivity, hyperpolarization was proposed. Hyperpolarization indicates that the polarization is no

J. H. Ardenkjaer-Larsen (✉)
Department of Health Technology, Technical University of Denmark, Lyngby, Denmark
e-mail: jhar@dtu.dk

T. Jue, D. Mayer (eds.), *Dynamic Hyperpolarized Nuclear Magnetic Resonance*, Handbook of Modern Biophysics, https://doi.org/10.1007/978-3-030-55043-1_1

longer determined by the static magnetic field of the scanner. A polarizer enhances the polarization of the nuclear spins outside the imaging system.

The prospect of such a sensitivity gain by many orders of magnitude has triggered the invention of several methods for enhancing the nuclear spin polarization. The theoretical potential for more than 10,000-fold enhancement of nuclear spin polarization at room temperature and typical magnetic field strengths would enhance signal sensitivity, would compensate for the low metabolite concentrations in tissue, and would enable noninvasive, real-time metabolic imaging of hyperpolarized agents.

Hyperpolarization can be based on several principles, of which three have successfully been applied to molecules in solution: Parahydrogen-induced polarization (PHIP), brute-force polarization, and dissolution dynamic nuclear polarization (dDNP). Optical pumping (OP) effectively polarizes noble gasses. The dDNP method has been particularly successful in making solutions of biologically interesting molecules with highly polarized nuclear spins. The method takes advantage of dynamic nuclear polarization (DNP) in the solid state followed by rapid dissolution in a suitable solvent. The dissolution step retains, almost completely, the nuclear spin polarization, thus creating a solution with a nonequilibrium nuclear polarization. It took a decade of research and development to translate the concept first published [1] in 2003 into a first-in-man study [2] in 2013. Since then, research groups have used dDNP in clinical studies and have published a number of papers [3–11]. The dDNP has continued to expand rapidly with publications of studies on the basic physics of DNP and hyperpolarized spin states, instrumentation, acquisition hardware and pulse sequences, and biomedical applications.

Optical pumping of the noble gases ^{3}He and ^{129}Xe was originally developed in atomic physics laboratories [12]. Noble gases are polarized by the method that won Alfred Kastler the Nobel Prize in 1966. It transfers polarization from photons to atoms by absorption of circularly polarized light. Two methods have been developed: the spin-exchange method, discovered at Princeton in 1960, and meta-stability exchange optical pumping. In the spin-exchange method, the unpaired spin of the valence electron of an alkali atom (e.g., Rb vapor) is polarized via the strong resonance line at 795 nm connecting the $^{2}S_{1/2}$ ground state of the alkali metal to the $^{2}P_{1/2}$ excited state. The alkali metal electron spin then transfers its polarization through magnetic hyperfine coupling to the nuclear spins of a noble gas, which is present at densities up to several bars. Due to the weakness of hyperfine coupling, this kind of spin exchange is a rare process occurring with a probability of order 10^{-4} for xenon and 10^{-7} for helium during a gas kinetic encounter of the partner atoms. As a result, it takes from minutes for ^{129}Xe up to several hours for ^{3}He to attain steady-state nuclear polarization.

The meta-stability spin-exchange process applies exclusively to ^{3}He. Meta-stable ^{3}He atoms are produced in the $^{3}S_{1}$ state at a relative population of 10^{-6} in a low-pressure plasma of about 1 mbar. The electron spin of these meta-stable atoms is then polarized by circularly polarized light at 1083 nm to the $^{3}P_{0}$ state. Exchange between a meta-stable atom and a ground state atom leaves the former in the atomic ground state with a polarized nucleus. The next atom is then ready for pumping, thus forming a fast, catalytic chain of successive energy transfers and pumping action. Within a few seconds, the ^{3}He plasma attains a nuclear polarization of more than 50%. Meta-stability spin exchange is much more efficient than alkali metal spin exchange.

Since the mid-1990s, hyperpolarized noble gases have been applied in MR imaging of the lungs [13] as well as in the dissolved state [14]. The optical pumping method is, however, practically limited to the noble gas isotopes bearing a nuclear spin of ½ (i.e., ^{3}He and ^{129}Xe) in order to achieve long-enough relaxation times. Noble gases with higher spin have short T_1 due to their quadrupolar moment. Since ^{3}He is rare and expensive (produced in radioactive decay of tritium), most effort has focused on optimizing polarization and producing ^{129}Xe, which is naturally occurring (26%) in the atmosphere. Attempts to transfer the ^{129}Xe nuclear polarization to other nuclei have been attempted with limited success [15].

In the late 1980s, Bowers and Weitekamp [16, 17] discovered, both theoretically and experimentally, that the hydrogenation of small organic molecules with parahydrogen led to a highly ordered spin state, which was manifest as very large MR signals for the corresponding protons as the symmetry of the two protons as broken, parahydrogen-induced polarization (PHIP). This phenomenon arises from the quantum statistical mechanical properties of dihydrogen. Of the four possible spin isomers of dihydrogen, the singlet state of the nuclear spins, called parahydrogen, has the lowest energy. At room temperature there is an even distribution of the four spin isomers, giving 25% of the para form. When the temperature is lowered, the fraction of parahydrogen increases and approaches unity at ~20 K. If a molecule of parahydrogen is allowed to react with another molecule, it will in many cases retain the spin correlation between the two protons. However, the symmetry of the hydrogen molecule will be broken due to magnetic inequivalence and couplings with other spins, and the spin order of the parahydrogen can be transferred to the other spin by various methods [18]. For imaging applications, complete hydrogenation at high concentrations in aqueous systems must be achieved. Several substrates have been demonstrated for angiographic and perfusion studies, but there has been limited success with biologically relevant molecules so far. However, recent work has shown that hydrogenation is not a necessary requirement [19]. With certain catalysts, the spin order is transferred in a transient binding without chemical reaction. Another recent approach has been the use of special "side-arm" moieties that can be cleaved off subsequently [20]. Finally, it should be mentioned that the instrumentation for PHIP is quite simple and inexpensive.

Brute-force polarization (i.e., the exposure of the sample to extremely low temperature in a high magnetic field) is not easily combined with liquid-state MR applications. This is primarily due to the extremely low temperatures and high magnetic fields that are required. The ultra-low temperature of millikelvin can be achieved by closed-cycle dilution refrigerators, but the polarization build-up times under brute-force conditions tend to be unacceptably long, days to weeks. Even though the use of so-called relaxation switches has been proposed [21], the brute-force method has not been made practical yet. Demonstrations of the brute method has only achieved less than a percent nuclear spin polarization [22–24].

The fourth hyperpolarization method, dDNP, is based on polarization transfer from unpaired electron spins (e.g., an organic free radical) that are added to the sample (e.g., a biological molecule enriched in ^{13}C in specific chemical positions). The electron spin has a higher magnetic moment and is fully polarized under less-challenging conditions such as temperatures close to a Kelvin and magnetic field strengths of some Tesla. In the solid state, microwave irradiation close to the resonance frequency of the electron spin, transfers, in part, the high electron spin polarization to the nuclear spins. The efficiency of this process depends not only on several parameters characterizing the various spin systems, but also on technical factors such as magnetic field strength, temperature and microwave frequency, and power. The DNP theory will only be described at a high level to allow the reader to understand the significance of the properties of the electron paramagnetic agent (EPA) and sample formulation for the efficiency of DNP.

1.2 Dynamic Nuclear Polarization

DNP was first described theoretically by Overhauser in 1953 [25] and a few months later was demonstrated by Carver and Slichter [26] in metallic lithium. Overhauser predicted that saturating the conduction electrons of a metal would lead to an enhancement of the polarization of the nuclear spins. This was a fundamental discovery causing disbelief at the time: that heating of one spin system could lead to the cooling of another. Abragam soon extended the prediction by Overhauser for metals to electron spins in solution [27], and most magnetic resonance scientists are today familiar with the nuclear and electronic Overhauser effect. Molecular dynamics average the electron–nuclear spin interactions and couple two-spin transitions to the lattice. Saturation of the electron spin transition (equalizing the populations of the electron spin eigenstates) causes a redistribution of the population of the nuclear spin states. The mechanism is most prominent in the liquid state but has also been reported in the solid state [28, 29].

Overhauser DNP in liquids with radicals has been investigated by several groups [30]. The Zeeman energy of unpaired electrons is at least 658 times larger than that of the nuclei, and the same for their spin polarizations at high temperatures. At low to moderate magnetic fields, dipolar interaction between the electron and nuclear spins leads to enhancements of the nuclear polarization up to the ratio of their respective gyromagnetic factors γ_e/γ_n, which is 658 for ^1H and 2,700 for ^{13}C. However, the efficiency of the mechanism decreases with increasing field strength. At typical nuclear magnetic resonance (NMR) field strengths used today (>9.4 T), Overhauser DNP enhancements are much reduced as is the sample size due to the high resonance frequency of the electron spins. Overhauser DNP at moderate magnetic field strength and subsequent shuttling of the sample into a higher field for NMR detection have also been proposed. In any case, Overhauser DNP yields moderate gains in polarization compared to thermal equilibrium at room temperature. Furthermore, combination of dipolar and scalar interactions, as well as three-spin interactions, yields a combination of positive and negative DNP enhancements. These effects cause severely distorted relative intensities in NMR spectra, which render this method unsuitable as a robust and universal enhancement tool for liquid-state NMR. Likewise, the polarization that can be achieved is not competitive to dDNP.

Soon after, Jeffries described the solid effect for spins in the solid state coupled by dipolar interactions [31]. Later, DNP in the solid state was extended mechanistically to processes involving several electron spins [32]. The irradiation of electron–nuclear (two-spin) or electron–electron–nuclear (three-spin) to induce a spin transition enhances the nuclear spin polarization above polarization at thermal equilibrium. The initial theoretical framework for DNP had limited success in quantitatively predicting the efficiency of DNP. Over the last few years, the theory has been extensively revisited, and new theoretical frameworks have been developed. Three mechanisms are generally described:

1. The solid effect (electron–nuclear two-spin interaction with the microwave magnetic field)
2. The cross-effect (electron–electron-nuclear three-spin interaction with the microwave magnetic field, inhomogeneous electron paramagnetic resonance [EPR])
3. Thermal mixing (electron–electron–nuclear three-spin effect, homogeneous EPR)

The grouping into these three mechanisms is primarily due to historical reliance on analytical models (primarily spin temperature theory) that make certain assumptions to derive expressions for the DNP effect. In this work, a spin temperature picture will be used to describe the main features of DNP. In more recent works, quantum mechanical treatments [33–37] have been developed that enable the handling of large spin systems in the solid state. It remains yet to be seen how these new approaches (enabled by increased computational abilities) agree with spin temperature models for realistic samples and experimental data. For the quantum mechanical treatments, the three mechanisms are naturally incorporated without distinction and will describe intermediate cases between (2) and (3) as well as both two- and three-spin transitions. The classical spin temperature models of DNP are known to fail in describing DNP quantitatively. However, these can be solved analytically with some assumptions. The spin temperature picture has been refined in recent years to consider properties of the electron spin system and thereby comes in closer agreement with experimental observations. However, these models must be solved numerically, and it is questionable if the models provide further physical insight into DNP than full quantum mechanical treatments. For a recent comprehensive description of DNP based on spin temperature concepts, see Refs. [38–40].

Polarization, for a spin ½, is given by the difference in populations of the two possible energy eigenstates

$$P = \frac{N^{+} - N^{-}}{N^{+} + N^{-}} = \tanh\left(\frac{\gamma \hbar B_0}{2k_B T}\right) \tag{1.1}$$

where N^+ and N^- denote the number of spins parallel (spin up) and antiparallel (spin down) to the external magnetic field, respectively. According to Eq. (1.1), a temperature can be assigned to the spins for any polarization. For example, $P = 0$, complete saturation, corresponds to an infinite spin temperature. Depending on the sign of the polarization, the spin temperature can be either negative or positive. It will approach zero (cooling) as the polarization approaches unity. As a point of caution, it is often stated that spins are either pointing up or down (i.e., they are in one of the two eigenstates). However, this is not correct. Quoting Slichter [41]: "We emphasize that an arbitrary orientation can be specified, since sometimes the belief is erroneously held that spins may only be found pointing either parallel or antiparallel to the quantizing field. One of the beauties of quantum theory is that it contains features of both discreteness and continuity. In terms of the two quantum states with $m=\pm\frac{1}{2}$ we can describe an expectation value of the magnetization which goes all the way from parallel to antiparallel, including all values in between…." The spin is therefore in a superposition of the eigenstates, and it is perfectly valid to speak of populations (probabilities) of the eigenstates. From Eq. (1.1) we get that the proton polarization in a 3T MR scanner becomes $10 \cdot 10^{-6}$, while the polarization for ^{13}C is only $2.5 \cdot 10^{-6}$, four times lower (Fig. 1.1). A theoretical enhancement of the nuclear spin polarization of 100,000 and 400,000 times, respectively, is therefore possible, if almost full polarization could be achieved.

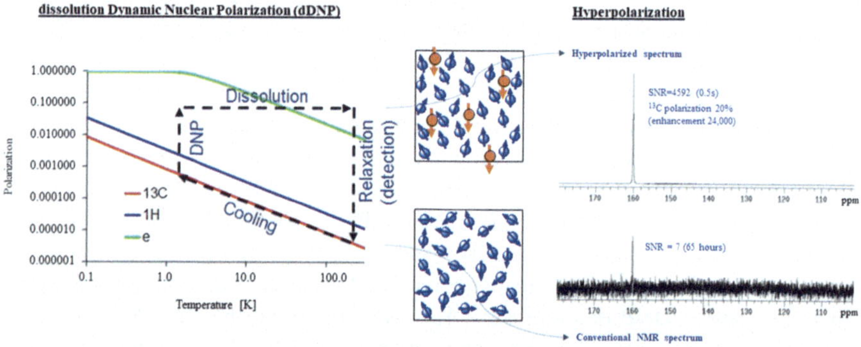

Fig. 1.1 The principle of dissolution dynamic nuclear polarization (dDNP) is illustrated. At room temperature and 3 T, the ^{13}C nuclear spins are only weakly polarized to about 2.5 ppm (left graph). However, electron spins have a 2,700 times stronger magnetic moment and are easily polarized. When the sample is cooled, the electron spin polarization reaches almost unity (97.8% at 3.35 T and 1 K). By irradiation with microwaves close to the resonance frequency of the electron spins, electron-electron-nuclear transitions are induced, and the nuclear spin polarization will be enhanced 100-fold by dynamic nuclear polarization (DNP) to several tens of percent. This process is slow at low temperatures and takes typically 15–60 min for ^{13}C. When the sample is polarized, superheated water or buffer dissolves the sample within seconds, and a room temperature solution of the hyperpolarized molecule is obtained. The hyperpolarized nuclear spins relax to thermal equilibrium (e.g., 2.5 ppm at 3 T and room temperature) with the longitudinal relaxation constant, T_1, of typically 40–80 s for carboxylic acids. A hyperpolarized ^{13}C spectrum of urea obtained in a single transient is shown to the right along with the thermal spectrum after 65 h of averaging

1.2.1 Solid Effect

The solid effect is explained most easily of the three mechanisms [32, 38]. A dilute EPA with spin $S = 1/2$ and Larmor frequency ω_e in a diamagnetic material with nuclear spins $I = 1/2$ with Larmor frequency ω_n is coupled by dipolar interactions. The dipolar interaction permits simultaneous flips of spins S and I (Fig. 1.2). In such spin reversals, the total energy of the spin system changes by an amount $\hbar(\omega_e \pm \omega_n)$ driven by the microwave field. This interaction is also a relaxation mechanism with the missing energy supplied by the lattice, usually in the form of one or several phonons. The rate (transition probability) of these processes is typically very small at low temperatures (typically 10^{-3} s^{-1} or less) due to the short electron spin T_{2e} and the relatively weak microwave field that can be applied. On the other hand, the relaxation of the electron spin alone, caused by lattice vibrations (phonons), occurs at a much faster rate, T_{1e}^{-1} (typically 1 s^{-1} to 10^3 s^{-1}).

Consider first an isolated electron–nucleus pair. In the absence of any mutual interaction, the eigenstates of this two-spin system are equal to the simple product states $|\alpha\alpha\rangle$, $|\alpha\beta\rangle$, $|\beta\alpha\rangle$, and $|\beta\beta\rangle$. We now consider that the spins are coupled by a dipolar interaction $H_{\text{dip}} = \frac{\mu_0}{4\pi}\frac{\gamma_e\gamma_n\hbar^2}{r^3}\left(\hat{I}\cdot\hat{S} - \frac{3}{r^2}\left(\hat{I}\cdot\hat{r}\right)\left(\hat{S}\cdot\hat{r}\right)\right)$. The dipolar interaction perturbs the pure product states and the resulting spin eigenstates will contain an admixture of states. For instance, one of the dipolar terms, S_zI_+, mixes into the original state $|\alpha\beta\rangle$ a contribution from the state $|\alpha\alpha\rangle$. The mixing coefficients p and q in

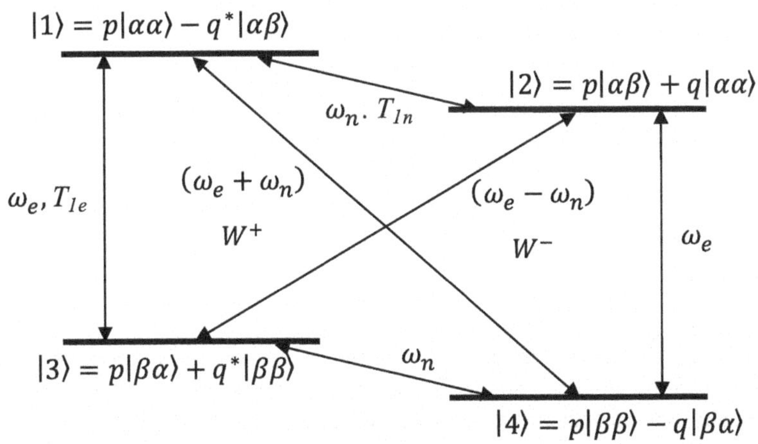

Fig. 1.2 The eigenstates for a dipolar coupled electron-nucleus spin pair. The electron paramagnetic resonance (EPR) transitions are 1 → 3 and 2 → 4 (degenerate) with resonance frequency ω_e. The nuclear magnetic resonance (NMR) transition is 1 → 2 and 3 → 4 (degenerate) with resonance frequency ω_e. Microwave irradiation at the sum or difference frequency of the electron and nuclear Larmor frequencies connect the diagonals and induce mutual electron-nuclear "flip-flips" or "flip-flops" with transition probabilities $\boldsymbol{W}^+$ and $\boldsymbol{W}^-$, respectively. Electron spin relaxation, T_{1e}, brings the electron spin back to the ground state, while leaves the nuclear spin in its flipped state

Fig. 1.2 can be calculated by perturbation theory, and it is found that $q = \frac{3}{2}\frac{\mu_0}{4\pi}\frac{\gamma_e \hbar}{B_0 r^3}\sin\theta\cos\theta\exp(i\varphi)$, and that $|q| \ll 1$. Because of the normalization ($|p|^2 + |q|^2 = 1$) we have $p \approx 1$. The transition probabilities between the states $|1\rangle \leftrightarrow |4\rangle$ (flip-flop) and $|2\rangle \leftrightarrow |3\rangle$ (flip-flip) are a factor $4|q|^2$ smaller than the transition probabilities between $|1\rangle \leftrightarrow |3\rangle$ and $|2\rangle \leftrightarrow |4\rangle$. Since the coefficient q is proportional to the inverse separation cubed, the transition probabilities are very sensitive to the distance between the electron and the nucleus. The electron spin transitions $|1\rangle \leftrightarrow |2\rangle$ and $|3\rangle \leftrightarrow |4\rangle$ can be induced by spin-phonon coupling [32]. In addition to the spin–lattice induced transitions, there can also be transitions induced by the microwave field. The frequency of the microwave field, corresponding to the difference in energy between the states, $\omega_e \pm \omega_n$, selectively drives either of the two transitions. The line width of the EPR, $\Delta\omega_e$, is assumed much smaller than the nuclear Larmor frequency, ω_n, such that when flip-flops occur, $\omega_e - \omega_n$, flip-flips, $\omega_e + \omega_n$, are impossible because they are off-resonance with the driving microwave field, and vice versa. For a microwave field with amplitude B_1 applied at a frequency $\omega = \omega_e - \Delta$, the transition probabilities, W^+ and W^-, for the flip-flop and flip-flip processes, respectively, are given by

$$W^{\pm} = 2|q|^2 \pi\gamma_e^2 B_1^2 g(\Delta \pm \omega_n) \tag{1.2}$$

where $g(\omega)$ is the normalized (by area) EPR. If the EPR is so broad that both transitions are induced simultaneously, the solid effect is said to be "unresolved." In the case of a resolved solid effect and a driven flip-flop transition, the following rate equations for the electron and nuclear spin polarizations are:

$$\frac{dP_e}{dt} = -C_n W^+ (P_e - P_n) - \frac{1}{T_{1e}}(P_e - P_e^0) \tag{1.3}$$

$$\frac{dP_n}{dt} = -W^+ (P_n - P_e) - \frac{1}{T_{1n}}(P_n - P_n^0) \tag{1.4}$$

Here $C_n = \frac{N_n}{N_e}$ is the number of nuclei that each electron has to "service," P_n^0 and P_e^0 are the equilibrium polarizations of the nuclear and electronic spins, and, finally, T_{1n} and T_{1e} are the nuclear and electronic relaxation time constants, respectively. If we assume that the electronic Zeeman relaxation rate constant dominates both cross-relaxation rates and the nuclear Zeeman relaxation rate, we can neglect the first term in Eq. (1.3). The steady-state electronic polarization will then be equal to P_e^0, and P_e can be replaced by P_e^0 in Eq. (1.4). The solution to the resulting equation, $dP_n / dt = -W^+(P_n - P_e^0) - T_{1n}^{-1}(P_n - P_n^0)$, is an exponential increase to the asymptotic limit $P_n^0((\gamma_e/\gamma_n)(W^+ + T_{1n}^{-1}))/(W^+ + T_{1n}^{-1})$ with a rate constant of $W^+ + T_{1n}^{-1}$. The build-up rate constant is thus always larger than the decay rate constant of the polarization. The maximum enhancement, γ_e / γ_n, is obtained when the nuclear relaxation rate is negligible compared to the transition probability of the microwave-driven transition.

Furthermore, if other relaxation mechanisms exist for the nuclear spins other than the coupling with the electrons, the enhancement is obviously further decreased, constituting a so-called "leakage." Another cause of reduced polarization enhancement occurs when the electron spin relaxation rate is no longer fast compared to the cross-relaxation terms. This is due to an inefficient coupling to the phonon lattice and is usually referred to as a "phonon bottleneck." Each electron spin must "service" a large number, N_n/N_e, of nuclear spins in order to be effective. The electron spin must return to the ground state before any of the N_n/N_e nuclear spins in its sphere of influence has flipped through a nuclear relaxation mechanism. This is true if the nuclear relaxation of the nuclear spins has no other origin than their couplings with the electron spins. If, however, other nuclear relaxation mechanisms, sometimes called leakage relaxation, are present, the condition may be violated, and the nuclear polarization would be smaller than P_e. This is expressed by the equation $N_n/N_e \ll T_{1n}/T_{1e}$.

Because of the r^{-6} dependence of the transitions rates, the dipolar coupling becomes increasingly less efficient at longer distances. It would thus seem that only spins in the vicinity of the paramagnetic centers would be polarized. However, spin diffusion among the nuclear spins tends to distribute the polarization more homogeneously. If nuclear spin diffusion is efficient, the nuclear polarization will be carried from the proximity of the electron spin to the bulk nuclei. For highly diluted nuclear spins, slow spin diffusion could also be a potential bottleneck. The nuclear spins in the immediate proximity of the paramagnetic centers experience a strong dipolar field, causing their Larmor frequencies to be shifted far off-resonance (nonobservable).

1.2.2 Cross-Effect and Thermal Mixing

The cross-effect and thermal mixing arises in the situation that the electron line width $\Delta\omega_s$ is larger than the nuclear frequency, ω_I, and involves two interacting electron spins and a nucleus. This situation arises when the electron spin g-tensor anisotropy is sufficiently large. The solid effect may still contribute, but it will in this case be unresolved and harder to drive. Thermal mixing is historically described using spin temperature formalism. When the EPR line is irradiated at a frequency $\omega = \omega_e + \Delta$, $\hbar\omega_e$ is absorbed by the electron Zeeman reservoir and the remaining $\hbar\Delta$ is absorbed or emitted by the electron spin–spin dipolar reservoir. The electron spin–spin dipolar reservoir is thereby cooled to a low spin temperature (positive or negative). The electron spin–spin dipolar reservoir is coupled to the nuclear Zeeman reservoir, and these two spin baths should eventually reach a common spin temperature (Fig. 1.3). For $\Delta < 0$, the electron spin–spin dipolar reservoir emits energy and

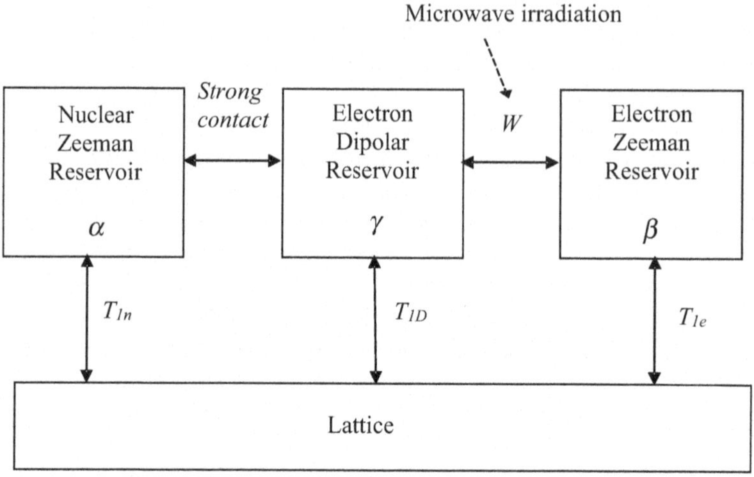

Fig. 1.3 Spin temperature model of thermal mixing. The electron spin Zeeman reservoir, α, the nuclear spin Zeeman reservoir, β, and the electron spin-spin non-Zeeman (dipolar) reservoir, γ. The three reservoirs are connected to the lattice by relaxation mechanisms (leakage) T_{1e}, T_{1n}, and T_{1D}, respectively. Microwave irradiation, W, heats the electron spin Zeeman reservoir and cools the non-Zeeman reservoir that is (strongly) connected to the nuclear Zeeman reservoir

is cooled by the irradiation of the EPR, and the nuclear polarization is enhanced by relaxation of the electron spin–spin bath to thermal equilibrium. For $\Delta > 0$, the electron–electron dipolar bath absorbs energy and will cool the nuclear Zeeman reservoir to a low negative spin temperature. This situation also forms the basis of the equal spin temperature theorem; that all nuclear spin systems (e.g., ^{13}C, ^{15}N, or ^{2}H) eventually reach the same spin temperature.

At low temperature the electron spin relaxation time, T_{1e}, becomes very long ($1\ s^{-1}$ to $10^3\ s^{-1}$) and spectral diffusion within the EPR effectively renders the EPR homogeneous. In this case it becomes meaningful to describe the spin–spin interaction by a temperature. The early spin temperature model (Borghini model) has failed to provide a quantitative description of thermal mixing, since the assumptions made seem invalid. Recent extensions of the theory provide better agreement with experiment and predict specific features of DNP.

At around 1 K, T_{1e} is expected to be dominated by the direct effect [32]

$$T_{1e}^{-1} = \frac{3\hbar}{2\pi\rho v^5}\frac{1}{\Delta^2}\left|V^{(1)}\right|^2 \omega^5 \coth\left(\frac{\hbar\omega}{2k_B T}\right) \tag{1.5}$$

that has a characteristic inverse temperature dependence in the high temperature approximation and is independent of temperature when $\hbar\omega \gg 2k_B T$. Other electron–phonon relaxation mechanisms have stronger temperature dependence. ρ is the density of the material, ν is the speed of sound in the material, Δ is the crystal field splitting, and $V^{(1)}$ is the first-order strain of the electric potential. At higher temperatures the electron T_{1e} becomes shorter than the electron–electron spectral diffusion and the EPR becomes inhomogeneous. The microwave irradiation saturates only a fraction of the EPR, denoted as hole-burning. Thermal mixing and cross-effect are distinguished by this characteristic, but the two effects have a continuous transition. For broad EPR (larger than the nuclear Larmor frequency), the effective mechanism is typically thermal mixing at very low temperatures and cross-effect at higher temperatures. An effective cross-effect therefore also relies on both a stronger microwave irradiation field (higher power) and a stronger electron–electron dipolar interaction. This is the reason that we have different requirements to the EPA for the two mechanisms and the two regimes (low and high temperature).

The nuclear longitudinal relaxation rate induced by the electron spin is given by [32]

$$\frac{1}{T_{1n}} \approx \left(\frac{\mu_0}{4\pi}\right)^2 \frac{8\pi}{5} \frac{S(S+1)}{3} \frac{N_e \gamma_n^2 \gamma_e^2 \hbar^2}{b^3} \frac{T_{2e}}{1+\left(\omega_I T_{2e}\right)^2}\left(1-P_{0e}^2\right) \tag{1.6}$$

where it has been assumed that spin diffusion among carbons is fast and the electron–electron bath is in fast thermal contact with the nuclear Zeeman bath. b is the radius of the diffusion barrier, ω_I is the nuclear resonance frequency times 2π, T_{2e} the electron transverse relaxation time, and P_{0e} the electron thermal equilibrium polarization. Additional relaxation mechanisms not originating from the added electron spin will also be present.

The dependence of the polarization on the spin temperature is given by (Eq. 1.1), where T is the actual spin temperature. We define the parameter β through $P = \tanh(\hbar\omega/2k_BT) \equiv \tanh(\omega\beta/2)$, which is often referred to as the "inverse spin temperature" although it actually has the dimension of time. In the following, the inverse temperatures α and β are assigned to electron and nuclear Zeeman reservoirs, respectively, and γ is the inverse temperature of the electron spin–spin (non-Zeeman) reservoir (γ should not to be confused with the nuclear and electron spin gyromagnetic ratios γ_n and γ_e, respectively). All three thermal reservoirs, the electron Zeeman reservoir α, the nuclear Zeeman reservoir β, and the electron non-Zeeman reservoir γ, are connected via relaxation to the lattice, with their respective relaxation times (Fig. 1.3). We now apply microwaves off-resonance with respect to the electron spin resonance, $\Delta = \omega_e - \omega$, by an amount of the order of the width of the EPR, D. The purpose of the microwaves is twofold: The application of microwaves will, in the rotating frame, result in a reduction of the effective field as seen by the electron spins, and will hence result in a cooling of the electron Zeeman reservoir according to the relation $\alpha \cdot \Delta = \gamma \cdot \omega_e$. At the same time, because of the matching of the effective field and the EPR width, the microwaves will establish a contact between the electron Zeeman and the non-Zeeman reservoirs, thus cooling the non-Zeeman reservoir. There is furthermore a direct coupling between the electron non-Zeeman reservoir and the nuclear Zeeman reservoir. This interaction exists also in the absence of the microwaves and can be understood as a simultaneous flip-flop within the EPR, providing the energy required for a nuclear transition. This will finally cool, that is, increase the polarization of, the nuclear spins.

Based on the Provotorov theory [42], rate equations for the spin temperatures α and γ can be written in the high temperature limit where the linearization of the hyperbolic tangent function for the electron spin polarization is valid. It is assumed that the nuclear Zeeman and the electron non-Zeeman reservoirs are in good contact, that is, always have the same temperature ($\gamma = \beta$) [10]:

$$\frac{d\alpha}{dt} = -W(\alpha - \beta) - \frac{1}{T_{1e}}(\alpha - \alpha_{L}) \tag{1.7}$$

$$\frac{d\beta}{dt} = W\frac{\Delta^2}{D'^2}(\alpha - \beta) - \frac{1}{T_{1n}}(\beta - \beta_{L}) \tag{1.8}$$

where $W = \pi\gamma_e^2 B_1^2 g(\Delta)$, $D'^2 = D^2 + \frac{N_n}{N_e}\frac{I(I+1)}{S(S+1)}\omega_n^2$, and $D^2 = \int_{-\infty}^{\infty} \Delta^2 g(\Delta) d\Delta$ is the second moment of the EPR. The steady-state solution ($d\alpha / dt = d\beta / dt = 0$) to Eqs. (1.7) and (1.8) gives the inverse spin temperature of the nuclear Zeeman reservoir:

$$\frac{\beta_{eq}}{\beta_L} = \frac{T_{1n}W\left(\frac{\Delta^2}{D'^2}\right)\left(\frac{\alpha_L}{\beta_L}\right) + T_{1e}W + 1}{1 + T_{1e}W + T_{1n}W\left(\frac{\Delta^2}{D'^2}\right)} \tag{1.9}$$

As long as the microwave field, ω_e, is on, the two thermal reservoirs converge toward different lattice temperatures according to $\alpha_L \cdot \Delta = \beta_L \cdot \omega_e$. Dividing by $\frac{WT_{1n}}{D'^2}$ and letting $W \rightarrow \infty$ (strong saturation) yield the equation:

$$\frac{\beta_{eq}}{\beta_L} = \frac{\omega_e\Delta + \Omega^2}{\Delta^2 + \Omega^2} \tag{1.10}$$

where $\Omega^2 = \frac{D'^2 T_{1e}}{T_{1n}}$. If we apply microwaves off-resonance by an amplitude larger than width of the EPR, that is, $\Delta \approx D$ and $\omega_e\Delta \gg \Omega^2$, then the second term in the numerator can be neglected. Close to resonance, however, the second term should be retained. The limiting cases are $\beta_{eq} = 0$ as $\Delta \rightarrow \pm\infty$ and $\beta_{eq} = \beta_L$ when $\Delta = 0$. The extreme values of Eq. (1.10) are obtained for $\Delta = -\Omega^2/2\omega_e \pm (\Omega^4/4\omega_e + \Omega^2)^{0.5}$. In the limit $\Omega \ll \omega_e$ (this is always the case at high field), the extrema are equal to $\pm\frac{\omega_e}{2\Omega}$ and are obtained for $\Delta = \pm\Omega$.

Since the frequency dependence of the transition probability W is given by the shape of the EPR, the strong saturation condition might not be satisfied in the wings of the line. Using a frequency-dependent W narrows the width of the polarization optima, and in the case of an asymmetric EPR introduces a deviation from the odd functionality of the polarization curve.

At low temperatures there exists no generalization of the Provotorov theory, but a specialized solution for the simplified case of inhomogeneous broadening of the EPR, caused by either a distribution of electron spin g-factors or by hyperfine interactions with neighboring nuclei, is given in Ref. [32], the Borghini model, Eq. 1.11. Assuming steady-state conditions, one obtains an integral equation for β that must be solved numerically:

$$-\Delta_0 P_{0e} + \frac{N_n}{N_e}\frac{T_{1e}}{T_{1n}} 2I\omega_n P_n = \int_{-\infty}^{\infty} (\Delta_0 - \Delta) g(\Delta) \tanh\left(\frac{1}{2}\beta(\Delta_0 - \Delta)\right) d\Delta \quad (1.11)$$

The function g is the normalized EPR with $\Delta = 0$ defined as the center of gravity

$$\int_{-\infty}^{\infty} g(\Delta) = 1 \qquad \int_{-\infty}^{\infty} \Delta g(\Delta) = 0 \quad (1.12)$$

$\omega_S - \Delta_0$ is the irradiation frequency, and P_{0e} is the electron spin thermal equilibrium polarization.

Microwave frequency or magnetic field modulation becomes effective when the conditions for thermal mixing breaks down, that is, spectral diffusion is too slow to establish well-defined spin temperature of the electron–electron interaction reservoir. In this situation, which arise for EPA that have large g-anisotropy relative to the electron spin–spin interaction, more electron spins can be "recruited" for DNP by sweeping the microwave frequency across several spin packets. The rate of the modulation has to be fast on T_{1e} time scale and the amplitude of the modulation will depend on the width of the EPR [43, 44].

It has recently been shown that higher nuclear polarization can be obtained for both nitroxides and trityls by increasing the magnetic field strength [45–48] or lowering the temperature [49]. Since DNP is typically not able to reach the theoretical limit, this appears obvious from the above equations. The thermal polarizations increase naturally, T_{1n} typically increases at higher magnetic field and T_{1e} typically gets shorter. For [1-^{13}C]pyruvic acid, the ^{13}C polarization improved from 27% at 3.35 T to 60% at 4.64 T and 70% at 6.7 T, all at 1–1.2 K.

1.2.3 Spectral Diffusion, Spin Diffusion, and Cross-Polarization

Dynamic nuclear polarization is effective near the paramagnetic centers. The distribution of the spins within the sample is critical for this efficiency. The average distance, d, between two electron spins in the simplest model is given by the inverse cube root of the density (radical concentration), c, that is, $d = \sqrt[-3]{1000 N_A c}$, where N_A is Avogadro's number. For typical radical concentrations of 15–60 mmol/L, this leads to distances of 48.0–30.2 Å. This is a simplistic model. Therefore, it seems more appropriate to consider the distribution of distances in order to calculate d, and the average electron–electron dipolar interaction, $D_{ee} \propto d^{-3}$. In the following calculation, only the dipolar coupling to the next nearest neighbor is considered. For a sample of randomly distributed radicals, the probability to find a radical in the distance k (discrete values) would seem to follow the geometric distribution with its probability density function $p_{geom}(k)$, cumulative distribution function $d_{geom}(k)$, mean, median, and variance given for the 1D case as:

$$p_{geom} = q(1-q)^{k-1}, \quad k = 1,2,3,\ldots$$

$$d_{geom} = 1-(1-q)^{k}$$

$$\mathrm{mean}_{geom} = q^{-1}; \quad \mathrm{median}_{geom} = \frac{-\log(2)}{\log(1-q)}; \quad \mathrm{Variance}_{geom} = \frac{1-q}{q^2}$$

Similarly, the continuous exponential distribution can be used to describe the probability to find a radical at the distance x. When extending to the 3D case, the equation for the probability density function of the geometric distribution changes to:

$$\mathrm{p}_{geom,3D}(n) = a_n q \prod_{k=1}^{n-1} (1 - a_k q), \quad n = 2,3,4,\ldots$$

$$\text{with } a_m = m^3 - (m-1)^3, \quad m = k,n$$

The probability of finding a radical in a spherical shell of thickness 1 a.u. and outer radius of n a.u. is $a_n q$ with the density of radicals q in $(\text{a.u.})^{-3}$. Additionally, the non-negligible size of the radical is considered by requiring $p_{geom,3D} = 0$ within twice the radius of the radical center located at the origin. Then the numerically calculated probability density function is normalized to give $\int_{n=2}^{\infty} p_{geom,3D}\, dn = 1$. The density of radicals, q, is calculated as $q = N_A c$ For $c = 15$ mmol/L, the result is $q = 9.033 \cdot 10^{-6}$ Å^{-3}. With a numerical spacing of 1 a.u. = 0.1 Å, $q = 9.033 \cdot 10^{-9}$ $(\text{a. u.})^{-3}$, and a trityl radius of 5.8 Å, the calculated probability function is shown in Fig. 1.4. The standard deviation is calculated as $\sigma_{geom} = \sqrt{\mathrm{Variance}_{geom}} = \sqrt{1-q/q^2} = 1.1 \cdot 10^5$ Å, which may betransformed to the standard deviation of the mean distance of Å. This probability function allows calculating the average distance

$$d_{geom} = \sum_{n=n_{min}}^{\infty} p_{geom,3D}\, n \Delta n = 43.4\ \text{Å} \quad \text{with } n_{min} = 11.6\ \text{Å}$$

and similarly, the dipolar coupling constant D_{ee} in frequency units

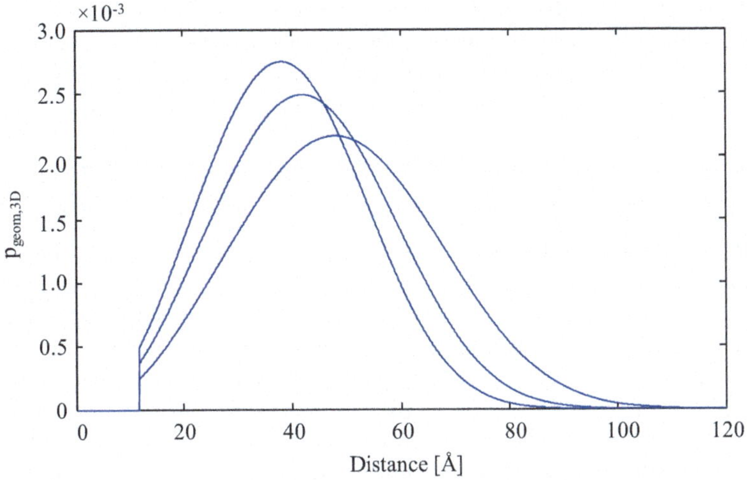

Fig. 1.4 Probability function for distribution of distances between pairs of electron spins based on the three-dimensional geometric distribution for radical concentrations of 10 mmol/L, 15 mmol/L, and 20 mmol/L, and a radical diameter of 11.6 Å (minimum distance of approach)

$$D_{\mathrm{ee}} = C\sum_{n=n_{\min}}^{\infty} p_{\mathrm{geom,3D}} n^{-3} \Delta n$$

$$C = \frac{\mu_0}{4\pi}\frac{\hbar\gamma_{\mathrm{e}}^2}{2\pi} \approx 5.20 \cdot 10^{-20}\,\mathrm{Hz} \cdot m^3$$

Numerical integration yields D_{ee}= 1.74 MHz, corresponding to the effective mean distance of 31.0 Å. As expected, the effective average electron spin–spin distance is shorter than d_{geom} since the dipolar coupling is weighting shorter distances more strongly. This homogeneous line width, D_{ee}, much smaller than the overall EPR width, determines the rate of spectral diffusion. We assume that such spectral diffusion is faster than all other processes up to the EPR width that can be much larger than the homogeneous width. This limits our treatment to lower temperatures, in which electron spin–lattice relaxation is sufficiently slow for this condition to hold. The process is denoted thermal mixing when spectral diffusion is so fast that a local equilibrium is established in the electron spin system, otherwise the cross-effect [39].

Nuclear spin diffusion carries the polarization from the proximity of the electron spins to the bulk of the sample. Homonuclear dipolar interactions are energy conserving, and nuclear spin flip-flops occur at a high rate. Lowe and Gade [50] have derived an expression for the spin diffusion constant

$$D_{\mathrm{SD}} = 0.15\frac{\mu_0\gamma^2\hbar}{4\pi a} \tag{1.13}$$

where a is the average distance between the nuclear spins. The equation tells us that spin diffusion scales with the square of the gyromagnetic ratio, that is, it is approximately 16 times faster for ^{1}H than ^{13}C, and the inverse of the spin-spin distance, which follows the cubic root of the spin concentration. For an order of magnitude calculation, we can use ^{13}C-pyruvic acid as example. Neat pyruvic acid is 14 mol/L, which, in a simplistic lattice model, leads to a spin-spin distance of 4.9 Å ($\sqrt[-3]{1000cN_{\mathrm{A}}}$). Eq. (1.13) then gives 1.16 10^{-18} m^2/s, which means that polarization will diffuse an average distance of 26 Å ($\sqrt{6D_{\mathrm{SD}}t}$) within a second. This is more than half the average distance between electron spins. It is however clear that for dilute (e.g., natural abundance isotopes or molecules with low solubility) samples of low gamma nuclei, spin diffusion can be the limiting factor in the build-up of polarization. Likewise, in heterogeneous samples, for example, nanoparticles, polymers, or silica with uneven distribution of EPA, spin diffusion also becomes a limiting factor. It has been assumed that spin diffusion within a certain radius of the EPA was impeded by the strong dipolar field of the electron spin. The diffusion barrier, b, can be estimated from [32]

$$\Delta\omega_{\mathrm{n}} = \gamma_{\mathrm{I}} a \left| \frac{\partial}{\partial r}\left(\frac{\mu_0}{4\pi}\frac{\gamma_S \hbar}{r^3}\right) \right|_{r=b} = \gamma_I \frac{\mu_0}{4\pi}\frac{3\gamma_S \hbar a}{b^4} \tag{1.14}$$

which defines the radius from the paramagnetic center where the dipolar gradient is equal to the width of the ^{13}C NMR, $\Delta\omega_{\mathrm{n}}$, which is determined from the full-width-at-half-height, and a is the ^{13}C–^{13}C average distance. This equation overestimates b, since the gradient is not linear. This is considered by using the difference instead of the derivative

$$\begin{aligned}\Delta\omega_{\mathrm{n}} &= \gamma_{\mathrm{n}} \frac{\mu_0}{4\pi}\overline{\left(1+3\cos^2\theta\right)}\gamma_{\mathrm{e}} S\hbar \left(\frac{1}{b^3} - \frac{1}{(b+a)^3}\right) \\ &= \frac{\mu_0}{4\pi}\gamma_{\mathrm{n}}\gamma_{\mathrm{e}}\hbar \frac{1}{b^3}\left(1 - \frac{1}{(1+a/b)^3}\right)\end{aligned} \tag{1.15}$$

where θ is the angle between the inter-spin vector and the static magnetic field. Similar result is obtained from Eqs. (1.14) and (1.15) for $a \ll b$. Data [51] and Eq. (1.15) suggest that the spin diffusion barrier is small and close to the radius of, for example, the trityl EPA. For an overview of spin diffusion in DNP, see Ref. [52].

For these types of samples, cross-polarization from protons to low gamma nuclei has proven to be very efficient. Dilute proton systems polarize fast and efficiently, and small samples are compatible with the technical requirements of large radiofrequency magnetic fields on the nuclear resonance frequencies [53–55]. At 6.7 T and 1.2 K, a ^{13}C polarization of 78% with a time constant of 470 s was achieved for [^{13}C]urea by cross-polarizing from the bulk ^{1}H to the ^{13}C of urea. However, efficient cross-polarization has yet to be demonstrated for large samples with high proton density. This is due to the strong dipolar broadening of the proton resonance (line width of 60–80 kHz) in fully protonated samples and the size of the sample (NMR coil) that easily leads to arcing (discharge in high electrical fields) in the helium atmosphere of the polarizer.

1.3 Electron Paramagnetic Agents

DNP requires the presence of unpaired electrons, typically in the form of an organic radical, but a few metal ions have high efficiency for DNP at low temperature, Cr(V) in particular. The magnetic moment of the electron is 658 times higher than that of the proton. This means that the electron spin will reach unity polarization at a moderate magnetic field strength and liquid helium temperature (Fig. 1.1). For example, at 3.35 T and 1 K, the electron spin polarization is already 98%. The choice of EPA will depend on several factors. Firstly, it is advantageous that the EPA is chemically stable and dissolves readily in the matrix of interest. Secondly, the electron paramagnetic resonance (EPR) spectrum of the radical should have a width that allows DNP to be effective for the nucleus of interest, that is, a width that exceeds the Larmor frequency of the nuclear spin of interest. Thirdly, the EPA should have low toxicity for any biological studies, and it should be removable from the hyperpolarized solution in the case of clinical applications. Finally, it should also have minimal effect on nuclear T_1 in the liquid state, that is, post dissolution. In practice the above criteria mean that two classes of EPA are available, namely nitroxides and trityls [56, 57]. The nitroxides belong to a class of molecules that have been studied extensively by EPR, and that have been used for DNP for many samples. Nitroxides are characterized by having a broad EPR spectrum. The EPR line width is approximately 4.0‰ of the EPR frequency, compared to the ^{1}H resonance frequency, which is 1.5‰ (1/658) of the EPR frequency. Some of them have reasonable chemical stability and come with different degrees of hydrophilicity. The trityl is another class of EPA with superior properties for direct polarization of low gamma nuclei such as ^{13}C, ^{15}N, and ^{2}H. These radicals have a line width that is only approximately 0.80‰ of the resonance frequency, much less than the proton resonance frequency, but perfectly matched for ^{13}C, which has a resonance frequency, which is 0.37‰ (1/2700) of the EPR frequency. Trityls exist with a range of hydrophilicities, and they are typically very stable chemically.

It has been shown that a small amount of gadolinium can positively affect the DNP enhancement with trityl in the solid state [58]. Other paramagnetic ions and molecules (Mn^{2+} and O_2) will also, in part, have the same effect. A shortening of the EPA relaxation times is the main explanation for the effect. Addition of 1–2 mmol/L chelated Gd^{3+} leads to a 50–100% improvement in the DNP enhancement factor for pyruvic acid. The effect seems to be present for most samples, but each sample must be optimized for the concentration of the EPA and Gd chelate. There is no direct DNP effect of the Gd^{3+} by itself under the conditions typically used (the Gd^{3+} ground state fully populated), as is otherwise the case at elevated temperatures [59]. Finally,

Gd^{3+} may enhance the solid-state DNP, but care should be taken in avoiding accelerated relaxation in the liquid state. Free Gd ions would cause detrimental liquid-state relaxation and pose an in vivo safety risk. After dissolution, the low concentration of trityl and chelated Gd will have negligible effect on T_1 in most cases. However, firstly, addition of Gd has not been shown to affect nitroxide DNP efficiency positively, and, secondly, nitroxides do have a significant effect on nuclear T_1 in liquid state post dissolution that should be considered.

A new class of EPA, with wider prospects, is ultraviolet (UV)-generated radicals [60, 61]. A precursor molecule, for example, pyruvic acid itself, can be excited with UV light to produce a radical species with good EPR properties for DNP. The radical is labile and must be generated at low temperature, for example, 77 K. In the dissolution process, the radical annihilates and decomposes into acetate and bicarbonate. However, the most exciting prospect for these radicals is the potential for generating radical-free, hyperpolarized solids that can be transported over longer distances. A centralized production site (within a city or region) for hyperpolarized solid samples could be imagined. Two approaches have been proposed [62, 63]. In both cases, long relaxation times (16–20 h) have been demonstrated in a permanent magnetic field at liquid helium temperature. This could enable during-the-day delivery of hyperpolarized solid samples, leaving only the dissolution and quality control at the local site. This would obviously not only interesting for medical applications, but could open up for wider access to hyperpolarization in other fields.

1.4 Imaging Agents

A key limitation of the method is the short half-life of hyperpolarization. Long longitudinal relaxation time constant, T_1, of tens of seconds to minutes can be found for ^{13}C that have no directly attached protons, such as carboxylic acids or carbonyls. An example is pyruvate enriched with ^{13}C in either the C-1 or C-2 positions, both having a T_1 of about 60 s. It is also possible to obtain sufficiently long T_1 by deuteration of directly attached protons, as in, for example, [U-^{13}C,U-^{2}H]glucose, however, in this case, obtaining a T_1 of only approx. 20 s. Therefore, the imaging agent, or bio-probe, must target metabolic pathways that have high uptake and metabolic rates on the time scale of T_1.

Another requirement for the imaging agent is that product and substrates have chemical shifts that are well separated to resolve the different metabolites. Chemical shift values for many in vivo metabolites can be found at the homepage of the Advanced Imaging Research Center, UTSouthwestern Medical Center (https://www.utsouthwestern.edu/education/medical-school/departments/airc/tools-references/chemical-shifts/).

Of the growing list of imaging agents studied in vivo by dDNP (for a comprehensive list, see Ref. [64]), the most studied is [1-^{13}C]pyruvate. Pyruvate can be reduced to lactate, transaminated to alanine, or will lose the C-1 carbon as $^{13}CO_2$ when it converts to acetylCoA in the Krebs cycle. This agent has shown great utility in oncology, as exemplified by the many studies summarized in other chapters.

[1-^{13}C]pyruvate was also the first agent to be used in a human study of hyperpolarized metabolic imaging.

1.5 Sample Preparation

The first step in hyperpolarizing a new imaging agent by dDNP is to find a formulation with high concentration of the imaging agent, and good solubility of the EPA. For the DNP process to be effective, the EPA must be homogeneously distributed within the sample. Many molecules will be crystalline or tend to crystallize as saturated aqueous solutions. This will cause the EPA to concentrate in domains and lead to a poor DNP effect. To prevent this, the imaging agent will have to be dissolved or mixed with a glassing solvent that makes the sample stay amorphous when frozen. Three examples of molecules that are liquids at room temperature and freeze as a glass are [1-^{13}C]pyruvic acid (or any other isotopic labeling), 2-keto-[1-^{13}C]isocaproic acid, and *bis*-1,1-(hydroxymethyl)-[1-^{13}C]cyclopropane-d_8 (HP001). All three molecules are liquids at room temperature.

For other compounds, it is necessary to prevent crystallization by mixing or dissolving the compound with a suitable solvent such as glycerol or dimethylsulfoxide (DMSO). For in vivo studies, it is necessary to formulate the molecule in a concentrated form in order to achieve a high concentration of the molecule after dissolution. To give an estimate of the requirements, a patient dose of 0.1 mmol/kg body weight can be assumed for an imaging agent with low toxicity, requiring approximately 10 mmol of compound. This means approximately 1 g of imaging agent, assuming a molecular weight of 100 g/mol. To keep the sample size reasonable requires a solubility of 30–50%. A good imaging agent requires a solvent mixture with high solubility for the molecule and EPA, has a low rate of sample crystallization, and exhibits good in vivo tolerance. An example of a biologically compatible formulation is fumaric acid (e.g., [1,4-$^{13}C_2$, 2,3-D_2]fumaric acid) in DMSO. DMSO is a widely used solvent for pharmaceuticals and has a good safety profile. As a saturated solution of fumaric acid with a molarity of 3.6 mol/L or 1:1.8 (w:w), the solution forms an amorphous solid when frozen, if the cooling rate is not too slow. Dry DMSO should be used as small amounts of water will decrease solubility and increase supersaturation.

Another means of improving the solubility involves changing the counter ion of the salt. Solubility typically increases with increasing size of the counter ion, and two examples of this can be mentioned: The cesium salt of bicarbonate, $CsH^{13}CO_3$, and the tris(hydroxymethyl)aminomethane (TRIS) salt of acetate [65]. Both salts have higher solubilities than their sodium counterpart. Finally, for amino acids (zwitterions at neutral pH), it has been shown that either high or low pH preparations increase the aqueous solubility by reducing the charge of the molecule to a point [65] that no or little glycerol is needed to form an amorphous sample.

Finally, Ji et al. [62] showed that microcrystals of solids can be polarized by impregnation with a water:glycerol solution of a nitroxide. The proton polarization of the microcrystals is enhanced by spin diffusion from the impregnation. However, reduced polarization and longer polarization time is observed.

Formulations for many of the most used imaging agents can be found on the homepage of the Hyperpolarized MRI Technology Resource Center at UCSF (https://radiology.ucsf.edu/research/labs/hyperpolarized-mri-tech/training).

1.6 Dissolution and Relaxation

To make the polarized solid sample useful for in vivo imaging, the sample is dissolved in a suitable buffer. Depending on the DNP sample, the dissolution may involve neutralization of the agent with acid or base, for example, as in the case of neat ^{13}C-pyruvic acid. Buffering of the solution may be required to maintain control of pH within the physiologic range of 6.8–8.1. Physiological buffers such as tris(hydroxymethyl)aminomethane (TRIS), 4-(2-hydroxyethyl)piperazine-1-ethanesulfonate (HEPES), or phosphate are commonly used. An isotonic formulation is desired. This may mean lowering the concentration of the imaging agent, if it is hypertonic, or adding sodium chloride if hypotonic. The dissolution must be efficient and fast compared to the nuclear T_1 in order to preserve the nuclear polarization during the phase transition. Formulating the solid sample as beads or powder may improve the dissolution in terms of polarization loss and recovery of the imaging agent. Optimizing the fluid dynamics, as well as providing the necessary heat, is essential for optimal performance of more difficult agents [66]. To minimize relaxation, dissolution is performed inside the cryostat in the high field of the polarizer (e.g., ~3 T in the case of a 3.35 T polarizer), but above the liquid helium surface. As an example, the shortest T_1 of the C-1 of neat [1-^{13}C]pyruvic acid at 9.4 T is ~1.6 s at 0 °C (unpublished data), and is attributed to dipolar relaxation by the methyl and acid protons. According to theory, the minimum T_1 scales with B_0, which means that a minimum T_1 of 0.7 s should be expected during the dissolution in the 3 T polarizer field. Thus, melting and dissolution should be fast on this time scale to avoid loss of polarization. The severity of the problem will depend on the nuclear spin and molecular structure.

Some parameters can be controlled, for example, distances to other spins (labeling position), the abundance of other spins (full or partial deuteration, as well as of the solvent), and the concentration of the EPA. Any paramagnetic impurities (originating, e.g., as leachable from the dissolution system or as a chemical impurity in the chemicals) that could increase the relaxation rate can be chelated by adding, for example, ethylenedinitrotetraacetic acid (EDTA) to the dissolution solvent. In most cases, the EPA or Gd chelate do not cause significant relaxation after dissolution and may be safe to inject into animals. However, for samples that require high concentration of nitroxide as EPA, T_1 of the nuclear spin may be shortened. If this is the case, it should be considered that relaxation is further accelerated at low magnetic field (at high magnetic field strengths the high electron spin resonance frequency ensures that relaxation terms disperse). For this reason, the background magnetic field should be considered from dissolution to detection of magnetic fields.

For pre-clinical imaging, it is not required to remove the EPA. The same applies to the Gd chelate in case it is used in the formulation. However, the solution may undergo a filtration or chromatography step to remove the EPA involved in the DNP process. In case a Gd chelate has been added, this agent may be removed as well. The filtration can either be in-line with the dissolution process or a subsequent step. In either case, the filtration is completed in a matter of a few seconds with insignificant loss of polarization or imaging agent concentration. As an example, in the case of clinical studies with pyruvate, the solid sample is neat pyruvic acid (^{13}C-labeled) with 15 trityl mmol/L radical (AH111501). The solid sample is dissolved in water-for-injection, which causes the trityl to precipitate under the acidic conditions. The solid EPA particles are then removed by filtration, and the pyruvic acid is neutralized post dissolution.

Most other relaxation mechanisms may play a role depending on imaging agent and formulation. Of most importance is chemical shift anisotropy that may contribute a significant relaxation contribution for carbonyls and carboxylic acids above approximately 7 T. Likewise, scalar relaxation of the second kind: scalar coupling to a fast relaxing nuclear spin, for example, ^{14}N in urea, will reduce T_1 in low (earth) magnetic field (and T_2 in high magnetic field) [67]. For all these relaxation mechanisms (except spin-rotational relaxation), it will be beneficial to keep the sample at a relatively high temperature, considering that eventually the temperature must be reduced to 37 °C for biological studies.

1.7 Summary

The DNP theory has received much attention in recent years with the renaissance of the method within NMR and magnetic resonance imaging (MRI). High field static and magic angle spinning dynamic nuclear polarization nuclear magnetic resonance (MAS-DNP-NMR) for sensitivity enhancement of solid-state NMR along with dDNP for in vivo biological and medical imaging has spawned renewed interest in revisiting the theory and developing new methods. The theoretical framework is on a much more solid ground, and the field is a great example of experiment and theory bringing scientists together.

Acknowledgment Funding support from the Danish National Research Foundation (DNRF124).

Disclosure I am an employee of GE Healthcare and the owner of the company Polarize.

Problems

1. Calculate the number of spins per cubic centimeter (spins/cm^3) for a 15 mmol/L EPA and a 40 m EPA concentration.

2. Calculate the average distance between electron spins based on a simple lattice model for a 40 mmol/L EPA concentration.
3. Calculate the average electron–electron dipolar interaction (MHz) for two trityl radicals at 10 Å distance.
4. Calculate the average electron–nuclear dipolar interaction (MHz) for a ^{13}C spin and an electron spin at 5 Å distance.
5. Calculate the ^{13}C spin diffusion rate for natural abundance pyruvic acid and calculate how long it takes to diffuse the average electron-electron distances calculated in problem 2.

References

1. Ardenkjær-Larsen, J.H., Fridlund, B., Gram, A., Hansson, G., Hansson, L., Lerche, M.H., Servin, R., Thaning, M., Golman, K.: Increase in signal-to-noise ratio of <10,000 times in liquid-state NMR. Proc. Natl. Acad. Sci. U. S. A. **100**, 10158–10163 (2003). https://doi.org/10.1073/pnas.1733835100
2. Nelson, S.J., Kurhanewicz, J., Vigneron, D.B., Larson, P.E.Z., Harzstark, A.L., Ferrone, M., van Criekinge, M., Chang, J.W., Bok, R., Park, I., Reed, G., Carvajal, L., Small, E.J., Munster, P., Weinberg, V.K., Ardenkjaer-Larsen, J.H.J.H., Chen, A.P., Hurd, R.E., Odegardstuen, L.-I., Robb, F.J., Tropp, J., Murray, J.A., Van Criekinge, M., Chang, J.W., Bok, R., Park, I., Reed, G., Carvajal, L., Small, E.J., Munster, P., Weinberg, V.K., Ardenkjaer-Larsen, J.H.J.H., Chen, A.P., Hurd, R.E., Odegardstuen, L.-I., Robb, F.J., Tropp, J., Murray, J.A.: Metabolic imaging of patients with prostate cancer using hyperpolarized [1-13C]pyruvate. Sci. Transl. Med. **5**, 198ra108 (2013). https://doi.org/10.1126/scitranslmed.3006070
3. Cunningham, C.H., Lau, J.Y., Chen, A.P., Geraghty, B.J., Perks, W.J., Roifman, I., Wright, G.A., Connelly, K.A.: Hyperpolarized 13C metabolic MRI of the human heart: initial experience. Circ. Res. **119**(11), 1177–1182 (2016). https://doi.org/10.1161/CIRCRESAHA.116.309769
4. Miloushev, V.Z., Granlund, K.L., Boltyanskiy, R., Lyashchenko, S.K., DeAngelis, L.M., Mellinghoff, I.K., Brennan, C.W., Tabar, V., Yang, T.J., Holodny, A.I., Sosa, R.E., Guo, Y.W., Chen, A.P., Tropp, J., Robb, F., Keshari, K.R.: Metabolic imaging of the human brain with hyperpolarized 13C pyruvate demonstrates 13C lactate production in brain tumor patients. Cancer Res. **78**(14), 3755–3760 (2018). https://doi.org/10.1158/0008-5472.CAN-18-0221
5. Aggarwal, R., Vigneron, D.B., Kurhanewicz, J.: Hyperpolarized 1-[13C]-pyruvate magnetic resonance imaging detects an early metabolic response to androgen ablation therapy in prostate Cancer. Eur. Urol. **72**, 1028–1029 (2017). https://doi.org/10.1016/J.EURURO.2017.07.022
6. Park, I., Larson, P.E.Z., Gordon, J.W., Carvajal, L., Chen, H.-Y., Bok, R., Van Criekinge, M., Ferrone, M., Slater, J.B., Xu, D., Kurhanewicz, J., Vigneron, D.B., Chang, S., Nelson, S.J.: Development of methods and feasibility of using hyperpolarized carbon-13 imaging data for evaluating brain metabolism in patient studies. Magn. Reson. Med. **80**, 864–873 (2018). https://doi.org/10.1002/mrm.27077
7. Grist, J.T., McLean, M.A., Riemer, F., Schulte, R.F., Deen, S.S., Zaccagna, F., Woitek, R., Daniels, C.J., Kaggie, J.D., Matyz, T., Patterson, I., Slough, R., Gill, A.B., Chhabra, A., Eichenberger, R., Laurent, M.-C., Comment, A., Gillard, J.H., Coles, A.J., Tyler, D.J., Wilkinson, I., Basu, B., Lomas, D.J., Graves, M.J., Brindle, K.M., Gallagher, F.A.: Quantifying normal human brain metabolism using hyperpolarized [1–13C]pyruvate and magnetic resonance imaging. NeuroImage. **189**, 171 (2019). https://doi.org/10.1016/J.NEUROIMAGE.2019.01.027
8. Mammoli, D., Gordon, J., Autry, A., Larson, P.E.Z., Li, Y., Chen, H.-Y., Chung, B., Shin, P., Van Criekinge, M., Carvajal, L., Slater, J.B., Bok, R., Crane, J., Xu, D., Chang, S., Vigneron,

D.B.: Kinetic modeling of hyperpolarized Carbon-13 pyruvate metabolism in the human brain. IEEE Trans. Med. Imaging. **39**, 320–327 (2019). https://doi.org/10.1109/TMI.2019.2926437
9. Lee, C.Y., Soliman, H., Geraghty, B.J., Chen, A.P., Connelly, K.A., Endre, R., Perks, W.J., Heyn, C., Black, S.E., Cunningham, C.H.: Lactate topography of the human brain using hyperpolarized 13C-MRI. NeuroImage. **2019**, 116202 (2019). https://doi.org/10.1016/J.NEUROIMAGE.2019.116202
10. Chung, B.T., Chen, H.-Y., Gordon, J., Mammoli, D., Sriram, R., Autry, A.W., Le Page, L.M., Chaumeil, M., Shin, P., Slater, J., Tan, C.T., Suszczynski, C., Chang, S., Li, Y., Bok, R.A., Ronen, S.M., Larson, P.E.Z., Kurhanewicz, J., Vigneron, D.B.: First hyperpolarized [2-13C]pyruvate MR studies of human brain metabolism. J. Magn. Reson. **2019**, 106617 (2019). https://doi.org/10.1016/J.JMR.2019.106617
11. Stødkilde-Jørgensen, H., Laustsen, C., Hansen, E.S.S., Schulte, R., Ardenkjaer-Larsen, J.H., Comment, A., Frøkiær, J., Ringgaard, S., Bertelsen, L.B., Ladekarl, M., Weber, B.: Pilot study experiences with hyperpolarized [1–13C]pyruvate MRI in pancreatic cancer patients. J. Magn. Reson. Imaging. **51**, 961–963 (2020). https://doi.org/10.1002/jmri.26888
12. Walker, T.G., Happer, W.: Spin-exchange optical pumping of noble-gas nuclei. Rev. Mod. Phys. **69**, 629–642 (1997). https://doi.org/10.1103/revmodphys.69.629
13. Albert, M.S., Cates, G.D., Driehuys, B., Happer, W., Saam, B., Springer, C.S., Wishnia, A.: Biological magnetic resonance imaging using laser-polarized 129Xe. Nature. **370**, 199–201 (1994). https://doi.org/10.1038/370199a0
14. Qing, K., Mugler, J.P., Altes, T.A., Jiang, Y., Mata, J.F., Miller, G.W., Ruset, I.C., Hersman, F.W., Ruppert, K.: Assessment of lung function in asthma and COPD using hyperpolarized 129Xe chemical shift saturation recovery spectroscopy and dissolved-phase MRI. NMR Biomed. **27**, 1490–1501 (2014). https://doi.org/10.1002/nbm.3179
15. Seydoux, R., Pines, A., Haake, M., Reimer, J.A.: NMR with a continuously circulating flow of laser-polarized 129Xe. J. Phys. Chem. B. **103**, 4629–4637 (1999). https://doi.org/10.1021/jp9821984
16. Bowers, C.R., Weitekamp, D.P.: Transformation of symmetrization order to nuclear-spin magnetization by chemical reaction and nuclear magnetic resonance. Phys. Rev. Lett. **57**, 2645–2648 (1986). https://doi.org/10.1103/PhysRevLett.57.2645
17. Bowers, C.R., Weitekamp, D.P.: Parahydrogen and synthesis allow dramatically enhanced nuclear alignment. J. Am. Chem. Soc. **109**, 5541–5542 (1987). https://doi.org/10.1021/ja00252a049
18. Hövener, J.B., Pravdivtsev, A.N., Kidd, B., Bowers, C.R., Glöggler, S., Kovtunov, K.V., Plaumann, M., Katz-Brull, R., Buckenmaier, K., Jerschow, A., Reineri, F., Theis, T., Shchepin, R.V., Wagner, S., Bhattacharya, P., Zacharias, N.M., Chekmenev, E.Y.: Parahydrogen-based hyperpolarization for biomedicine. Angew. Chem. Int. Ed. **57**, 11140–11162 (2018). https://doi.org/10.1002/anie.201711842
19. Adams, R.W., Aguilar, J.A., Atkinson, K.D., Cowley, M.J., Elliott, P.I.P., Duckett, S.B., Green, G.G.R., Khazal, I.G., López-Serrano, J., Williamson, D.C.: Reversible interactions with parahydrogen enhance NMR sensitivity by polarization transfer. Science. **323**, 1708–1711 (2009). https://doi.org/10.1126/science.1168877
20. Cavallari, E., Carrera, C., Aime, S., Reineri, F.: Studies to enhance the hyperpolarization level in PHIP-SAH-produced C13-pyruvate. J. Magn. Reson. **289**, 12–17 (2018). https://doi.org/10.1016/j.jmr.2018.01.019
21. Peat, D.T.: Achievement of high nuclear spin polarization using lanthanides as low-temperature NMR relaxation agents. Phys. Chem. Chem. Phys. **15**, 7586–7591 (2013). https://doi.org/10.1039/c3cp00103b
22. Hirsch, M.L., Smith, B.A., Mattingly, M., Goloshevsky, A.G., Rosay, M., Kempf, J.G.: Transport and imaging of brute-force 13 C hyperpolarization. J. Magn. Reson. **261**, 87–94 (2015). https://doi.org/10.1016/j.jmr.2015.09.017
23. Hirsch, M.L., Kalechofsky, N., Belzer, A., Rosay, M., Kempf, J.G.: Brute-force hyperpolarization for NMR and MRI. J. Am. Chem. Soc. **137**, 8428–8434 (2015). https://doi.org/10.1021/jacs.5b01252

24. Peat, D.T., Hirsch, M.L., Gadian, D.G., Horsewill, A.J., Owers-Bradley, J.R., Kempf, J.G.: Low-field thermal mixing in [1- 13 C] pyruvic acid for brute-force hyperpolarization. Phys. Chem. Chem. Phys. **18**, 19173–19182 (2016). https://doi.org/10.1039/C6CP02853E
25. Overhauser, A.W.: Polarization of nuclei in metals. Phys. Rev. **92**, 411 (1953). https://doi.org/10.1103/PhysRev.92.411
26. Carver, T.R.R., Slichter, C.P.P.: Polarization of nuclear spins in metals. Phys. Rev. **92**, 212 (1953). https://doi.org/10.1103/PhysRev.92.212.2
27. Abragam, A., Proctor, W.G.: Experiments on spin temperature. Phys. Rev. **106**, 160–161 (1957). https://doi.org/10.1103/PhysRev.106.160
28. Can, T.V., Caporini, M.A., Mentink-Vigier, F., Corzilius, B., Walish, J.J., Rosay, M., Maas, W.E., Baldus, M., Vega, S., Swager, T.M., Griffin, R.G.: Overhauser effects in insulating solids. J. Chem. Phys. **141**, 064202 (2014). https://doi.org/10.1063/1.4891866
29. Ji, X., Can, T.V., Mentink-Vigier, F., Bornet, A., Milani, J., Vuichoud, B., Caporini, M.A., Griffin, R.G., Jannin, S., Goldman, M., Bodenhausen, G.: Overhauser effects in non-conducting solids at 1.2 K. J. Magn. Reson. **286**, 138–142 (2018). https://doi.org/10.1016/j.jmr.2017.11.017
30. Orlando, T., Dervişoğlu, R., Levien, M., Tkach, I., Prisner, T.F., Andreas, L.B., Denysenkov, V.P., Bennati, M.: Dynamic nuclear polarization of 13 C nuclei in the liquid state over a 10 tesla field range. Angew. Chem. Int. Ed. **58**, 1402–1406 (2019). https://doi.org/10.1002/anie.201811892
31. Jeffries, C.D.: Polarization of nuclei by resonance saturation in paramagnetic crystals. Phys. Rev. **106**, 164–165 (1957). https://doi.org/10.1103/PhysRev.106.164
32. Abragam, A., Goldman, M.M.: Nuclear magnetism: order and disorder. Clarendon Press, Oxford (1982). http://findit.dtu.dk/en/catalog/2300286932
33. Hovav, Y., Feintuch, A., Vega, S.: Theoretical aspects of dynamic nuclear polarization in the solid state—the solid effect. J. Magn. Reson. **207**, 176–189 (2010). https://doi.org/10.1016/j.jmr.2010.10.016
34. Hovav, Y., Feintuch, A., Vega, S.: Theoretical aspects of dynamic nuclear polarization in the solid state—the cross effect. J. Magn. Reson. **214**, 29–41 (2012). https://doi.org/10.1016/j.jmr.2011.09.047
35. Hovav, Y., Feintuch, A., Vega, S.: Theoretical aspects of dynamic nuclear polarization in the solid state—spin temperature and thermal mixing. Phys. Chem. Chem. Phys. **15**, 188–203 (2012). https://doi.org/10.1039/C2CP42897K
36. Karabanov, A., Wiśniewski, D., Raimondi, F., Lesanovsky, I., Köckenberger, W.: Many-body kinetics of dynamic nuclear polarization by the cross effect. Phys. Rev. A. **97**, 031404 (2018). https://doi.org/10.1103/PhysRevA.97.031404
37. Wiśniewski, D., Karabanov, A., Lesanovsky, I., Köckenberger, W.: Solid effect DNP polarization dynamics in a system of many spins. J. Magn. Reson. **264**, 30–38 (2016). https://doi.org/10.1016/j.jmr.2016.01.016
38. W.T. Wenckebach, Essentials of dynamic nuclear polarization, http://www.wenckebach.net/html/dnp-book.html.
39. Wenckebach, W.T.: Spectral diffusion and dynamic nuclear polarization: beyond the high temperature approximation. J. Magn. Reson. **284**, 104–114 (2017). https://doi.org/10.1016/J.JMR.2017.10.001
40. Wenckebach, W.T.: Dynamic nuclear polarization via thermal mixing: beyond the high temperature approximation. J. Magn. Reson. **277**, 68–78 (2017). https://doi.org/10.1016/J.JMR.2017.01.020
41. Slichter, C.P.: Principles of magnetic resonance. Springer, Berlin (1990). Page 15. https://doi.org/10.1007/978-3-662-09441-9
42. B.N. Provotorov, A quantum-statistical theory of cross relaxation "'±1" '±1, 1962
43. Adeva, B., Arik, E., Ahmad, S., Arvidson, A., Badelek, B., Ballintijn, M.K., Bardin, G., Baum, G., Berglund, P., Betev, L., Bird, I.G., Birsa, R., Björkholm, P., Bonner, B.E., de Botton, N., Boutemeur, M., Bradamante, F., Bressan, A., Brüll, A., Buchanan, J., Bültmann, S., Burtin, E., Cavata, C., Chen, J.P., Clement, J., Clocchiatti, M., Corcoran, M.D., Crabb, D., Cranshaw, J., Çuhadar, J., Torre, S.D., Deshpande, A., van Dantzig, R., Day, D., Dhawan, S., Dulya, C., Dyring, A., Eichblatt, S., Faivre, J.C., Fasching, D., Feinstein, F., Fernandez, C., Frois, B.,

Garabatos, C., Garzon, J.A., Gaussiran, T., Giorgi, M., von Goeler, E., Golutvin, I.A., Gomez, A., Gracia, G., de Groot, N., Perdekamp, M.G., Gülmez, E., von Harrach, D., Hasegawa, T., Hautle, P., Hayashi, N., Heusch, C.A., Horikawa, N., Hughes, V.W., Igo, G., Ishimoto, S., Iwata, T., de Jong, M., Kabuß, E.M., Kageya, T., Kaiser, R., Karev, A., Kessler, H.J., Ketel, T.J., Kiryushin, I., Kishi, A., Kisselev, Y., Klostermann, L., Krämer, D., Krivokhijine, V., Kukhtin, V., Kyynäräinen, J., Lamanna, M., Landgraf, U., Lau, K., Layda, T., Le Goff, J.M., Lehar, F., de Lesquen, A., Lichtenstadt, J., Lindqvist, T., Litmaath, M., Lopez-Ponte, S., Lowe, M., Magnon, A., Mallot, G.K., Marie, F., Martin, A., Martino, J., Matsuda, T., Mayes, B., McCarthy, J.S., Medved, K., van Middelkoop, G., Miller, D., Mitchell, J., Mori, K., Moromisato, J., Mutchler, G.S., Nagaitsev, A., Nassalski, J., Naumann, L., Neganov, B., Niinikoski, T.O., Oberski, J.E.J., Ogawa, A., Okumi, S., Özben, C.S., Penzo, A., Perez, C.A., Perrot-Kunne, F., Peshekhonov, D., Piegaia, R., Pinsky, L., Platchkov, S., Plo, M., Pose, D., Postma, H., Pretz, J., Pussieux, T., Pyrlik, J., Reyhancan, I., Rieubland, J.M., Rijllart, A., Roberts, J.B., Rock, S.E., Rodriguez, M., Rondio, E., Rondon, O., Ropelewski, L., Rosado, A., Sabo, I., Saborido, J., Salvato, G., Sandacz, A., Sanders, D., Savin, I., Schiavon, P., Schüler, K.P., Segel, R., Seitz, R., Semertzidis, Y., Sergeev, S., Sever, F., Shanahan, P., Sichtermann, E., Smirnov, G., Staude, A., Steinmetz, A., Stuhrmann, H., Teichert, K.M., Tessarotto, F., Thiel, W., Velasco, M., Vogt, J., Voss, R., Weinstein, R., Whitten, C., Willumeit, R., Windmolders, R., Wislicki, W., Witzmann, A., Yañez, A., Zamiatin, N.I., Zanetti, A.M., Zhao, J.: Large enhancement of deuteron polarization with frequency modulated microwaves. Nucl. Instr. Method. Phys. Res. Sect. A Accel. Spectromet. Detect. Assoc. Equip. **372**, 339–343 (1996). https://doi.org/10.1016/0168-9002(95)01376-8
44. Thurber, K.R., Yau, W.-M., Tycko, R.: Low-temperature dynamic nuclear polarization at 9.4 T with a 30 mW microwave source. J. Magn. Reson. **204**, 303–313 (2010). https://doi.org/10.1016/j.jmr.2010.03.016
45. Jóhannesson, H., Macholl, S., Ardenkjaer-Larsen, J.H.: Dynamic nuclear polarization of [1-13C]pyruvic acid at 4.6 tesla. J. Magn. Reson. **197**, 167–175 (2009). https://doi.org/10.1016/j.jmr.2008.12.016
46. Cheng, T., Capozzi, A., Takado, Y., Balzan, R., Comment, A.: Over 35% liquid-state 13C polarization obtained via dissolution dynamic nuclear polarization at 7 T and 1 K using ubiquitous nitroxyl radicals. Phys. Chem. Chem. Phys. **15**, 20819–20822 (2013). https://doi.org/10.1039/C3CP53022A
47. Yoshihara, H.A.I., Can, E., Karlsson, M., Lerche, M.H., Schwitter, J., Comment, A.: High-field dissolution dynamic nuclear polarization of [1- 13 C]pyruvic acid. Phys. Chem. Chem. Phys. **18**, 12409–12413 (2016). https://doi.org/10.1039/C6CP00589F
48. Ardenkjaer-Larsen, J.H., Bowen, S., Petersen, J.R.J.R., Rybalko, O., Vinding, M.S.M.S., Ullisch, M., Nielsen, N.C.N.C., Ardenkjær-Larsen, J.H., Bowen, S., Petersen, J.R.J.R., Rybalko, O., Vinding, M.S.M.S., Ullisch, M., Nielsen, N.C.N.C., Ardenkjaer-Larsen, J.H., Bowen, S., Petersen, J.R.J.R., Rybalko, O., Vinding, M.S.M.S., Ullisch, M., Nielsen, N.C.N.C.: Cryogen-free dissolution dynamic nuclear polarization polarizer operating at 3.35 T, 6.70 T, and 10.1 T. Magn. Reson. Med. **81**, 2184–2194 (2018). https://doi.org/10.1002/mrm.27537
49. Ardenkjaer-Larsen, J.H., Leach, A.M., Clarke, N., Urbahn, J., Anderson, D., Skloss, T.W.: Dynamic nuclear polarization polarizer for sterile use intent. NMR Biomed. **24**, 927–932 (2011). https://doi.org/10.1002/nbm.1682
50. Lowe, I.J., Gade, S.: Density-matrix derivation of the spin-diffusion equation. Phys. Rev. **156**, 817–825 (1967). https://doi.org/10.1103/PhysRev.156.817
51. Tan, K.O., Mardini, M., Yang, C., Ardenkjær-Larsen, J.H., Griffin, R.G.: Three-spin solid effect and the spin diffusion barrier in amorphous solids. Sci. Adv. **5**, eaax2743 (2019). https://doi.org/10.1126/sciadv.aax2743
52. Ramanathan, C.: Dynamic nuclear polarization and spin diffusion in nonconducting solids. Appl. Magn. Reson. **34**, 409–421 (2008). https://doi.org/10.1007/s00723-008-0123-7
53. Bornet, A., Pinon, A., Jhajharia, A., Baudin, M., Ji, X., Emsley, L., Bodenhausen, G., Ardenkjaer-Larsen, J.H., Jannin, S.: Microwave-gated dynamic nuclear polarization. Phys. Chem. Chem. Phys. **18**, 30530–30535 (2016). https://doi.org/10.1039/C6CP05587G
54. Vinther, J.M.O., Zhurbenko, V., Albannay, M.M., Ardenkjær-Larsen, J.H.: Design of a local quasi-distributed tuning and matching circuit for dissolution DNP cross polarization. Solid State Nucl. Magn. Reson. **102**, 12–20 (2019). https://doi.org/10.1016/J.SSNMR.2019.04.006

55. Aghelnejad, B., Bodenhausen, G., Marhabaie, S.: A low-temperature broadband NMR probe for multinuclear cross-polarization. ChemPhysChem. **20**, 2830–2835 (2019). https://doi.org/10.1002/cphc.201900723
56. Hu, K.-N.: Polarizing agents and mechanisms for high-field dynamic nuclear polarization of frozen dielectric solids. Solid State Nucl. Magn. Reson. **40**, 31–41 (2011). https://doi.org/10.1016/j.ssnmr.2011.08.001
57. Lilly Thankamony, A.S., Wittmann, J.J., Kaushik, M., Corzilius, B.: Dynamic nuclear polarization for sensitivity enhancement in modern solid-state NMR. Prog. Nucl. Magn. Reson. Spectrosc. **102–103**, 120–195 (2017). https://doi.org/10.1016/j.pnmrs.2017.06.002
58. Ardenkjaer-Larsen, J.H., MacHoll, S., Jóhannesson, H.: Dynamic nuclear polarization with trityls at 1.2 K. Appl. Magn. Reson. **34**, 509–522 (2008). https://doi.org/10.1007/s00723-008-0134-4
59. Corzilius, B., Smith, A.A., Barnes, A.B., Luchinat, C., Bertini, I., Griffin, R.G.: High-field dynamic nuclear polarization with high-spin transition metal ions. J. Am. Chem. Soc. **133**, 5648–5651 (2011). https://doi.org/10.1021/ja1109002
60. Capozzi, A., Karlsson, M., Petersen, J.R., Lerche, M.H., Ardenkjaer-Larsen, J.H.: Liquid-state 13C polarization of 30% through photo-induced non-persistent radicals. J. Phys. Chem. C. **122**, 7432–7443 (2018). https://doi.org/10.1021/acs.jpcc.8b01482
61. Capozzi, A., Patel, S., Gunnarsson, C.P., Marco-Rius, I., Comment, A., Karlsson, M., Lerche, M.H., Ouari, O., Ardenkjær-Larsen, J.H.: Efficient hyperpolarization of U-13C-glucose using narrow-line UV-generated labile free radicals. Angew. Chem. Int. Ed. **58**, 1334–1339 (2019). https://doi.org/10.1002/anie.201810522
62. Ji, X., Bornet, A., Vuichoud, B., Milani, J., Gajan, D., Rossini, A.J., Emsley, L., Bodenhausen, G., Jannin, S.: Transportable hyperpolarized metabolites. Nat. Commun. **8**, 13975 (2017). https://doi.org/10.1038/ncomms13975
63. Capozzi, A., Cheng, T., Boero, G., Roussel, C., Comment, A.: Thermal annihilation of photo-induced radicals following dynamic nuclear polarization to produce transportable frozen hyperpolarized 13C-substrates. Nat. Commun. **8**, 15757 (2017). https://doi.org/10.1038/ncomms15757
64. Keshari, K.R., Wilson, D.M.: Chemistry and biochemistry of 13C hyperpolarized magnetic resonance using dynamic nuclear polarization. Chem. Soc. Rev. **43**, 1627–1659 (2014). https://doi.org/10.1039/c3cs60124b
65. Karlsson, M., Jensen, P.R., Duus, J.Ø., Meier, S., Lerche, M.H.: Development of dissolution DNP-MR substrates for metabolic research. Appl. Magn. Reson. **43**, 223–236 (2012). https://doi.org/10.1007/s00723-012-0336-7
66. Bowen, S., Ardenkjaer-Larsen, J.H.J.H.: Formulation and utilization of choline based samples for dissolution dynamic nuclear polarization. J. Magn. Reson. **236**, 26–30 (2013). https://doi.org/10.1016/j.jmr.2013.08.007
67. Chiavazza, E., Kubala, E., Gringeri, C.V., Düwel, S., Durst, M., Schulte, R.F., Menzel, M.I.: Earth's magnetic field enabled scalar coupling relaxation of 13C nuclei bound to fast-relaxing quadrupolar 14N in amide groups. J. Magn. Reson. **227**, 35–38 (2013). https://doi.org/10.1016/j.jmr.2012.11.016

Further Reading

Abragam, A., Goldman, M.M.: Nuclear magnetism: order and disorder. Clarendon Press, Oxford (1982). http://findit.dtu.dk/en/catalog/2300286932

W.T. Wenckebach, Essentials of dynamic nuclear polarization, n.d. http://www.wenckebach.net/html/dnp-book.html. Accessed 5 Feb 2018.

Chapter 2
Design Considerations for Implementing a Hyperpolarizer

Arnaud Comment

Abbreviations

BDPA	1,3-bisdiphenylene-2-phenylallyl
DNP	Dynamic nuclear polarization
ESR	Electron spin resonance
MR	Magnetic resonance
NMR	Nuclear magnetic resonance
UV–Vis	Ultraviolet–visible
VTI	Variable temperature insert

2.1 Introduction

Historically, Thomas Carver and Charles P. Slichter were the first ones to experimentally demonstrate what Albert W. Overhauser had predicted [1, 2]: electron spins can be used to polarize nuclear spins, theoretically up to the quotient between their respective gyromagnetic ratio. In their initial work, Carver and Slichter polarized ^{7}Li spins using the inherent conduction electrons of a metallic matrix, before they demonstrated that it was also possible to polarize ^{1}H in liquid ammonia [3]. A historical description of this discovery has been recently published by Slichter [4]. These experiments were performed in a magnetic field of 1–5 mT, at which the electron spins resonate below 125 MHz. The choice of working at very low field was motivated by the fact that, first, high-frequency irradiation has a limited penetration inside a metallic sample, and second, microwave sources powerful enough to saturate electron spins at higher frequency were not widely available at that time. One of the major difficulties was therefore to detect the nuclear magnetic resonance

A. Comment (✉)
Cancer Research UK Cambridge Institute, Li Ka Shing Centre, Cambridge, UK

General Electric Healthcare, Buckinghamshire, UK
e-mail: arnaud.comment@ge.com

T. Jue, D. Mayer (eds.), *Dynamic Hyperpolarized Nuclear Magnetic Resonance*, Handbook of Modern Biophysics, https://doi.org/10.1007/978-3-030-55043-1_2

(NMR) signal at such a low field, that is, at frequencies around 50 kHz. Even today, this experiment would be quite challenging!

Starting from 1959, several research groups showed that the concept can be extended to electron spins that are fixed within a solid matrix [5]. Since then, a large number of apparatuses to perform solid-state dynamic nuclear polarization (DNP) have been designed by the community working in the realm of particle physics [6]. These systems are used to polarize the nuclear spins of samples acting as targets for particle beams. Each apparatus consists of a cryostat placed in the center of a magnetic field of several Tesla and capable of holding liquid helium. The sample is placed inside a microwave cavity connected to a source via a waveguide. A radiofrequency coil surrounds the sample to monitor the nuclear polarization by NMR. Such an apparatus is known as a polarizer or a DNP polarizer and for the beam experiments typically performed with these systems, the sample environment is kept closed during the entire experiment and the polarization speed is usually not a major issue.

This type of apparatus was first modified for what is nowadays known as "dissolution DNP" by a team working at Amersham Health [7]. The challenge was to rapidly transform the cold sample into a liquid and extract it from the cryostat for injection into a living system. This meant that the instrument was supposed to operate under drastically different environmental constraints than a classical DNP polarizer: it needed to be a sort of stand-alone production apparatus that provides a highly controlled liquid-state solution in a highly reproducible manner so that magnetic resonance (MR) experiments can be subsequently performed. This type of apparatus will therefore never be part of the set of main analytical instruments in an MR facility but rather the machine we would like to "hide" in a corner of a back room. It should therefore ideally be a small, low-maintenance, highly automated apparatus in order to make sure that it can be operated without interfering with the surrounding MR systems.

For a given sample, the maximum nuclear spin polarization that can be reached using a DNP polarizer will depend on its environment, namely, the temperature T, the static magnetic field B_0, as well as the frequency and power of the microwaves delivered to the sample. In practice, these parameters are set by the hardware design, and once the system has been assembled, the challenges lie in the optimization of the sample to get the best polarization for a given temperature, field, and microwave source. To date, only two dissolution DNP polarizers have been made commercially available, namely, the HyperSense™ and the SPINlab™, but the few prototype systems that were built in various labs provide additional information as to what could be the ideal system for specific research.

This chapter will review the specifics of all these systems. The content of this chapter should provide enough background information to evaluate what performances can be expected from a system with specified characteristics.

2.2 The Quest for Optimal System Parameters

The goal of hyperpolarization is to increase the orientation of the nuclear spins along the magnetic field. It may seem obvious from the formula defining the polarization P for a ½-spin particle, namely,

$$P = \tanh \frac{\frac{h}{2\pi}\gamma B_0}{2k_B T}, \tag{2.1}$$

with h, the Planck constant, γ, the gyromagnetic ratio of the spins under investigation, and k_B, the Boltzmann constant, that this can simply be achieved by increasing the magnetic field and reducing the temperature. However, when looking at the orders of magnitude, one quickly realizes that for any standard achievable magnetic field, the temperature must be decreased to milli-Kelvin (mK) range to reach any reasonably large nuclear spin polarization (see Fig. 2.1). This could be done, but two prominent difficulties would have to be overcome. First, the complexity of the cryogenic apparatus would most likely render the experiment both expensive, possibly requiring the use of the scarce ^{3}He isotope, and extremely complex, especially the implementation of a rapid method to extract the polarized samples from the cryostat; second, the nuclear spin longitudinal relaxation time $T_{1,n}$ at such low

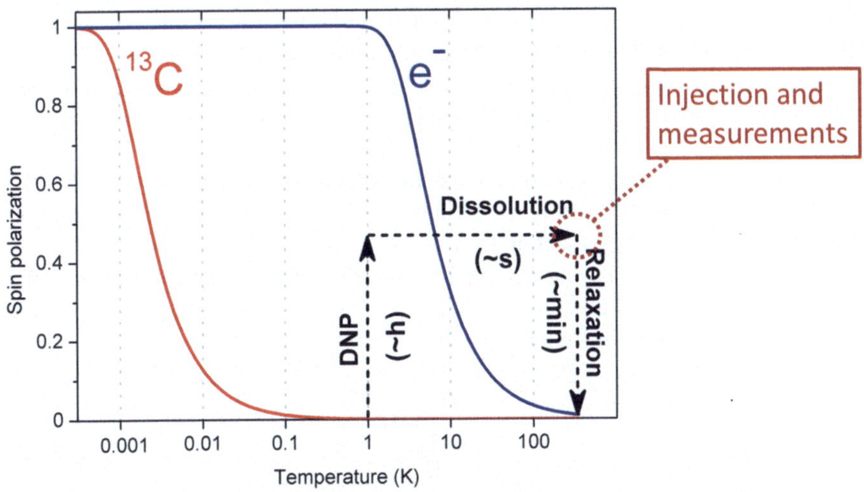

Fig. 2.1 Theoretical electron (blue) and ^{13}C (red) spin polarization at 5 T as a function of temperature calculated from Equation (2.1). The dashed arrows show the evolution of the ^{13}C spins during a hyperpolarized MR experiment, starting with solid-state DNP, which typically lasts a few hours. Dissolution takes 1–10 s depending on the sample size. Longitudinal relaxation to room-temperature thermal equilibrium ^{13}C polarization occurs within the few minutes following dissolution. If the hyperpolarized ^{13}C-molecules are rapidly injected into a rodent or other living organism after dissolution, in vivo measurements can be performed with a highly enhanced ^{13}C MR signal

temperatures becomes prohibitively long and reaching thermal equilibrium nuclear spin polarization would become terribly slow unless the sample is doped with adequate relaxing agents [8]. An implementation of a hyperpolarizer based on this so-called "brute force" approach has been proposed. As expected, this apparatus was not operating below pumped ^{4}He temperature. The ^{13}C polarization was too low to be useful for biomedical applications [9].

Instead of attempting to reach directly a thermal equilibrium state with a high nuclear spin polarization, DNP follows a three-step process: (a) dope the sample with unpaired electron spins, (b) reach a thermal equilibrium state with a high electron spin polarization, and (c) use microwave irradiation at a frequency slightly off-resonance from the unpaired electron spins, to drive the electron and nuclear spins in an out-of-equilibrium steady state in which the nuclear spins are highly polarized. From Equation (2.1), we can deduce that unpaired electron spins are essentially fully polarized at any temperature below 1.5 K if the magnetic field is 3.35 T or above (see Fig. 2.1). Step (c) essentially consists of irradiating the doped sample with microwaves while maintaining it at low temperature and high field to force all interacting spins in the sample to acquire the same polarization, which in this context is also referred to as "inverse spin temperature" [10]. As was highlighted by Comment et al. [11], if we neglect the subtleties related to the shape of the electron spin resonance (ESR) linewidth of the radical used as source of unpaired electron spins, the equation for calculating the minimum achievable inverse spin temperature derived by Redfield is proportional to B_0/T [12]. This means that if we assume that microwave power is unlimited, the maximum achievable nuclear spin polarization in a given sample will be a function of only two parameters: temperature and magnetic field. This crude first approximation is corroborated by the maximum ^{13}C polarization values reported in the literature for the most common sample, namely, neat pyruvic acid doped with trityl radicals, at various fields and temperatures (see Fig. 2.2). Although the optimal trityl concentration slightly increases (from 15 to 20 mM) with increasing magnetic field, it is clear that at least within the 3–7 T field range and 0.7–1.5 K temperature range, the maximum ^{13}C polarization is essentially linearly dependent on the ratio B_0/T. This linear dependence is merely empirical as no model has yet fully explained this trend. It must, however, be noted that it was demonstrated that the fact that the ^{13}C polarization is proportional to the thermal polarization in the high temperature approximation, that is,

$$P \cong \frac{\frac{h}{2\pi}\gamma B_0}{2k_B T}, \tag{2.2}$$

was in agreement with the model proposed by Borghini [11, 13].

From Fig. 2.2, one might think that it would be beneficial to further push the magnetic field up and the temperature down to increase B_0/T. This purely static approach, however, does not take into account the dynamics of the process. In a simple picture, there is a competition between the rate at which the nuclear spin

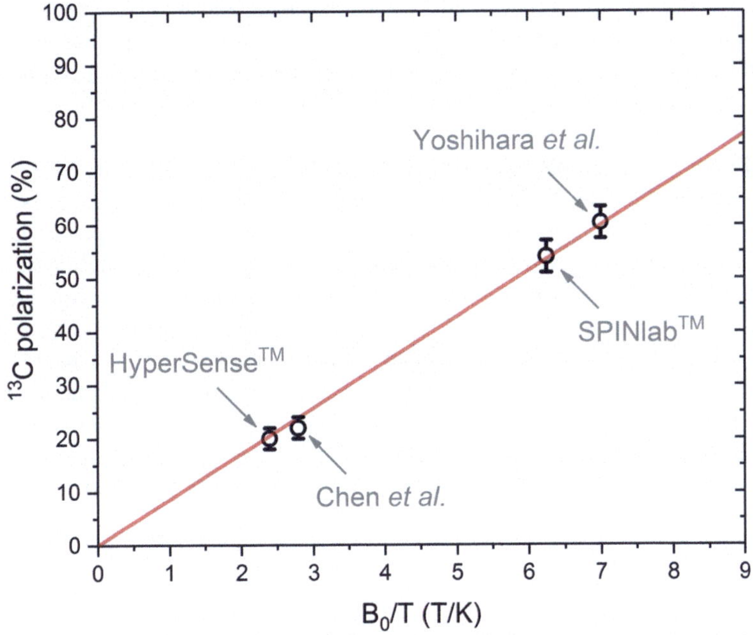

Fig. 2.2 Liquid-state ^{13}C polarization of [1-^{13}C] pyruvate obtained from hyperpolarized [1-^{13}C]pyruvic acid (neat) doped with trityl radical (15–20 mM) as a function of B_0/T. Four data points taken from the literature are shown on the plot: (1) estimated mean ^{13}C polarization obtained with HyperSense™ (3.35 T, 1.4 K); (2) reported ^{13}C polarization obtained using the original hyperpolarizer by Chen et al. (3.35 T, 1.2 K) [54]; (3) estimated mean ^{13}C polarization obtained with SPINlab™ (5 T, 0.8 K); (4) reported ^{13}C polarization obtained by Yoshihara et al. (7 T, 1 K) [15]. The red line is a linear fit obtained using the least square method (the resulting slope is 8.56 with a standard error of 0.12 and R^2 of 0.9992)

polarization builds up, $1/\tau_{\mathrm{Build}}$, and the rate at which it is lost through relaxation processes, which is characterized by $1/T_{1,n}$. So, ideally, we would want a $T_{1,n}$ as long as possible and a short τ_{Build}. In reality, both are tightly correlated since they both depend on the electron spin longitudinal relaxation time $T_{1,\mathrm{e}}$, the hyperfine interaction being the vector of both the nuclear polarization and the nuclear relaxation.

At this point, it is quite clear that one of the key goals in determining the ideal conditions for DNP is to figure out what is the optimal $T_{1,\mathrm{e}}$. Heuristically, a short $T_{1,\mathrm{e}}$ is beneficial because each electron spin can be used more frequently to polarize nuclear spins, but if it is too short, it becomes more difficult to saturate the electron spins, and the probability of effectively polarizing nuclear spins is reduced. Since to date no quantitative theoretical model can predict the optimal $T_{1,\mathrm{e}}$ value, estimates have relied on available set of data published in the literature and extrapolated to the expected temperature and field dependence of $T_{1,\mathrm{e}}$. Obviously, the nature and concentration of the radical used in a specific sample will strongly impact $T_{1,\mathrm{e}}$. However, because this chapter is dedicated to hardware, we will not discuss the properties of the various radicals that can be used for DNP (see Chap.

1 for more details on this topic). The chapter assumes, unless stated otherwise, that the sample to be polarized is [1-^{13}C]pyruvic acid doped with an optimal concentration of trityl radical, which will vary with magnetic field [14, 15]. The reported $T_{1,e}$ value for such sample at 3.35 T and 1.2 K is about 1 s [16]. As mentioned earlier, we will also always assume that the hardware design will provide sufficient microwave power to the sample to maximize the nuclear spin polarization.

Let us examine how $T_{1,e}$ depends on T and B_0. In a standard glassy matrix at low temperature, $1/T_{1,e}$ is proportional to the temperature because the relaxation is dominated by the direct process [17]. The field dependence is more complicated and if we still assume that the direct process is the main relaxation mechanism, we shall expect a dependence in B_0^4 [18]. Although the experiments Lumata et al. performed at 9.5, 95, 240, and 336 GHz, were done around 20 K instead of at 1 K [19], their results seem to confirm a high-power dependence of $T_{1,e}$ on B_0. The authors indeed observed that $T_{1,e}$ is very similar at 9.5 and 95 GHz but drastically increases at 240 and 336 GHz. In conclusion, within the 0.7–1.5 K temperature range where the direct process is expected to drive $1/T_{1,e}$, we can expect a field and temperature dependence comparable to the following:

$$1/T_{1,e} \propto TB_0^4. \tag{2.3}$$

Equation (2.3) means that increasing or decreasing the field cannot directly compensate an increase or decrease in temperature, because although the final nuclear spin polarization might scale with B_0/T (see Fig. 2.2), DNP dynamics scales with TB_0^4. Together, with practical considerations, these dynamics will be key in determining the optimal operating field and temperature of a hyperpolarizer.

2.3 Selecting a Magnet

Since the polarization seems to linearly depend on B_0/T and because within the temperature-field parameter space used in DNP, increasing the field by 20% is more straightforward than decreasing the temperature by 20%, it is tempting to propose a design that operates at higher field to simplify the cryogenic system. However, as mentioned in the previous section, the caveat of working at greater field is that $T_{1,e}$ will become prohibitively short because of its B_0^4 field dependence, dramatically slowing down the DNP process because of the difficulties in saturating the electron spins. So, the design goal should generally start by trying to reach the lowest possible temperature within an acceptable budget and then determine the optimal field at this temperature. The selection of the optimal field is perhaps the most critical and debated point among the researchers in the DNP community.

Obviously, the field strength that is of interest for a hyperpolarizer requires working with a superconducting magnet. The two crucial parameters when considering the integration of a superconducting magnet in a system are (1) the bore size and (2) the magnetic field homogeneity. The bore size is especially important when one considers developing a system capable of polarizing several samples in parallel. Comment et al. have shown that a sample as big as 10 mL could be polarized and subsequently dissolved in a system built around a cryostat placed in the center of a room-temperature (88 mm diameter) bore magnet [20], and the sample space of this system (about 40 mm diameter) could accommodate several samples of 1–2 mL, which would be sufficient to polarize the current human doses. Two systems were developed using ultrawide bore room-temperature (150 mm diameter) superconducting magnets [21, 22], but they were not used to hyperpolarize multiple samples. The inner diameter of the magnet integrated in the commercial clinical polarizer (SPINlab™) that can simultaneously polarize four samples is also around 150 mm, but it is not a room-temperature bore. The homogeneity over the entire sample space is important to make sure that the microwave frequency is optimal for all electron spins contained inside the sample(s), especially for narrow ESR line such as trityl radicals. However, the requirements are far less than for a high-resolution NMR system and the typical homogeneity required is around 100 ppm across the sample space.

Recent developments in superconducting magnet technology allow reaching fields of 10 T or above with a cryogen-free or "dry" magnet of appropriate bore size and homogeneity, at least for a single sample system. This has the advantage of not requiring the input of liquid helium, which is clearly a big asset in a clinical environment and helps conserve the limited world supply of helium. Several prototype DNP systems have already been developed with this technology [23–26]. The challenge with this type of magnet is that, since there is no bath of liquid helium, there is only a limited thermal capacity to dissipate the heat load brought to the system during dissolution and this can lead to a spike in temperature of the magnet and hence a quench. A dissolution apparatus such as the one proposed in the original paper, which requires the insertion of a warm and fairly bulky dissolution insert or "wand" [7], is therefore ill adapted to a polarizer designed around a dry magnet. However, using the concept of fluid path originally developed for sterile intent (see Sect. 2.6), it is possible to perform dissolution experiments without having to introduce a dissolution insert inside the cryogenic environment, hence reducing the heat load [23, 25].

2.4 Cryogenic System

The original hyperpolarizer was based on a so-called "variable temperature insert (VTI)" consisting of an insulated sample space connected to an external bath from which liquid helium is drawn into the sample space (Fig. 2.3a) [7]. This external liquid helium bath was also designed such as to embed the superconducting magnet. A needle valve controls the flow from the external bath. As long as the external liquid helium bath is refilled on a regular basis, the sample space has a constant supply of liquid helium. To lower the temperature to 1.5 K or below, the sample space

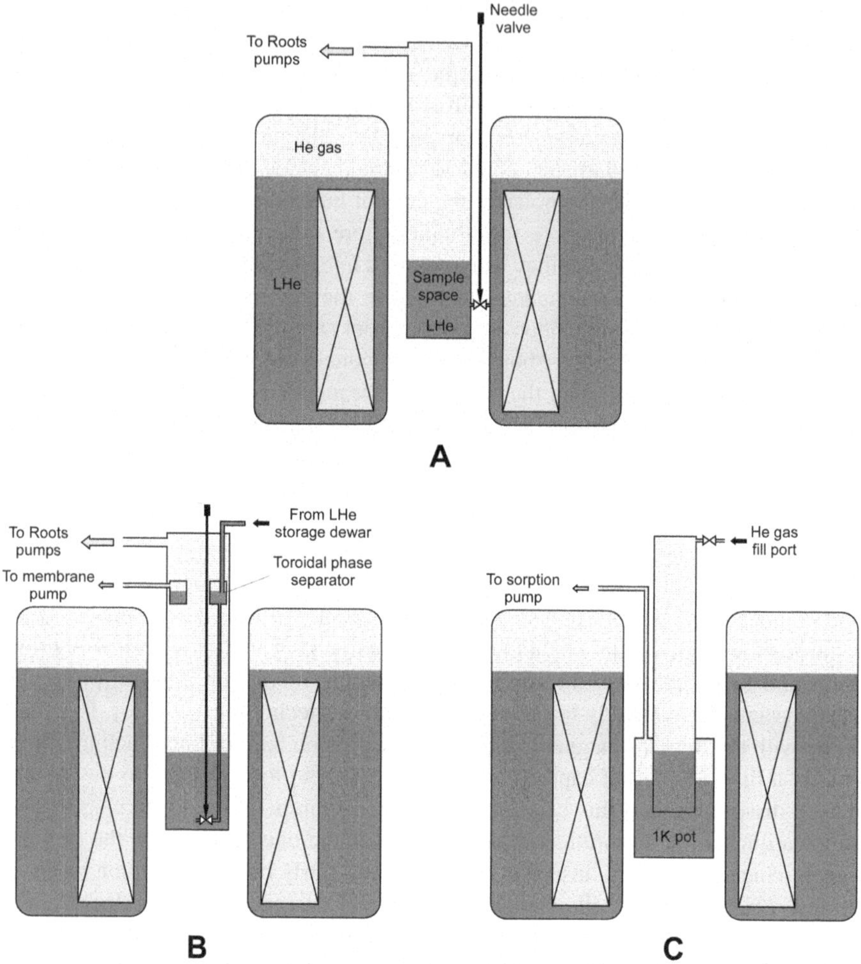

Fig. 2.3 Cryogenic designs for hyperpolarizers. (**a**) The sample space is cooled by pumping with mechanical (Roots) pumps and it is continuously fed with liquid helium (LHe) drawn from the magnet bath through an adjustable needle valve; (**b**) the sample space is also cooled by pumping with Roots pumps but in this design, it is fed with LHe drawn from a phase separator, which is itself filled from an external LHe storage dewar using a low-loss transfer line and a membrane pump; (**c**) the sample space is cooled by conduction through the copper wall between the sample space and the "1 K pot." The 1 K pot is cooled by pumping with a charcoal sorption pump. The helium gas absorbed by the pump can be recondensed inside the 1 K pot by heating the charcoal. The sample space only needs to be filled once (during installation) with He gas since there is no direct pumping on the sample space

is pumped on with a high-flow vacuum pump (Roots pump) while maintaining a near constant inflow of liquid helium by adjusting the needle valve. Practically, the temperature is limited to around 1 K, even when large Roots pumps are used. This system was the basis on which the commercial HyperSense™ system was designed.

A variant in which the VTI is replaced by a stand-alone cryostat that fits into a room-temperature bore magnet and is fed by an external liquid helium storage dewar was later proposed by Comment et al. (Fig. 2.3b) [11]. In this alternative system, the liquid helium inflow is also controlled by a needle valve, but a phase separator is added between the external dewar and the sample space to extract the helium gas resulting from the evaporation during the transfer of liquid helium from the storage dewar to the cryostat. A few similar systems were later implemented by Batel et al., Lumata et al., and Cremilleux et al. [21, 22, 27].

Although these systems perform well and the first in-human hyperpolarized MR scan was done using this type of hyperpolarizer [28], it quickly became obvious that such cryogenic system is not adequate for routine clinical applications. Indeed, the necessity to fill the system with liquid helium on a weekly basis is not adapted to the clinical environment. In addition, because of the worldwide scarcity of helium, there is a general trend to move to cryogenic system requiring little or no external supply of cryogenic liquids. Thanks to the recent developments in cryocooler technology, it has become possible to reach very low temperatures in a quite efficient manner using relatively small compressors. Current Gifford–McMahon and Pulse Tube cryocoolers will typically maintain a temperature between 2.5 and 4.2 K with up to 0.5–2 W cooling power at 4.2 K, so that helium can be liquefied inside the volume thermally anchored to the second-stage of the cryocooler.

A team at General Electric developed a hyperpolarizer with a closed-cycle cryogenic system based on a sorption pump that can bring the temperature of the sample space to 0.75–0.8 K [14], which is essentially as low as any cryogenic system based on ^{4}He will go (Fig. 2.3c). The design requirement was to have enough liquid helium in a "1 K pot" to cool the sample space during a full day of operation. The helium gas that was pumped into the sorption pump is then retransferred and liquefied into the 1 K pot overnight. A particularity of this system is that there is no transfer of fluid (helium) between the 1 K pot and the sample space, which is separately prefilled with ^{4}He gas and cooled by conduction through the copper walls separating the two baths. There is also no need to mechanically adjust a needle valve to cool the sample space. The current clinical hyperpolarizer (SPINlab™) is based on this technology.

Other simpler and cheaper closed-cycle cryogenic systems were also more recently proposed to replace the HyperSense™ design. One is fairly similar to the cryostat of the original hyperpolarizer since it is based on a pumped helium bath inside a VTI fed through a needle valve but, unlike for the HyperSense™, helium gas is recondensed through an integrated closed cycle, and the sample space has its own helium bath that is cooled by conduction inside the VTI [23, 24]. Another approach more similar to the SPINlab™ was developed by Cheng et al. [25]. In this system, there is no VTI or needle valve, but instead of a 1 K pot pumped on with a sorption pump, the sample space is cooled by conduction through a "1 K stage" consisting of a cooper plate anchored to a closed-cycle ^{4}He circulator such as the one described by Uhlig [29].

Another important feature of the SPINlab™ is that it can polarize four samples simultaneously. This is crucial for routine biomedical applications because, in addition to the quite long polarization time characteristic of DNP, loading a sample into a hyperpolarizer is not instantaneous (typically 15–30 min) since it must be cooled from room temperature to 1 K while avoiding evaporating all the liquid helium from the sample space. For this reason, even with schemes aiming at boosting the polarization speed such as cross-polarization [30], the delay between consecutive dissolutions will always be limited by the loading time. In other words, with a single channel system, it is hardly conceivable to perform two consecutive dissolutions in less than 1 h, which is definitely too long in the case where two doses need to be prepared for a single patient or even for an animal under anesthesia. In the author's opinion, preclinical systems should therefore also be designed such as to have multiple samples like the system developed by Batel et al. [27].

2.5 Microwave Setup

One of the most important points to consider when designing the microwave setup for a hyperpolarizer is that the sample size is typically much larger than the wavelength of the microwaves required to perform DNP, especially at larger magnetic fields. This means that, unlike ESR experiments, the sample will not be placed inside a resonant cavity but instead in a "nonresonant" sample space only aimed at confining the microwaves. Rather than exciting the electron spins in a coherent manner, the goal is to create a B_1 field with an optimal power density in a relatively homogenous manner across the entire sample. The key is to deliver a sufficiently large B_1 to the electron spins for DNP to be efficient without warming up the sample space. It has been shown that the required input microwave power is proportional to the volume of the sample space since the power density linearly decreases with increasing volume [20].

To transmit the microwaves from the source located outside the cryogenic environment to the sample space, it is necessary to use a waveguide. Because DNP does not require the use of a specific electromagnetic mode or a controlled phase, it is best to use an oversized waveguide, which limits the losses. The losses indeed become increasingly large when the frequency is increased from W-band (75–110 GHz) toward microwaves with sub-millimeter wavelengths. The microwave setup design including the waveguide becomes more critical when the magnetic field of the polarizer is increased. Circular waveguides are better adapted and easier to manufacture. Gold-platting will also help reduce the losses and has been shown to improve the power density leading to more performant DNP setup [31]. Corrugated waveguides can also be used to improve the transmission efficiency but they are expensive and difficult to adapt to the large temperature gradient between the sample space and the top of the cryostat without adding additional unwanted heat leaks [32].

To enter the subatmospheric pressure environment of the cryogenic system, the microwaves must travel through a leak-tight window. A sealed quartz window is the most common implementation but a thin film of polytetrafluoroethylene (PTFE), which is transparent at these microwave frequencies (50–300 GHz), can simply be clamped between two sections of waveguides. Attempts to optimize the microwave setup using computational methods have been proposed but their usefulness is limited by the fact that the sample position and dielectric properties are not usually well defined. In addition, the coupling of the waveguide to the sample space is ill defined.

Microwave frequency modulation has been shown to help ^{1}H polarization when wide-ESR line width radicals are used in DNP experiments performed at 3.35 T or below [33]. This is generally not the case for X nuclei such as ^{13}C, but it has recently been shown that when the field is increased to 6.7 T, the situation is different. Indeed, fairly broad-ESR line width radicals were shown to better polarize ^{13}C when the microwave field was modulated [34]. Microwave modulation can be fairly easily implemented by applying a sawtooth or sinusoidal voltage modulation to the low-frequency (typically X-band) voltage-control oscillator from which the high-frequency microwaves are generated using frequency doublers. The useful range of frequency modulation is 0.1–10 kHz, and the typical modulation amplitude, which depends on the ESR line width, is 1–100 MHz [34].

2.6 How to Obtain a Hyperpolarized Solution

Hyperpolarized nuclear spins are obtained following low-temperature DNP by transforming the frozen sample into a liquid-state solution through a fast heating process. To maintain the high nuclear spin polarization built up in the solid state, it is necessary to rapidly go through the solid-to-liquid phase transition. This is because the longitudinal nuclear spin relaxation times become very short in the solid state, when the temperature increases above 20 K and especially at low magnetic field [35]. This fast relaxation is exacerbated by the presence of free radicals inside the frozen samples which precludes the extraction of the sample in its solid state from the polarizer [36]. For a sample doped with persistent radicals, it is therefore necessary to implement a rapid melting procedure that takes place within the cryogenic and high-magnetic field environment of the polarizer. For a small sample, this can be done using various techniques, for instance using a laser [37], or heated nitrogen gas [38]. However, if the free radicals are persistent, the nuclear spin $T_{1,n}$ will still be quite short in the liquid-state solution obtained following the heating process because of the high concentration of paramagnetic agents, typically tenth of mM. For this reason, a dissolution step using a substantial amount of superheated water has been shown to be the most adequate implementation to obtain a solution with a diluted concentration of radicals and this hyperpolarization method is commonly called “dissolution DNP” [7]. The heat capacity of water, being large, also allows rapid melting of fairly large samples, up to several mL [14, 20].

The dissolution process is a critical step and it requires the implementation of a hardware design providing an efficient way to drive superheated water onto the frozen sample without contaminating the cryogenic environment. This is done by creating a leak-tight fluid path which can either be connected a few seconds prior to dissolution [7, 11], or assembled as a kit prior to the introduction of the sample inside the polarizer [14]. The latter has the advantage of minimizing the heat load to the cryogenic environment since there is no need for introducing a dissolution insert or "wand" into the polarizer to direct the water onto the sample. This last point might not be important in a "wet" system containing tens of liters of liquid helium because of the large cooling capacity of the helium bath. However, it will be a concern in a "dry" system in which the cryocooler has to compensate the sudden heat load while maintaining the low temperature of the system, and in particular keep the superconducting magnet below its critical temperature.

In the original hyperpolarizer developed by Amersham Health and most of the later implementations, the dissolved liquid is pushed out of the polarizer using a burst of helium gas as chase gas. This allows recovering the entire volume of liquid introduced inside the sample cup and fluid path, which is important when small samples are polarized because a small variation in volume can lead to a large discrepancy in concentration and/or pH. In the clinical polarizer SPINlab™, it was decided to avoid using a chase gas because of potential risks of contaminating the injectable solution through the introduction of nonsterile gas into the fluid path. However, recent developments in preclinical polarizers showed that a chase gas is compatible with the preassembled fluid paths and that the gas can be sterilized by running it through a 0.1 μm filter. Because the use of a chase gas leads to a better recovery and a reduced dead volume, it could allow reducing the dilution factor and therefore possibly allow preparing doses with higher concentration, which would be especially desirable for substrates other than pyruvic acid that cannot be polarized at high concentration. For these reasons, the use of a chase gas for the preparation of clinical doses might be revisited in the near future.

2.7 Transfer Strategies

Once the solid sample has been dissolved, the hyperpolarized nuclear spins immediately start relaxing toward their room-temperature thermal equilibrium polarization. This means that from this point the clock is ticking and that, whether it will be in a tube placed inside an NMR magnet or in vivo in a MR scanner, it is necessary to minimize the delay until the MR measurements of interest take place. This is usually even more critical when the stray field between the hyperpolarizer and the MR system is very low. In most cases, it is indeed advantageous to maintain the solution at a moderately large field, say above 0.1 T, to reduce the nuclear spin relaxation. This can be done by keeping the solution in either a permanent magnet or an electromagnet during transfer [39], or by very rapidly driving the solution into the bore of the MR system. For the latter, one of the great advantages of using a chase gas is that the hyperpolarized molecules can be directly and rapidly transferred into the bore of the MR scanner. We will examine the various strategies that have been proposed so far to transfer the hyperpolarized spins from the polarizer to the MR system. Since the requirements and constraints are quite different for the various types of MR experiments, we will treat them separately.

2.7.1 Fast Transfer into a Tube Placed Inside a Vertical-Bore MR System

To perform a successful high-resolution NMR experiment, it is of the uttermost importance to not only have a homogenous solution across the entire field of view of the coil, especially if a vertical gradient ("z-gradient") is applied during the experiments, but also to maintain a highly homogenous field across the sample. This means that the solution should be well mixed and that no gas bubbles can be present inside the sample. The development of such fast transfer techniques was initiated by Frydman and Blazina for ultrafast two-dimensional NMR measurements of hyperpolarized solutions [40]. But the system that has arguably given the most spectacular results was developed by Bowen and Hilty, with a subsecond delay allowing measuring fast molecular dynamics [41]. The key point of the implemented hardware is that a computer-controlled valve rapidly pressurizes the NMR tube with nitrogen gas once the hyperpolarized solution has been transferred to prevent outgassing of the solution and hence the formation of gas bubbles. Note that Leggett et al. have designed a DNP system based on a dual-isocenter superconductive magnet allowing liquid-state NMR measurements only 700 ms after dissolution but this system was most likely too complex and expensive to be disseminated to a wider community [42].

2.7.2 Fast Transfer for In Vivo Animal Studies

For in vivo studies, the challenge is to decouple the fast transfer between the hyperpolarizer and the MR scanner performed using a chase gas from the rate-limited intravenous injection of the solution that should obviously not contain any substantial amount of gas bubbles. Comment et al. developed a dedicated injection pump containing a liquid-gas phase separator (see Fig. 2.4) [11]. With this apparatus, the delay between dissolution and injection can be as low as 3 s [43]. This device also has the advantage of reducing the length of the injection line to the animal, hence increasing the dose of hyperpolarized solution in the limited-volume bolus injection by reducing the volume of prefilled saline. This is especially important when the MR scanner has a relatively long and narrow bore and the animal is already placed at the isocenter prior to injection for optimizing position and shimming. An experimental setup based on this injection pump allowed the detection of glucose metabolism in a healthy mouse brain following the injection of hyperpolarized ^{13}C-glucose [44]. No such device has yet been developed for clinical experiments because some

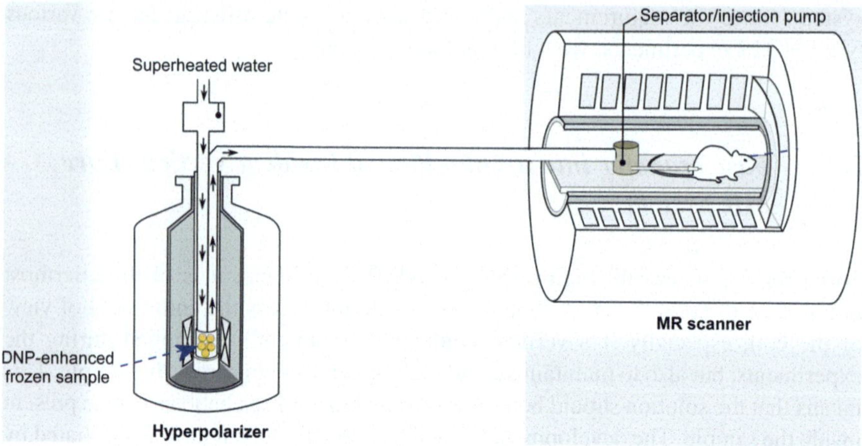

Fig. 2.4 Illustration of a dedicated setup implemented for in vivo hyperpolarized MR experiments. Following DNP in the hyperpolarizer, the frozen sample is quickly dissolved in superheated water and automatically transferred into a separator/injection pump located inside the bore of an MR magnet. MR measurements can be performed while the hyperpolarized solution is inside the separator/injection pump and/or after the solution has been injected into the animal

additional care must be taken prior to injecting the hyperpolarized solution into a human (see Sect. 2.9) and the transfer from the clinical polarizer to the MR room is still done manually. However, it is likely that someday an automated transfer procedure will be developed to reduce the polarization losses.

2.8 Monitoring the Polarization by NMR

Since the nuclear spin polarization is driven out of its thermal equilibrium, both in the solid state during DNP and in the liquid state after dissolution, it is useful to monitor its value at various stages during the hyperpolarization process to assess the adequacy of the system parameters, in particular, the microwave frequency, and the quality of the sample.

2.8.1 Solid-State Polarization Measurement

In most hyperpolarizers, monitoring the nuclear spin polarization build-up during DNP is done either by continuous-wave or pulsed NMR using a simple probe consisting of a copper coil placed inside the sample cup that is connected through a coaxial line to a room-temperature matching and tuning circuit placed outside the cryostat. Such a configuration is usually sensitive enough to measure the polarization growth since the DNP-enhanced signal is large. The intensity of this signal is in arbitrary units but it can be calibrated by correlating it with the liquid-state signal of a reference sample measured after dissolution to provide a rough estimate of the solid-state polarization of similar samples placed inside the hyperpolarizer. However, if one wishes to directly measure the solid-state polarization of a particular sample, it is necessary to measure its thermal equilibrium signal inside the cryostat, at the temperature at which DNP is performed. To do so, a better NMR probe is typically needed. Comment et al. have shown that by using a set of fixed capacitors placed next to the coil and a second remote fine tuning and matching circuit, it is possible to greatly improve the sensitivity and hence measure the solid-state thermal equilibrium signal with a good precision [45]. It is nevertheless still fastidious to record the thermal equilibrium signal at around 1 K because of the long $T_{1,n}$ that is typically of several hours. For such measurement, it is also necessary to be able to stabilize the temperature at around 1 K for many hours. In addition, since the solid-state polarization is based on the comparison of two single-shot measurements, one after DNP and the other at thermal equilibrium, the precision will rarely exceed ±10–20%. It is still an important value that provides a way to understand the DNP efficiency and allows one to estimate the expected subsequent liquid-state polarization.

2.8.2 Liquid-State Polarization Measurement

Liquid-state polarization can be straightforwardly measured after dissolution using an NMR spectrometer by comparing the hyperpolarized signal measured immediately after the sample has been placed inside the NMR magnet with the signal obtained once the nuclear spins have relaxed to room-temperature thermal equilibrium. The latter will often require some signal averaging and since the $T_{1,n}$ of the spins of interest are usually quite long, typically of the order of a minute, the addition of a lanthanide-based contrast agent is advised to reduce the experimental time. Although this measurement might seem trivial, the fact that the hyperpolarized signal is up to 5–6 orders of magnitude larger than the thermal equilibrium, care must be taken with the linearity of all hardware components (preamplifier, in particular) across such a large dynamic range, and, if the flip angle is not identical for both measurements, it must be very well calibrated. The temperature of the sample might also vary during the experiment since it is typically above room-temperature just after dissolution.

When performing in vivo experiments, the polarization cannot be measured once the solution has been injected but an aliquot can be isolated prior to injection in order to measure the thermal equilibrium on a separate dedicated NMR spectrometer, called a "polarimeter" in this context. Cheng et al. have demonstrated that it is also possible to perform this measurement inside the separator/injection pump described in Sect. 2.7.2, in line with the transfer of the solution from the hyperpolarizer to the MR scanner (see Fig. 2.5) [43].

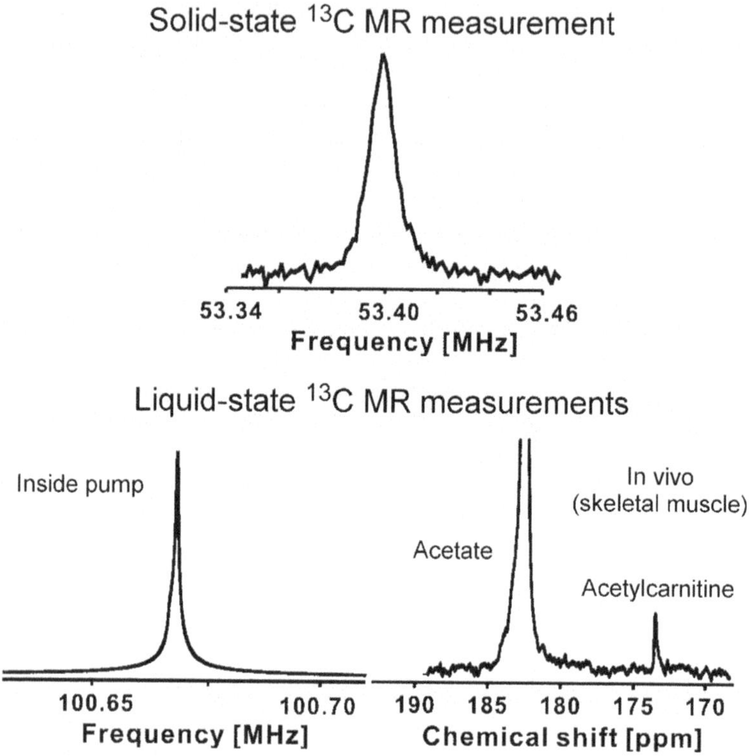

Fig. 2.5 Three consecutive single-shot low-flip-angle ^{13}C MR measurements during a metabolic study with hyperpolarized [1-^{13}C]acetate: solid-state signal inside the hyperpolarizer operating at 5 T and 1 K (top), liquid-state signal inside the injection pump at the time the infusion started (bottom left), and in vivo [1-^{13}C]acetate and [1-^{13}C]acetylcarnitine signals in the rat skeletal muscle (bottom right)

2.9 Radical Filtration and Quality Control

Following the precautionary principle, the trityl radical used to prepare hyperpolarized solutions for clinical use is filtered out prior to injection into humans, despite the fact that this molecule is stable and likely harmless. Indeed, since no complete toxicology study has been completed to date, the medical regulatory bodies require the filtration of these radicals down to a residual concentration of a few μM (typically less than 5 μM). This can be achieved using an inline mechanical filter in conjunction with a specific form of trityl radical designed to precipitate and form clusters upon dissolution in water [14].

For the preparation of clinical doses of hyperpolarized molecules, additional hardware is required to ensure that the solution has the adequate properties and is safe for injection. The main parameters that need to be checked are the following: (1) concentration of the molecule(s) of interest (e.g., ^{13}C-pyruvate in the current clinical studies); (2) residual concentration of radicals; (3) pH of the final solution; (4) temperature; and (5) volume of injectable dose. Because time is critical to avoid losing polarization prior to injection, all concentration measurements are performed using optical methods. Many of the interesting molecules thought to have potential clinical applications exhibit a strong photoabsorption in the ultraviolet–visible (UV–Vis) light spectral range. Upon proper calibration at different wavelengths and for a set of reference solutions of known concentration, a linear prediction of the light absorption can provide an accurate measurement in less than 1 s. If the molecule of interest does not have a UV–Vis absorption that is strong enough to accurately measure its concentration directly, a dye that chemically reacts with the molecule of interest can be added to the solution. To date, ^{13}C-pyruvate is the only hyperpolarized molecule that has been injected into humans and the typical concentration for pyruvate dose is 250 ± 20 mM.

A pH-sensitive dye can also be used to measure the pH of the hyperpolarized solution through optical methods. This is the fastest method to assess pH and it is the one currently implemented in the clinical polarizer. However, because pH is arguably the most critical parameter in the preparation of clinical doses due to the fact that very acidic and basic preparations are used to generate the final solution, additional measurements are sometimes implemented, for instance using a pH strip or a fast electrode. The temperature is also monitored because the doses are prepared with superheated water (130 °C in the current implementation) and it is necessary to ensure that the solution is not above body temperature once it is released. It must be highlighted though that most of the heat passively dissipates along the tubing of the fluid path during the dissolution process, and after the hot acidic solution has been mixed with the room-temperature neutralization medium and transferred inside the injection syringe, the final solution is typically cooler than 30 °C. Finally, the volume of hyperpolarized solution transferred into the injection syringe is important because it helps assess whether or not the dissolution process has correctly taken place and that all substances (in particular acid and base) have been recovered.

Liquid-state polarization can be measured as a complementary quality control check to ensure that the dissolution process did not lead to an important loss of nuclear spin polarization (see Sect. 2.8.2). An integrated 0.3 T polarimeter was provided in the first version of the quality control hardware that is coupled to the SPINlab™ but commercially available benchtop NMR spectrometers operating at 1–1.5 T are best suited for this optional measurement.

2.10 Photogenerated Radicals for Hyperpolarization Technology

A method to photogenerate nonpersistent radicals for DNP was introduced by Eichhorn et al. and it was demonstrated that pyruvic acid itself can be used as radical precursor [46]. The radical produced after photoirradiation of pyruvic acid does not have an ESR line width as narrow as trityl radicals [16] or 1,3-bisdiphenylene-2-phenylallyl (BDPA) [47] but, as mentioned in Sect. 2.6, when the polarizer field is increased to or above 5 T, wider ESR line width radicals start becoming competitive if microwave frequency modulation is applied. Capozzi et al. showed that 30% liquid-state ^{13}C polarization could be obtained with photoirradiated ^{13}C-pyruvic acid using a hyperpolarizer operating at 7 T and 1 K [34]. Preclinical studies demonstrated that photogenerated nonpersistent radicals are suitable for in vivo applications, and although initially developed with pyruvic acid and mixtures containing pyruvic acid [46, 48, 49], the technique has been extended to other keto acids, in particular, molecules that are not directly involved in mammalian metabolism [50, 51]. These latest articles indeed show that if ^{13}C-labeled metabolic substrates such as glucose and dihydroxyacetone are mixed with specific nonlabeled keto acids and that the resulting solution is photoirradiated at liquid nitrogen temperature, high ^{13}C-polarization can be obtained even with a low-field (3.35 T) polarizer.

The most straightforward benefit of using nonpersistent radicals for in vivo hyperpolarized MR scans is that no filtration is required [52]. An added advantage is that the liquid-state nuclear spin relaxation time is extended because paramagnetic relaxation is reduced. However, the most attractive feature of these nonpersistent radicals is that they can be annihilated in the solid state without having to dissolve the frozen sample [36]. As a consequence, the solid-state nuclear spin relaxation time is dramatically extended, and the sample can be extracted from the polarizer as a frozen solution that can be subsequently stored and transported, allowing remote production of hyperpolarized samples. From a hardware perspective, the dissolution equipment should be replaced by a "thermalization" apparatus as well as an extraction device to collect the frozen solid samples as presented in the original paper by Capozzi et al. [36]. This is not completely trivial, and no commercial solution has been proposed to date.

2.11 Summary and Perspectives

Hyperpolarizers were initially designed by integrating a dissolution apparatus into a modified DNP polarizer similar to those used by the polarized target community. Because hyperpolarized MR has found many applications in biology and medicine, the instrumentation has started to be more adapted to the biomedical environment rather than physics laboratories. Although, to date, two commercial hyperpolarizers have been produced; both these systems are expected to undergo some major redesign to better fit the needs of the biomedical community. It is the author's opinion that the ultimate goal would be to design an instrument allowing quasi-continuous production of hyperpolarized ^{13}C-molecules so that repeated or nearly continuous injection would be made possible. This would require the production of small hyperpolarized ^{13}C frozen aliquots at adequate concentration that can be melted on demand next to the MR system. We could, for instance, imagine going from the single-shot 30 s hyperpolarized ^{13}C-glucose experiment presented by Mishkovsky et al. [44] to a ~5 min experiment with hundreds of acquisitions that could be averaged to observe metabolic intermediates that are today invisible, and eventually compare these novel hyperpolarized ^{13}C experiments with the long infusions of thermally polarized ^{13}C-glucose experiments [53]. This type of experiments could potentially be performed in patients with essentially no side effect due to the non-ionizing nature of MR and the use of solutions exclusively containing endogenous molecules. We can certainly expect a lot of exciting new developments in this field over the coming years.

Acknowledgements This work is part of a project that has received funding from the European Union's Horizon 2020 European Research Council (ERC Consolidator Grant) under grant agreement no. 682574 (ASSIMILES). The author would like to thank Dr. Adam Gaunt for his suggestions and for proof reading this chapter.

Problems

1. A ^{13}C MR signal enhancement ε of 55,000 was recorded in a phantom containing a solution of hyperpolarized [1-^{13}C]pyruvate placed inside a 3 T MR scanner. The solution was prepared from neat [1-^{13}C]pyruvic acid doped with trityl radicals and placed inside the bore of the scanner 20 s before the hyperpolarized ^{13}C MR measurement. Polarization, dissolution, and neutralization took place in a 7 T hyperpolarizer located in an adjacent room and it took 40 s to transport the solution through the stray field of the scanner. Knowing that the room-temperature $T_{1,\,13C}$ of [1-^{13}C]pyruvate in solution is 65 s at 3 T and 50 s in the stray field of the scanner, determine at what temperature the pyruvic acid sample was polarized (hint: use the linear fit from Fig. 2.2).
2. During a conference, a scientist presented a model showing that the optimal $T_{1,\,e}$ leading to the highest ^{13}C polarization with trityl radical is 300 ms whatever the magnetic field as long as the temperature is in the range 1–1.5 K. You decide to find out experimentally if this is true using your polarizer operating at 1.5 K by measuring the ^{13}C polarization of [1-^{13}C]pyruvic acid as a function of magnetic field knowing that you can set the field anywhere between 0 and 5 T (you also have access to a large set of microwave sources with adjustable frequency that can match the ESR frequency of trityl radical over this whole range of magnetic field). Considering that the $T_{1,\,e}$ of trityl in [1-^{13}C] pyruvic acid has been measured to be approximately 1 s at 3.35 T and 1.2 K (see Sect. 2.2), do you think that you can test this theoretical model with your polarizer? Estimate the largest ^{13}C polarization you could expect measuring in your polarizer if this model was correct.

References

1. Carver, T.R., Slichter, C.P.: Polarization of nuclear spins in metals. Phys. Rev. **92**(1), 212–213 (1953)
2. Overhauser, A.W.: Polarization of nuclei in metals. Phys. Rev. **92**(2), 411–415 (1953)
3. Carver, T.R., Slichter, C.P.: Experimental verification of the Overhauser nuclear polarization effect. Phys. Rev. **102**(4), 975–980 (1956)
4. Slichter, C.P.: The discovery and renaissance of dynamic nuclear polarization. Rep Prog Phys. **77**, 7 (2014)
5. Abraham, M., Mccausland, M.A.H., Robinson, F.N.H.: Dynamic nuclear polarization. Phys Rev Lett. **2**(11), 449–451 (1959)
6. Jeffries, C.D.: History of the development of polarized targets. In: High energy spin physics, pp. 3–19. Springer, Berlin, Heidelberg (1991)
7. Wolber, J., Ellner, F., Fridlund, B., Gram, A., Johannesson, H., Hansson, G., Hansson, L.H., Lerche, M.H., Mansson, S., Servin, R., Thaning, M., Golman, K., Ardenkjaer-Larsen, J.H.: Generating highly polarized nuclear spins in solution using dynamic nuclear polarization. Nucl Instrum Meth A. **526**(1–2), 173–181 (2004)
8. Krjukov, E.V., O'Neill, J.D., Owers-Bradley, J.R.: Brute force polarization of Xe-129. J. Low Temp. Phys. **140**(5–6), 397–408 (2005)
9. Hirsch, M.L., Kalechofsky, N., Belzer, A., Rosay, M., Kempf, J.G.: Brute-force hyperpolarization for NMR and MRI. J. Am. Chem. Soc. **137**(26), 8428–8434 (2015)
10. Abragam, A., Goldman, M.: Principles of dynamic nuclear-polarization. Rep. Prog. Phys. **41**(3), 395–467 (1978)
11. Comment, A., van den Brandt, B., Uffmann, K., Kurdzesau, F., Jannin, S., Konter, J.A., Hautle, P., Wenckebach, W.T., Gruetter, R., van der Klink, J.J.: Design and performance of a DNP prepolarizer coupled to a rodent MRI scanner. Concept Magn Reson. **31B**(4), 255–269 (2007)
12. Redfield, A.G.: Nuclear magnetic resonance saturation and rotary saturation in solids. Phys. Rev. **98**(6), 1787–1809 (1955)
13. Borghini, M.: Spin-temperature model of nuclear dynamic polarization using free radicals. Phys. Rev. Lett. **20**(9), 419–421 (1968)
14. Ardenkjaer-Larsen, J.H., Leach, A.M., Clarke, N., Urbahn, J., Anderson, D., Skloss, T.W.: Dynamic nuclear polarization polarizer for sterile use intent. NMR Biomed. **24**(8), 927–932 (2011)
15. Yoshihara, H.A., Can, E., Karlsson, M., Lerche, M.H., Schwitter, J., Comment, A.: High-field dissolution dynamic nuclear polarization of [1-(13)C]pyruvic acid. Phys. Chem. Chem. Phys. **18**(18), 12409–12413 (2016)
16. Ardenkjaer-Larsen, J.H., Macholl, S., Johannesson, H.: Dynamic nuclear polarization with trityls at 1.2 K. Appl Magn Reson. **34**(3–4), 509–522 (2008)
17. Comment, A., Merritt, M.E.: Hyperpolarized magnetic resonance as a sensitive detector of metabolic function. Biochemist. **53**(47), 7333–7357 (2014)
18. Wenckebach, W.T., Swanenburg, T.J.B., Poulis, N.J.: Thermodynamics of spin systems in paramagnetic crystals. Phys. Rep. **14**(5), 181–255 (1974)
19. Lumata, L., Kovacs, Z., Sherry, A.D., Malloy, C., Hill, S., van Tol, J., Yu, L., Song, L.K., Merritt, M.E.: Electron spin resonance studies of trityl OX063 at a concentration optimal for DNP. Phys. Chem. Chem. Phys. **15**(24), 9800–9807 (2013)
20. Comment, A., Rentsch, J., Kurdzesau, F., Jannin, S., Uffmann, K., van Heeswijk, R.B., Hautle, P., Konter, J.A., van den Brandt, B., van der Klink, J.J.: Producing over 100 ml of highly concentrated hyperpolarized solution by means of dissolution DNP. J. Magn. Reson. **194**(1), 152–155 (2008)
21. Cremillieux, Y., Goutailler, F., Montcel, B., Grand, D., Vermeulen, G., Wolf, P.E.: A superwide bore DNP system for multiple sample polarization: cryogenic performance and polarization at low temperature. Appl. Magn. Reson. **43**(1–2), 167–180 (2012)

22. Lumata, L., Martin, R., Jindal, A., Kovacs, Z., Conradi, M., Merritt, M.: Development and performance of a 129-GHz dynamic nuclear polarizer in an ultra-wide bore superconducting magnet. MAGMA. **28**, 195–205 (2014)
23. Ardenkjaer-Larsen, J.H., Bowen, S., Petersen, J.R., Rybalko, O., Vinding, M.S., Ullisch, M., Nielsen, N.C.: Cryogen-free dissolution dynamic nuclear polarization polarizer operating at 3.35 T, 6.70 T, and 10.1 T. Magn Reson Med. **81**(3), 2184–2194 (2019)
24. Baudin, M., Vuichoud, B., Bornet, A., Bodenhausen, G., Jannin, S.: A cryogen-consumption-free system for dynamic nuclear polarization at 9.4 T. J. Magn. Reson. **294**, 115–121 (2018)
25. Cheng, T., Gaunt, A.P., Marco-Rius, I., Gehrung, M., Chen, A.P., van der Klink, J.J., Comment, A.: A multisample 7 T dynamic nuclear polarization polarizer for preclinical hyperpolarized MR. NMR Biomed. **33**(5), e4264 (2020)
26. Kiswandhi, A., Niedbalski, P., Parish, C., Wang, Q., Lumata, L.: Assembly and performance of a 6.4T cryogen-free dynamic nuclear polarization system. Magn Reson Chem. **55**(9), 846–852 (2017)
27. Batel, M., Krajewski, M., Weiss, K., With, O., Dapp, A., Hunkeler, A., Gimersky, M., Pruessmann, K.P., Boesiger, P., Meier, B.H., Kozerke, S., Ernst, M.: A multi-sample 94 GHz dissolution dynamic-nuclear-polarization system. J. Magn. Reson. **214**, 166–174 (2012)
28. Nelson, S.J., Kurhanewicz, J., Vigneron, D.B., Larson, P.E., Harzstark, A.L., Ferrone, M., van Criekinge, M., Chang, J.W., Bok, R., Park, I., Reed, G., Carvajal, L., Small, E.J., Munster, P., Weinberg, V.K., Ardenkjaer-Larsen, J.H., Chen, A.P., Hurd, R.E., Odegardstuen, L.I., Robb, F.J., Tropp, J., Murray, J.A.: Metabolic imaging of patients with prostate cancer using hyperpolarized [1-13C]pyruvate. Sci Transl Med. **5**(198), 198ra108 (2013)
29. Uhlig, K.: Dry dilution refrigerator with he-4-1 K-loop. Cryogenics. **66**, 6–12 (2015)
30. Bornet, A., Melzi, R., Linde, A.J.P., Hautle, P., van den Brandt, B., Jannin, S., Bodenhausen, G.: Boosting dissolution dynamic nuclear polarization by cross polarization. J. Phys. Chem. Lett. **4**(1), 111–114 (2013)
31. Cheng, T., Capozzi, A., Takado, Y., Balzan, R., Comment, A.: Over 35% liquid-state 13C polarization obtained via dissolution dynamic nuclear polarization at 7 T and 1 K using ubiquitous nitroxyl radicals. Phys. Chem. Chem. Phys. **15**(48), 20819–20822 (2013)
32. Macor, A., de Rijk, E., Alberti, S., Comment, A., Ansermet, J.P.: Universal polarizer at 140 for DNP-enhanced NMR experiments. Int Conf Infra Milli. (2011)
33. Kisselev, Y.F.: The modulation effect on the dynamic polarization of nuclear spins. Nucl Instrum Meth A. **356**(1), 99–101 (1995)
34. Capozzi, A., Karlsson, M., Petersen, J.R., Lerche, M.H., Ardenkjaer-Larsen, J.H.: Liquid-state 13C polarization of 30% through Photoinduced nonpersistent radicals. J. Phys. Chem. C. **122**(13), 7432–7443 (2018)
35. Hirsch, M.L., Smith, B.A., Mattingly, M., Goloshevsky, A.G., Rosay, M., Kempf, J.G.: Transport and imaging of brute-force (13)C hyperpolarization. J. Magn. Reson. **261**, 87–94 (2015)
36. Capozzi, A., Cheng, T., Boero, G., Roussel, C., Comment, A.: Thermal annihilation of photo-induced radicals following dynamic nuclear polarization to produce transportable frozen hyperpolarized 13C-substrates. Nat. Commun. **8**, 15757 (2017)
37. Joo, C.G., Hu, K.N., Bryant, J.A., Griffin, R.G.: In situ temperature jump high-frequency dynamic nuclear polarization experiments: enhanced sensitivity in liquid-state NMR spectroscopy. J. Am. Chem. Soc. **128**(29), 9428–9432 (2006)
38. Sharma, M., Janssen, G., Leggett, J., Kentgens, A.P.M., van Bentum, P.J.M.: Rapid-melt dynamic nuclear polarization. J. Magn. Reson. **258**, 40–48 (2015)
39. Shang, H., Skloss, T., von Morze, C., Carvajal, L., Van Criekinge, M., Milshteyn, E., Larson, P.E.Z., Hurd, R.E., Vigneron, D.B.: Handheld electromagnet carrier for transfer of hyperpolarized Carbon-13 samples. Magn. Reson. Med. **75**(2), 917–922 (2016)
40. Frydman, L., Blazina, D.: Ultrafast two-dimensional nuclear magnetic resonance spectroscopy of hyperpolarized solutions. Nat. Phys. **3**(6), 415–419 (2007)

41. Bowen, S., Hilty, C.: Rapid sample injection for hyperpolarized NMR spectroscopy. Phys. Chem. Chem. Phys. **12**(22), 5766–5770 (2010)
42. Leggett, J., Hunter, R., Granwehr, J., Panek, R., Perez-Linde, A.J., Horsewill, A.J., McMaster, J., Smith, G., Kockenberger, W.: A dedicated spectrometer for dissolution DNP NMR spectroscopy. Phys. Chem. Chem. Phys. **12**(22), 5883–5892 (2010)
43. Cheng, T., Mishkovsky, M., Bastiaansen, J.A.M., Ouari, O., Hautle, P., Tordo, P., van den Brandt, B., Comment, A.: Automated transfer and injection of hyperpolarized molecules with polarization measurement prior to in vivo NMR. NMR Biomed. **26**(11), 1582–1588 (2013)
44. Mishkovsky, M., Anderson, B., Karlsson, M., Lerche, M.H., Sherry, A.D., Gruetter, R., Kovacs, Z., Comment, A.: Measuring glucose cerebral metabolism in the healthy mouse using hyperpolarized 13C magnetic resonance. Sci. Rep. **7**, 11719 (2017)
45. Comment, A., van den Brandt, B., Uffmann, K., Kurdzesau, F., Jannin, S., Konter, J.A., Hautle, P., Wenckebach, W.T., Gruetter, R., van der Klink, J.J.: Principles of operation of a DNP prepolarizer coupled to a rodent MRI scanner. Appl. Magn. Reson. **34**(3–4), 313–319 (2008)
46. Eichhorn, T.R., Takado, Y., Salameh, N., Capozzi, A., Cheng, T., Hyacinthe, J.N., Mishkovsky, M., Roussel, C., Comment, A.: Hyperpolarization without persistent radicals for in vivo real-time metabolic imaging. Proc. Natl. Acad. Sci. U. S. A. **110**(45), 18064–18069 (2013)
47. Lumata, L., Ratnakar, S.J., Jindal, A., Merritt, M., Comment, A., Malloy, C., Sherry, A.D., Kovacs, Z.: BDPA: an efficient polarizing agent for fast dissolution dynamic nuclear polarization NMR spectroscopy. Chem. **17**(39), 10825–10827 (2011)
48. Bastiaansen, J.A.M., Yoshihara, H.A.I., Capozzi, A., Schwitter, J., Gruetter, R., Merritt, M.E., Comment, A.: Probing cardiac metabolism by hyperpolarized 13C MR using an exclusively endogenous substrate mixture and photo-induced nonpersistent radicals. Magn. Reson. Med. **79**(5), 2451–2459 (2018)
49. Capozzi, A., Hyacinthe, J.N., Cheng, T., Eichhorn, T.R., Boero, G., Roussel, C., van der Klink, J.J., Comment, A.: Photoinduced nonpersistent radicals as polarizing agents for X-nuclei dissolution dynamic nuclear polarization. J. Phys. Chem. C. **119**(39), 22632–22639 (2015)
50. Capozzi, A., Patel, S., Gunnarsson, C.P., Marco-Rius, I., Comment, A., Karlsson, M., Lerche, M.H., Ouari, O., Ardenkjaer-Larsen, J.H.: Efficient hyperpolarization of U-(13) C-glucose using narrow-line UV-generated labile free radicals. Angew. Chem. Int. Ed. Engl. **58**(5), 1334–1339 (2019)
51. Marco-Rius, I., Cheng, T., Gaunt, A.P., Patel, S., Kreis, F., Capozzi, A., Wright, A.J., Brindle, K.M., Ouari, O., Comment, A.: Photogenerated radical in Phenylglyoxylic acid for in vivo hyperpolarized (13)C MR with photosensitive metabolic substrates. J. Am. Chem. Soc. **140**(43), 14455–14463 (2018)
52. Comment, A.: The benefits of not using exogenous substances to prepare substrates for hyperpolarized MRI. Imag Med. **6**(1), 1–3 (2014)
53. Duarte, J.M., Lanz, B., Gruetter, R.: Compartmentalized cerebral metabolism of [1,6-C]glucose determined by in vivo C NMR spectroscopy at 14.1 T. Front Neuroenerget. **3**, 3 (2011)
54. Chen, A.P., Albers, M.J., Cunningham, C.H., Kohler, S.J., Yen, Y.-F., Hurd, R.E., Tropp, J., Bok, R., Pauly, J.M., Nelson, S.J., Kurhanewicz, J., Vigneron, D.B.: Hyperpolarized C-13 spectroscopic imaging of the TRAMP mouse at 3T—initial experience. Magn. Reson. Med. **58**(6), 1099–1106 (2007)

Further Reading

Ardenkjaer-Larsen, J.H., Leach, A.M., Clarke, N., Urbahn, J., Anderson, D., Skloss, T.W.: Dynamic nuclear polarization polarizer for sterile use intent. NMR Biomed. **24**(8), 927–932 (2011)

Ardenkjaer-Larsen, J.H., Bowen, S., Petersen, J.R., Rybalko, O., Vinding, M., Ullisch, M., Nielsen, N.C.: Cryogen-free dissolution dynamic nuclear polarization polarizer operating at 3.35 T, 6.70 T and 10.1 T. Magn Reson Med. **81**(3), 2184–2194 (2019)

Baudin, M., Vuichoud, B., Bornet, A., Bodenhausen, G., Jannin, S.: A cryogen-consumption-free system for dynamic nuclear polarization at 9.4 T. J Magn Reson. **294**, 115–121 (2018)
Cheng, T., Gaunt, A.P., Marco-Rius, I., Gehrung, M., Chen, A.P., van der Klink, J.J., Comment, A.: A multisample 7 T dynamic nuclear polarization polarizer for preclinical hyperpolarized MR. NMR Biomed. **33**(5), e4264 (2020)
Comment, A., van den Brandt, B., Uffmann, K., Kurdzesau, F., Jannin, S., Konter, J.A., Hautle, P., Wenckebach, W.T., Gruetter, R., van der Klink, J.J.: Design and performance of a DNP prepolarizer coupled to a rodent MRI scanner. Concepts Magn Reson. **31B**(4), 255–269 (2007)
Wolber, J., Ellner, F., Fridlund, B., Gram, A., Johannesson, H., Hansson, G., Hansson, L.H., Lerche, M.H., Mansson, S., Servin, R., Thaning, M., Golman, K., Ardenkjaer-Larsen, J.H.: Generating highly polarized nuclear spins in solution using dynamic nuclear polarization. Nucl Instrum Meth A. **526**(1–2), 173–181 (2004)

Chapter 3
Metabolic Imaging with Dynamic Hyperpolarized NMR

Dirk Mayer and Stephen J. DeVience

Abbreviations

CSDA	Chemical shift displacement artifact
CSI	Chemical shift imaging
DNP	Dynamic nuclear polarization
EPI	Echo planar imaging
EPSI	Echo planar spectroscopic imaging
FFT	Fast Fourier transform
FOV	Field of view
FWHM	Full width at half maximum
HP	Hyperpolarized
iNUFFT	Inverse nonuniform fast Fourier transform
LRMC	Low-rank matrix completion
MRI	Magnetic resonance imaging
MRS	Magnetic resonance spectroscopy
NSA	Number of signal averages
NUFFT	Nonuniform fast Fourier transform
PE	Phase encoding
PSF	Point spread function
RF	Radiofrequency
SNR	Signal-to-noise ratio
SW	Spectral width
TR	Repetition time

D. Mayer (✉)
Diagnostic Radiology and Nuclear Medicine, University of Maryland, Baltimore, Baltimore, MD, USA

Fischell Department of Bioengineering, University of Maryland, College Park, MD, USA
e-mail: dmayer@som.umaryland.edu

S. J. DeVience
Department of Biomedical Engineering, Case Western Reserve University, Cleveland, OH, USA

T. Jue, D. Mayer (eds.), *Dynamic Hyperpolarized Nuclear Magnetic Resonance*, Handbook of Modern Biophysics, https://doi.org/10.1007/978-3-030-55043-1_3

3.1 Introduction

The large amount of signal amplification provided by dynamic nuclear polarization (DNP) and the ability of magnetic resonance spectroscopy (MRS) to differentiate the product from the substrate signal based on their chemical shift, that is, their resonance frequency, permits the noninvasive real-time measurement of metabolism in vivo. Given that virtually no background signal is present (^{13}C has a very low 1.1% natural abundance signal), it is possible to collect valuable time-resolved data predominantly localized to a specific organ or tissue type, for example, liver, heart, muscle, or flank tumor, simply by exploiting the sensitivity profile of a surface coil placed in proximity of the organ or region of interest. However, localization techniques are critical when targeting deeper organs or observing differences within organs, in particular when translating this technology into clinical application. But due to the nonrecoverable nature of the hyperpolarized (HP) magnetization, performing a 4D data acquisition (3 spatial dimensions +1 spectral dimension) within a few seconds in order to capture the distribution and metabolic conversion of the injected substrate poses a technical challenge. This chapter will describe the basic concepts of spatial information encoding in magnetic resonance imaging (MRI)/MRS and current acquisition methods that permit the spatially and temporally resolved measurement of the metabolite signals within the limited lifetime of the amplified HP signal. As it is still the most widely used HP compound the examples will focus on [1-^{13}C]pyruvate and its metabolic products. However, the basic principles are applicable to imaging other HP substrates.

3.2 Localization Techniques and *k*-Space

3.2.1 *Slice-Selective Excitation*

The most widely used methods of spatial localization in MRI are based on the application of linear gradient fields ("gradients") that are superimposed on the static B_0 field

$$\vec{G} = \frac{\partial B_0}{\partial x}\vec{e}_x + \frac{\partial B_0}{\partial y}\vec{e}_y + \frac{\partial B_0}{\partial z}\vec{e}_z = G_x\vec{e}_x + G_y\vec{e}_y + G_z\vec{e}_z \tag{3.1}$$

where $\vec{e}_j$ are the unit vectors of the Cartesian coordinate system in the laboratory frame. By convention the direction of the static field is aligned along z. It is important to note that the direction of these additional fields are parallel to B_0 irrespective of the direction of the gradient, that is, $\vec{B} = \left(B_0 + G_x x + G_y y + G_z z\right)\vec{e}_z$. The point in the magnet where all three gradient coils generate zero additional magnetic field defines the isocenter of the gradient system. When applying these gradients, the resonance frequency of a spin becomes spatially dependent according to

$$\omega(\vec{r}) = \gamma\left(B_0 + \vec{G}\cdot\vec{r}\right) = \omega_0 + \gamma\vec{G}\cdot\vec{r} \tag{3.2}$$

where γ is the nucleus-specific gyromagnetic ratio and ω_0 is the *Larmor frequency*. Hence, the spin's location relative to the isocenter determines its resonance frequency. The larger the distance of a spin from isocenter, the larger the shift in resonance frequency. Note that the effect of the gradients is directly proportional to γ. As a consequence, the performance of a gradient system for ^{13}C applications is approximately four times lower compared to the case of ^1H-MRI. Most clinical MRI scanners are equipped with gradient systems on the order of 30–50 mT/m and a slew rate, that is, the rate of changing the gradient amplitude, on the order of 100–200 mT/m/ms. The gradient specifications for dedicated small animal MR scanners are about an order of magnitude higher. Due to the linear relationship between resonance frequency and spatial location, a frequency-selective radiofrequency (RF) pulse affects only spins within a slice with its surface normal parallel to the gradient direction. Position and slice profile are determined by the frequency offset and profile of the applied RF pulse, respectively (Fig. 3.1). Depending on the RF pulse the slice-selection gradient typically leads to a dispersion of the phase of the magnetization across the selected slice. As this results in signal loss, this phase dispersion is reversed by the application of a *slice-refocusing* gradient pulse applied along the slice direction, but with the opposite polarity (see also Fig. 3.5).

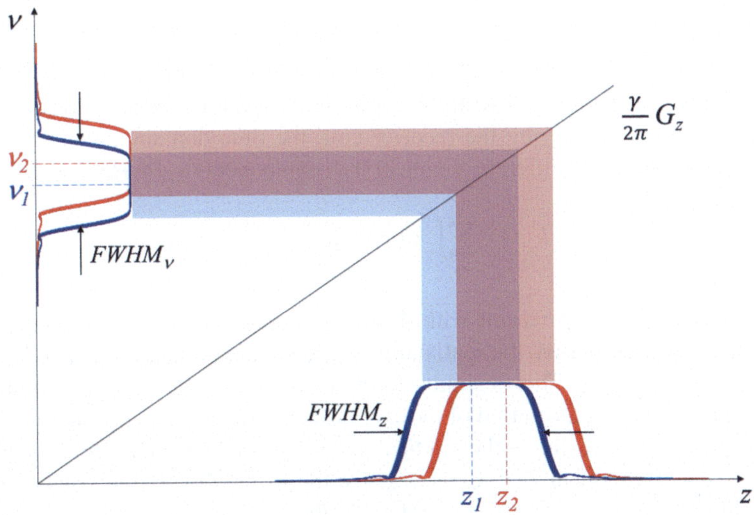

Fig. 3.1 RF slice selection and chemical shift displacement artifact. When using a frequency-selective RF pulse in the presence of a constant gradient G_z the spectral selectivity translates into spatial selectivity. The distance of the selected slice from the gradient isocenter is proportional to the offset frequency of the RF pulse. Consequentially, the slice location is different for spins from different chemical shift species as they possess different frequency offsets

It also follows from Eq. (3.2) that the chemical shift and the gradient-induced frequency offset interfere. This leads to an offset in slice position between resonances with different chemical shift (chemical shift displacement artifact, CSDA) as illustrated in Fig. 3.1. This artifact can be reduced by increasing the strength of the slice-selection gradient. However, for a given slice thickness the pulse width of the RF pulse needs to be increased proportionally, which leads to an increase in required RF power. Therefore, the reduction in CSDA is limited by the capabilities of both the gradient and the RF components of the MR scanner. The CSDA is in particular a problem in imaging of HP [2-^{13}C]pyruvate due to the large chemical shift dispersion in the corresponding in vivo spectra (>160 ppm as compared to 25 ppm for [1-^{13}C]pyruvate and its products).

3.2.2 *Phase Encoding*

Linear field gradients can also encode spatial information after excitation when the magnetization has been transferred into the transverse plane, that is, perpendicular to the main B_0 field. If a *phase encoding* (PE) gradient pulse in the y-direction of strength G and duration τ is applied between excitation and data acquisition, the magnetization will accrue an additional phase that depends on the y location

$$\Delta\phi(y) = \gamma \int_0^\tau G_y \cdot y dt = k_y \cdot y. \tag{3.3}$$

This equation defines the y-component of the *spatial frequency k*, which has the unit of an inverse length (e.g., cycles/cm) and will be further discussed in Sect. 3.2.4. Without relaxation and with the assumption of a homogeneous receive profile, that is, no weighting by the receiver coil, the time domain signal of a one-dimensional object with different chemical shift components ν_m (in Hz) has the form:

$$s(k_y, t) = \sum_m \int_V \rho(y, \nu_m) e^{-ik_y \cdot y - 2\pi i \nu_m \cdot t} dy \tag{3.4}$$

By repeating the experiment with different amplitudes of the phase encoding gradient for each excitation, the spatial information in the y direction is encoded into the phase of the signal. This is similar to the increase in phase in a free induction decay signal due to the evolution of the chemical shift during data acquisition. Note that the phase increase due to the gradient encoding is independent from the phase evolution due to chemical shift. Consequently, the spatial distribution can be reconstructed by an inverse Fourier transform along the k_y dimension.

3.2.3 *Frequency Encoding*

An alternative way to encode the spatial information is the *frequency encoding* technique, where the data are acquired while a gradient pulse is applied. Assuming a constant gradient along x with amplitude G_x during the readout, and again without considering relaxation, the acquired time domain signal is

$$s(t) = \sum_m \int_V \rho(x, \nu_m) e^{-i\gamma G_x \cdot x \cdot t - 2\pi i \nu_m \cdot t} dx \tag{3.5}$$

As in the case of the slice-selective excitation, the applied gradient leads to a spatially dependent resonance frequency. If only one chemical shift species is present its spatial distribution can be reconstructed directly via Fourier transform. Because the chemical shift also evolves during the data acquisition, frequency, and spatial information get mixed and analogous to the CSDA during slice-selective excitation, the spatial distribution for the individual chemical shift components are shifted proportionally to their frequency offset. However, if the frequency offset is known as it is the case of a chemical shift experiment, this artifact can be corrected in post-processing.

3.2.4 K-*Space Formalism and Fourier Reconstruction*

Extending the definition of k to all three spatial dimensions

$$\vec{k}(t) = \gamma \int_0^t \vec{G}(t')dt' \tag{3.6}$$

and neglecting relaxation, the time domain signal of an object with transverse magnetization $\rho(\vec{r})$ and a single chemical shift component on-resonance with the scanner's center frequency, that is, $\nu = 0$, is

$$s(\vec{k}(t)) = \int_V \rho(\vec{r}) e^{-i\vec{k}(t)\cdot\vec{r}} d\vec{r}. \tag{3.7}$$

In the k-space interpretation of MRI, the data acquisition corresponds to sampling in the space of spatial frequencies associated with the object [1, 2]. Accordingly, $\rho(\vec{r})$ can be reconstructed by the inverse Fourier transform of the sampled k-space data. The k-space concept is helpful in analyzing and designing advanced imaging techniques, especially as they apply time-varying gradients such as echo planar imaging (EPI) and spiral imaging. The k-space trajectory (i.e., the location in k-space of the acquired data points) is given by the time integral of the applied gradient waveforms (see Eq. (3.6)). Figure 3.2 illustrates the concept for the example of an EPI acquisition scheme assuming ideal gradients that can instantly be switched on or off.

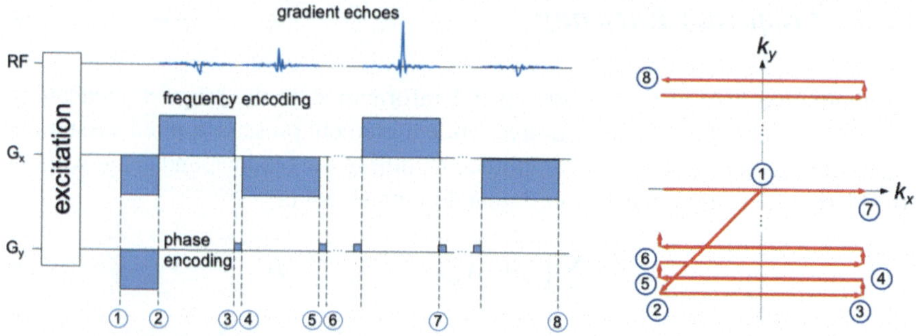

Fig. 3.2 Gradient waveforms and corresponding k-space trajectory for an EPI readout scheme. After preparation of transverse magnetization, which could be as simple as slice-selective excitation pulse, the initial negative gradient pulses ("prefocusing gradients") along x and y move the k-space trajectory from the k-space center to the corner of the k-space area to be sampled. Frequency-encoding gradient pulses on the x-axis with alternating polarity traverse the k-space trajectory along k_x in a back and forth pattern while the interspersed small "blip" gradient pulses on the y-axis act as phase encoding gradients that move the k-space trajectory further along k_y. Data acquired during the duration of the frequency-encoding gradients will show a maximum each time the cumulative area of the x-gradient equals 0, that is, $k_x = 0$, defining the so-called *gradient echo*

Analogous to how the temporal sampling rate and duration in a spectroscopy experiment determine the spectral width (SW) and nominal spectral resolution, respectively, the sampling interval in k-space and the extent of sampled k-space determine the spatial field of view (FOV) (i.e., the distance in image space that can be encoded) and the nominal spatial resolution. If k-space is sampled with N points and an increment Δk_x equidistantly along k_x, the FOV and resolution along x-dimension of the image are FOV = $2\pi/\Delta k_x$ and $\Delta x = 2\pi/(N \cdot \Delta k_x)$, respectively. If k-space sampling violates the Nyquist criterion and the increment Δk is too large (i.e., the corresponding FOV is smaller than the object to be imaged) *aliasing* will occur. Signal from outside the FOV will fold into the reconstructed image (Fig. 3.3). The effect of the k-space sampling pattern on the image can also be illustrated by the so-called *point spread function* (PSF), which is calculated as the Fourier transform of the sampling support in k-space (Fig. 3.4). Aliasing does not occur in the frequency-encoding direction if the bandwidth of the acquisition filter matches the readout bandwidth of the acquisition as it is usually the case in MR imaging ("anti-aliasing filter").

For the description of chemical shift imaging pulse sequences, it is useful to combine the 3D spatial frequency k-space and the time dimension to a 4D space in which the time coordinate t is interpreted as k_ω (i.e., the conjugated Fourier domain of the frequency or chemical shift dimension).

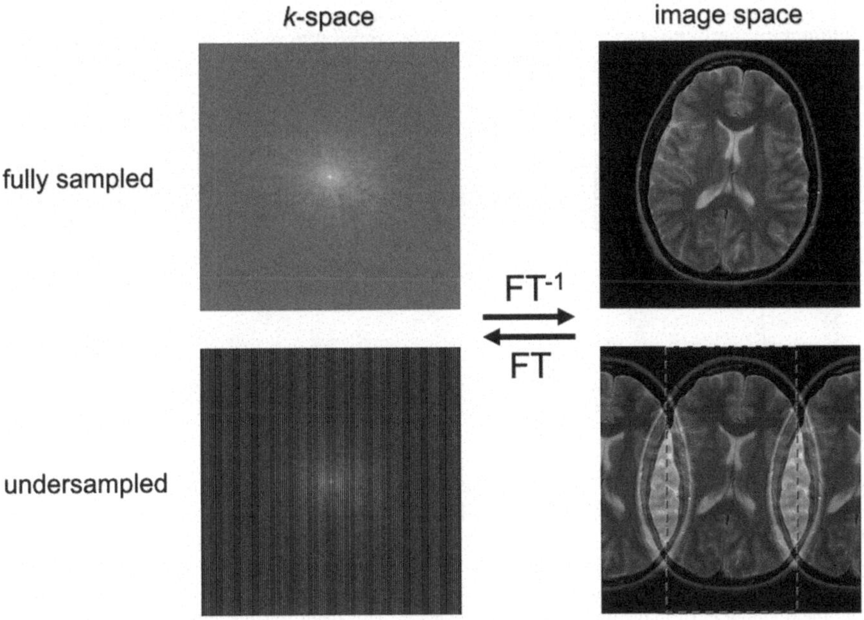

Fig. 3.3 Fourier relationship between *k*-space and image space in MRI. Utilizing linear field gradients, MRI data are acquired in the space of spatial frequencies, that is, *k*-space, that is related to the image space through Fourier transform. The extent of the object in image space determines how finely the data need to be sampled in *k*-space. For the case of equidistant sampling the image FOV is inversely proportional to the *k*-space sampling interval. If every other line in *k*-space is omitted the FOV in the corresponding dimension is cut in half (indicated by the dashed red lines). Signal from outside the reduced FOV will then "wrap around" and appear in the wrong location ("aliasing")

3.3 Chemical Shift Imaging (CSI)

3.3.1 Phase-Encoded CSI

The most straightforward two-dimensional CSI pulse sequence consists of a slice-selective excitation RF pulse followed by phase-encoding gradients applied along the remaining two orthogonal directions [3, 4]. The data are acquired without the application of a gradient, hence, only the chemical shift information is encoded. In the *k*-space interpretation of the experiment a single line in *k*-space along k_ω is sampled with each excitation. The *k*-space in the two spatial frequency dimensions is sampled by changing the amplitude of the phase-encoding gradients (Fig. 3.5). Applying a 3D Fourier transform results in a 2D map of spectra where the individual resonances can be quantified to yield the spatial distribution of the corresponding metabolites ("metabolic maps"). An extension to 3D CSI is possible simply by adding another phase-encoding gradient in the third dimension and either increasing the slice thickness ("slab selection") or replacing the slice-selective excitation pulse

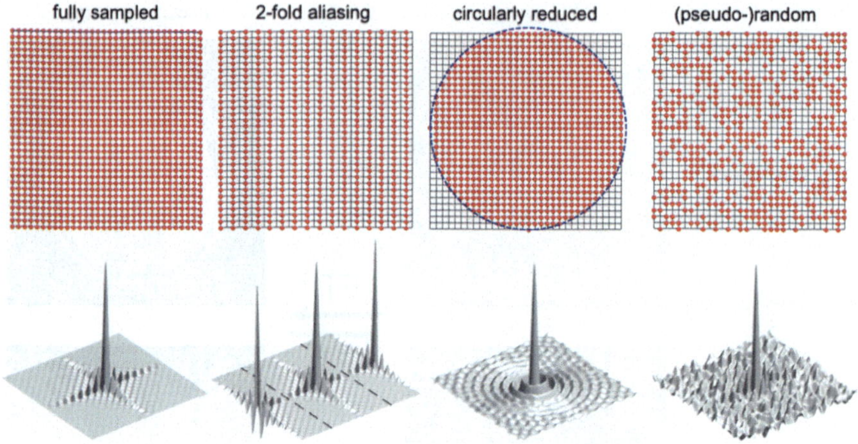

Fig. 3.4 Sampling patterns in k-space and the corresponding point spread functions for a nominal 32×32 Cartesian sampling scheme. (**a**) In the fully sampled case the limited extent in k-space leads to the broadening of the central peak of the PSF as well as sidelobes of lesser intensity primarily along the two principal sampling dimensions ("Gibbs ringing"). (**b**) When omitting every other line in k-space the point spread function consists of two peaks of equal intensity separated by half the nominal FOV in the undersampled dimension. This leads to signal aliasing if the objects is larger than the reduced FOV. (**c**) The data acquisition can be sped up by only sampling at k-space points within a circular region of k-space. As a consequence, the Gibbs ringing of the corresponding PSF is more radially symmetrical. At the same time, the main lobe has a slightly larger FWHM while the amplitude of the sidelobes is slightly reduced compared to the PSF of the fully sampled acquisition. (**d**) When the data are (pseudo-)randomly undersampled the PSF still has a single main peak and the artifact signal appears as incoherent noise. Under certain conditions an image without artifacts can be restored using a nonlinear reconstruction scheme that is presented in Sect. 3.6

with a nonselective pulse. However, as the total acquisition time is given by the product of the repetition time (TR) and the total number of phase encoding steps, 3D phase-encoded CSI is usually too time consuming. A simple modification to reduce the measurement time is to restrict k-space sampling to a circular (or spherical in the case of 3D-CSI) region around the k-space center [5]. Omitting the corners of k-space has only a minor effect on the spatial resolution (see also Fig. 3.4). Another strategy to reduce the number of PE steps without sacrificing FOV or resolution is to (pseudo-)randomly undersample k-space and use a nonlinear reconstruction scheme that exploits data sparsity [6] (see also Sect. 3.6). While CSI entirely based on phase encoding is the slowest acquisition scheme, this pulse sequence has the advantage that it is the least demanding on the gradient system. For example, it is possible even on a clinical MRI scanner to perform ^{13}C-CSI with a nominal in-plane resolution of 2.5 mm for metabolic imaging of the rat brain. Low sensitivity to gradient imperfections and ease of reconstruction are additional advantages of this method.

A pulse sequence for phase-encoded CSI is usually offered by the scanner manufacturer. However, the imaging parameters need to be modified for the special conditions in imaging HP magnetization. The large dispersion of the chemical shifts in

^{13}C-MRS (compared to ^{1}H-MRS) permits short data acquisition windows in the order of 50 ms with no or only minimal spectral overlap. With total scan times on the order of 20 s for common imaging matrix sizes of 16 × 16, the application of this technique is generally limited to single time-point measurements with the start of acquisition chosen near the maximum in the time curve of the metabolite of interest. An important aspect of HP MRI is the fact that T_1 relaxation does not replenish the magnetization between excitations, but instead leads to a decay toward the thermal-equilibrium level, which is usually below the detection limit. Therefore, low excitation flip angles (5°–10°) or variable flip-angle schemes that progressively increase the flip angle [7, 8] are typically used to conserve the HP magnetization for the necessary PE steps. For conventional CSI applications, that is, applications involving magnetization levels at thermal equilibrium, this acquisition scheme is considered the gold standard with respect to signal-to-noise ratio (SNR). However, considering the decay of longitudinal magnetization during the total acquisition time, faster CSI techniques as discussed in the next section can actually exceed the achievable SNR [9]. An example of 2D-CSI from a slice through the kidneys of a healthy rat is shown in Fig. 3.5c.

3.3.2 *Echo Planar Spectroscopic Imaging (EPSI)*

Whereas only a single line in *k*-space (along k_{ω}) is acquired with each excitation in phase-encoded CSI the methods described in the following permit the simultaneous sampling of *k*-space in two or more dimensions. This is achieved by applying time-varying gradients during data acquisition that can reduce the total acquisition time by more than an order of magnitude.

In EPSI the data are acquired while a periodically inverted frequency encoding gradient is applied. As the chemical shift evolves during the duration of the time-varying gradient waveform, a plane of *k*-space in k_x and k_{ω} is sampled with each TR [10]. The corresponding *k*-space trajectory is shown in Fig. 3.6. The remaining dimension in 2D-EPSI is then sampled by repeated excitations and phase encoding. If a complete plane in k_x-k_{ω} can be sampled with a single excitation and using conventional phase encoding for the second spatial dimension, the total acquisition time for a 2D-CSI experiment can be reduced to TR times number of PE steps. The reduction in scan time then permits time-resolved acquisitions to measure metabolite kinetics or extensions to volumetric imaging. The spectral width of an EPSI sequence is given by the inverse of the time between consecutive sampling of $k_x = 0$, that is, the inverse of the duration of the frequency-encoding gradient. Therefore, the increase in imaging speed is at the expense of the SW as the maximum achievable SW is limited by the speed and strength of the gradient system. In fact, Mansfield already proposed the EPSI approach in the mid1980s [11], but could not implement it because the gradient capabilities of the clinical MR scanners at that time were inadequate. Additionally, eddy-current effects or timing errors with respect to synchronization between data acquisition and gradient waveform can

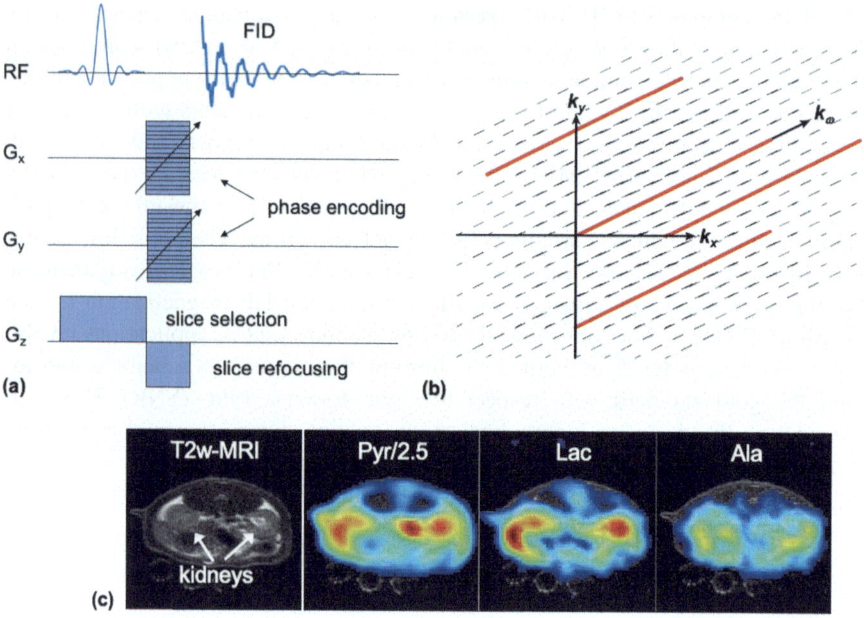

Fig. 3.5 (**a**) Pulse sequence diagram for 2D phase-encoded CSI and (**b**) the corresponding *k*-space sampling scheme. After each slice-selective excitation spatial information is encoded via phase-encoding gradients and a single line of *k*-space along k_ω is acquired. (**c**) T2-weighted ^{1}H-MRI from a slice through the kidneys of a rat as well as the corresponding metabolic maps of pyruvate, lactate, and alanine superimposed onto the MRI. The ^{13}C-CSI data were acquired for a FOV of 72 mm and an 18 × 18 matrix with a progressively increasing excitation flip scheme ending in 90°. With a TR of 75 ms and circularly reduced *k*-space sampling the total acquisition time was 24 s

lead to alternating modulations in k_ω for data acquired during positive and negative encoding gradients causing aliasing artifacts. Therefore, the data usually are reconstructed separately and reduce the achievable SW by a factor 2.

A variant that is also less sensitive to eddy currents or timing issues while permitting a larger SW is the *asymmetric* or *fly-back* EPSI approach: the data are only acquired during the readout gradient with one polarity while a refocusing gradient of opposite polarity, usually operating at the maximum gradient specifications, is used to quickly reverse *k*-space [12, 13]. Data acquisition in EPSI is often limited to the duration of the gradient plateau for easier reconstruction as *k*-space is sampled equidistantly, hence, fast Fourier transform (FFT) can be used. However, this approach leads to reduced SNR efficiency as the SNR is proportional to the square root of total acquisition time [14]. Another strategy of increasing the SW is performing multiple excitations (for each PE step) and increasing the delay between excitation and start of readout gradient/data acquisition by an integer fraction of the frequency encoding gradient ("interleaved EPSI"). As the chemical shift continues to evolve during data acquisition in the presence of a readout gradient, the spatial and spectral information are mixed. This leads to a displacement in readout direction

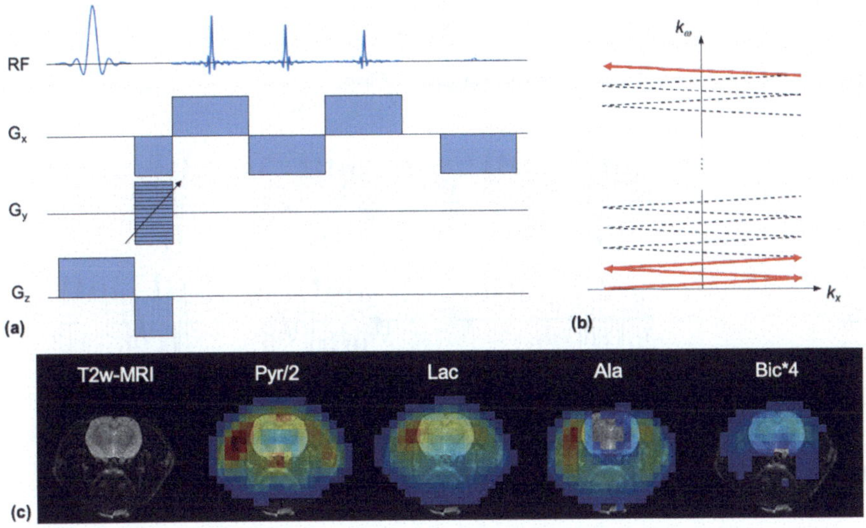

Fig. 3.6 (**a**) Pulse sequence diagram for 2D EPSI and (**b**) the corresponding *k*-space sampling scheme. After each slice-selective excitation data are acquired in a k_x-k_ω-plane of *k*-space for combined spatial-spectral encoding. Often data acquired during gradients of opposite polarity are processed separately, in which case the corresponding SW is the inverse of the duration for two consecutive readout gradients. In the asymmetric or fly-back variant of EPSI data are only acquired during the readout gradient with one polarity while a refocusing gradient of opposite polarity, usually operating at the maximum gradient specifications, is used to quickly reverse *k*-space. This scheme permits larger SWs at the expense of lower SNR efficiency as data are only acquired during a fraction of time. (**c**) T2-weighted ^{1}H-MRI from a slice through the brain of a rat as well as the time-averaged metabolic maps of pyruvate, lactate, alanine, and bicarbonate superimposed onto the MRI. The ^{13}C-CSI data were acquired with dynamic fly-back EPSI that allowed a SW of 833 Hz for a FOV of 60 mm and a 16 × 16 matrix (only the central 40 mm are displayed). A constant flip angle of 5° was used for the slice-selective excitation pulse. The metabolic maps were calculated as the average of 32 time points sampled every 2.6 s

proportional to the amount of off-resonance of the respective metabolite peak. However, this chemical shift artifact can easily be removed by applying a frequency-dependent linear phase correction prior to the FT in readout direction.

3.3.3 Spiral CSI

Spiral CSI extends the concept of combined spatial-spectral encoding to sampling of a 3D volume in k_x-k_y-k_ω space by applying a series of time-varying gradients simultaneously on two axes during the data acquisition [15] (Fig. 3.7). This potentially allows the complete acquisition of a given slice after a single excitation. However, for most applications, multiple excitations are necessary to achieve the desired SW due to limitations in gradient performance. Similar to EPSI, the SW can

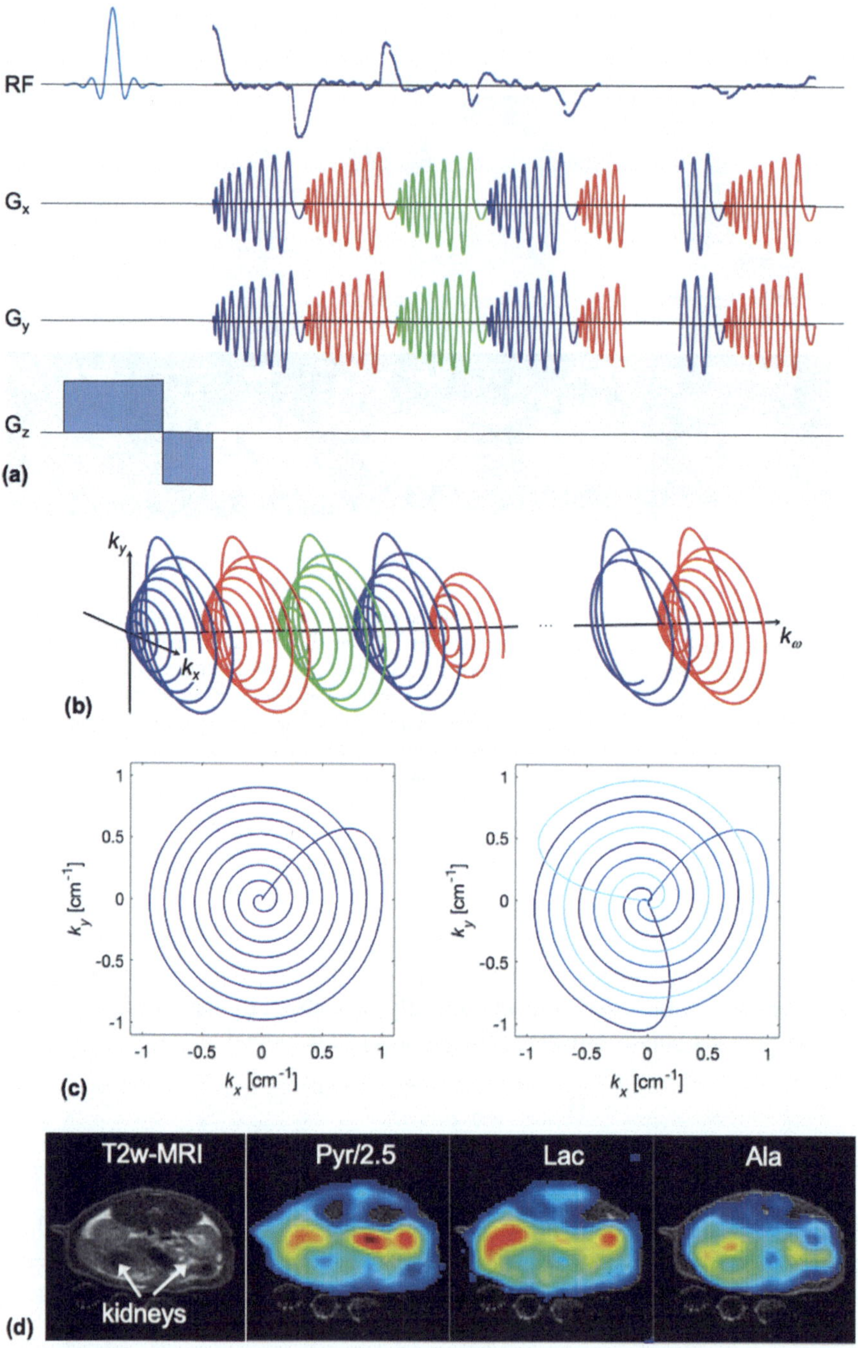

Fig. 3.7 (**a**) Pulse sequence diagram for 2D spiral CSI and (**b**) the corresponding *k*-space sampling scheme. After each slice-selective excitation data are acquired in a 3D volume of *k*-space in k_x, k_y,

be increased by performing additional acquisitions in which the start of the acquisition is shifted with respect to the excitation pulse ("spectral interleaving"). Alternatively, and more commonly, the spiral *k*-space trajectory can be split up into several subsections that can be covered using shorter gradient waveforms, hence, increasing the spectral resolution ("spatial interleaving", Fig. 3.7c).

A simple way to reduce the number of interleaves is by exploiting the sparsity of the HP ^{13}C spectrum and using a smaller SW chosen in a way that avoids spectral overlap. For example, for a FOV of 24 cm and a 1-cm nominal resolution it would require 16 spatial interleaves using a standard clinical gradient system to achieve a SW large enough to cover the spectral range from [1-^{13}C]pyruvate and ^{13}C-bicarbonate. However, only three interleaves are necessary for a 280-Hz SW that allows detecting the peaks from pyruvate, lactate, alanine, and bicarbonate without spectral overlap at 3 T [9]. As the data points along the spiral *k*-space trajectory are not uniformly sampled, a straightforward application of the FFT is not possible and the reconstruction of the image data is slightly more complicated.

Although an exact solution to the reconstruction problem is possible using the nonuniform discrete FT it is often impractical due to long computational times. Therefore, the most common construction strategies use approximate algorithms such as first interpolating the acquired *k*-space data onto a Cartesian grid followed by FFT ("gridding") [16] or by applying so-called *nonuniform FFT* (NUFFT) algorithms (see, e.g., [17]) that do not require uniform sampling. While spiral CSI is a very efficient way to sample *k*-space it also places higher demands on the gradient hardware. As the reconstruction requires knowledge of the *k*-space trajectory the designed trajectory can be used in case of sufficient gradient fidelity, that is, the applied gradient waveforms correspond to a *k*-space trajectory with negligible differences to the original design. In case of larger deviations, the actual *k*-space trajectory obtained from independent trajectory measurements should be used in the reconstruction to avoid artifacts, such as image blurring or rotation [18].

Compared to EPSI, spiral CSI is also more sensitive to the synchronization of the acquisition window and the spiral gradient waveforms. A misalignment leads only to a linear phase along the frequency direction in EPSI whereas it produces a rotation of the data and consequently the metabolic maps in spiral CSI. Another disadvantage

Fig. 3.7 (continued) and k_ω. With a sufficiently fast gradient system, this approach permits single-shot 2D-CSI. Otherwise the *k*-space trajectory can be split into several, shorter interleaves that allow for higher SW at the cost of multiple excitations to acquire the full *k*-space data. (**c**) The *k*-space trajectories for an acquisition with a single interleaf (left) and with three spatial interleaves (right) for a FOV of 80 mm and a 16 × 16 matrix that can be implemented on a clinical gradient system (40 mT/m, 150 mT/m/ms). Whereas the single-interleaf acquisition has a SW of 109 Hz the acquisition scheme with three spatial interleaves permits a SW of 275 Hz that is sufficient for imaging of HP [1-^{13}C]pyruvate and its main metabolic products at 3 T. (**d**) T2-weighted ^{1}H-MRI from a slice through the kidneys of a rat (same as in Fig. 3.5c) as well as the metabolic maps of pyruvate, lactate, and alanine acquired with single-shot spiral CSI on a clinical MRI scanner equipped with a high-performance gradient insert. The single-shot spiral CSI data were acquired in 125 ms for the same spatial parameters as the phase-encoded CSI in Fig. 3.5c. Using a 90° excitation, there is no reduction in SNR despite the almost 200-fold reduction in acquisition time

of spiral CSI is the fact that, unlike in EPSI, it is not possible to use the anti-aliasing filter for the readout bandwidth to eliminate potential signal contributions from outside the FOV as the readout direction is effectively rotated during the data acquisition. Therefore, the prescribed FOV must be large enough to cover the whole object.

3.4 Metabolic Imaging Using Model-Based Image Reconstruction

The metabolic imaging techniques described in the previous sections were all based on acquiring the spatially and spectrally encoded *k*-space data, applying a Fourier transform to yield a map of spectra that are then quantified to produce the metabolic maps. An alternative strategy to determine the spatial distribution of the metabolites is using a model-based image reconstruction formalism [19]. By explicitly incorporating a priori knowledge of the imaging object, in particular the metabolites' resonance frequencies, the number of data points and, hence, the necessary acquisition time can be reduced [20–23]. The basic idea is to formulate the reconstruction problem in matrix form and directly solve for the metabolite distribution. Extending Eq. (3.4) the *k*-space signal from *L* metabolites can be written as:

$$s(t)=\sum_{m=1}^{L}\int_{V}\rho_m(\vec{r})\,P_m(\vec{r},t)\,\mathrm{e}^{-i\vec{k}(t)\cdot\vec{r}-2\pi i\left(\nu_m+\Phi(\vec{r})\right)\cdot t}\,d\vec{r} \tag{3.8}$$

with ρ_m as the spatial distribution of the metabolite with frequency ν_m. The function $P_m(\vec{r},t)$ represents signal modulation due to transverse relaxation and J-coupling while $\Phi(\vec{r})$ denotes the so-called *field map* that takes into account local deviations of the field strength from its nominal value. After discretizing *k*-space sampling (*K* sampling points) and spatial location in the image domain ($M \times N$ voxels), the signal equation can be expressed in matrix–vector notation as

$$\boldsymbol{y}=\boldsymbol{E}\rho+\varepsilon \tag{3.9}$$

Here the *K*-by-($L \times M \times N$) *encoding matrix* $\boldsymbol{E}$ maps the vector $\boldsymbol{\rho}$ of the spatial distribution of the individual metabolites (length $L \times M \times N$) to the vector $\boldsymbol{y}$ of *K* acquired data points. The vector $\boldsymbol{\varepsilon}$ represents the measurement error. The reconstruction problem now corresponds to an optimization process of finding $\bar{\rho}$ so that $\boldsymbol{E}\bar{\rho}$ best matches the measured data $\boldsymbol{y}$. Assuming an uncorrelated noise distribution with uniform variance for the measurement error, a reasonable approach is to estimate $\bar{\rho}$ by a least-squares solution using the Euclidian norm $\|\bullet\|_2$:

$$\bar{\rho}=\underset{\rho}{\operatorname{argmin}}\,\|\,\boldsymbol{E}\rho-\boldsymbol{y}\,\|_2 \tag{3.10}$$

If B_0 inhomogeneities and relaxation are ignored in the model or determined independently this is a linear problem and the solution can be found using the Penrose–Moore pseudoinverse

$$\bar{\rho} = \left(\boldsymbol{E}^H \boldsymbol{E}\right)^{-1} \boldsymbol{E}^H \boldsymbol{y} \tag{3.11}$$

If the matrix $\boldsymbol{E}^H\boldsymbol{E}$ is too large for its inverse to be computed directly, a solution can be found using iterative approaches, for example, the conjugate gradient method [23].

An acquisition scheme for 2D metabolic imaging that utilizes the model-based reconstruction is illustrated in Fig. 3.8. It consists of multiple slice-selective excitations followed by a spiral readout gradient acquisition to encode the spatial information. With each excitation the start of the acquisition module with respect to the excitation is shifted by ΔTE to encode the spectral information. The noise performance of the estimation process depends on how well-conditioned the matrix $\boldsymbol{E}$ is. The number of excitations has to be equal to or exceed the number of metabolite

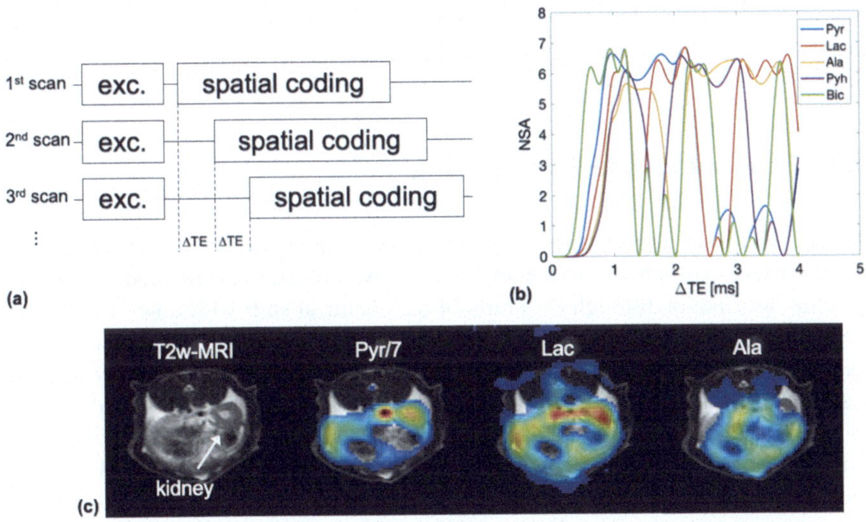

Fig. 3.8 (**a**) Acquisition scheme for 2D metabolic imaging that utilizes the model-based reconstruction. The spatial encoding module could be an EPI or spiral readout. The spectral information is encoded by incrementing the delay from excitation to the data acquisition module. (**b**) NSA calculated as a function ΔTE for [1-^{13}C]pyruvate and its metabolic products at 3 T for an acquisition with seven excitations. The choice of ΔTE can be optimized by maximizing the NSA for the individual metabolites. (**c**) T2-weighted ^{1}H-MRI from a slice through the kidneys of a rat as well as the metabolic maps of pyruvate, lactate, and alanine. The ^{13}C data were acquired with a pulse sequence scheme as shown in (**a**) using a single-shot spiral readout (80 mm FOV, 32 × 32 matrix) for the spatial encoding module. A ΔTE of 1.12 ms was used based on the NSA analysis shown in (**b**). The seven excitations for encoding the chemical information were preceded by an additional excitation without spatial encoding to provide prior knowledge of about the absolute resonance frequencies that it used during the model-based reconstruction [22]

peaks, and the noise amplification for the individual metabolites also depends on the value of ΔTE. Besides calculating the condition number for $\boldsymbol{E}$, another useful metric is the *number of signal averages* (NSA) [24], which is calculated for metabolite m as

$$\mathrm{NSA}_m = \frac{1}{\left(\left(\boldsymbol{E}^H \boldsymbol{E}\right)^{-1}\right)_{m,m}} \tag{3.12}$$

Compared to the condition number, the NSA is calculated for each metabolite individually. This has the advantage that ΔTE can be optimized for a subset of the metabolites. Figure 3.8b shows the dependence of NSA on ΔTE for the main metabolites observed in vivo after the injection of HP [1-^{13}C]pyruvate for such an acquisition scheme at 3 T using 7 echoes. An NSA value equal to the number of acquired echoes represents the optimum noise performance for the estimation of the respective metabolite. In contrast, an NSA of 0 corresponds to a case where it is not possible to accurately estimate the metabolite's contribution to the measured signal. For example, this is the case when a ΔTE was chosen so that the difference in phase increments for two metabolite signals at consecutive acquisitions is a multiple of 2π, that is, the corresponding SW is not large enough to cover the spectrum and one metabolite signal is aliased on top of another peak.

3.5 Metabolite-Selective Imaging

All the previously described methods for metabolic imaging simultaneously excite the resonances from multiple metabolites followed by subsequent encoding of the spectral information through evolution of the chemical shift with time. A different strategy is applied in metabolite-specific imaging [25] where the individual resonances are selectively excited using *spectral-spatial* RF pulses [26] followed by a fast imaging module, such as EPI or spiral imaging. This approach takes advantage of the large chemical shift dispersion of the metabolite peaks in the HP ^{13}C spectrum. An example for this technique is shown in Fig. 3.9. This method permits reducing the overall acquisition time by limiting the number of metabolites that are acquired. For example, in applications of HP ^{13}C MRI in cancer studies, it can often be sufficient to just detect pyruvate and lactate in order to measure changes in glycolytic metabolism due to the Warburg effect, that is, the shift from oxidative to glycolytic metabolism even in the presence of oxygen. The method has a downside as it is more sensitive to B_0 inhomogeneities. Ideally the resonance for the metabolite of interest should be excited with the same flip angle throughout the imaging volume, while the signal from the remaining resonances is sufficiently suppressed. However, frequency offsets due to B_0 inhomogeneities can shift the target resonance out of the spectral passband or reduce the suppression of the resonances. Therefore, in addition to the resonance present in the HP ^{13}C spectrum, the expected frequency variation has to be taken into account when designing the spectral-spatial RF pulses.

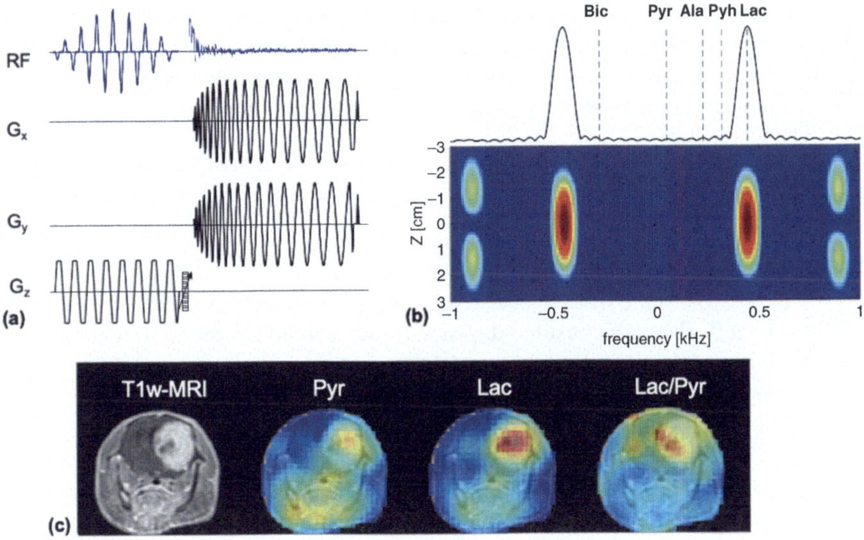

Fig. 3.9 (**a**) Pulse sequence diagram for 3D metabolite-selective imaging. A spectral-spatial RF pulse excites a slab for a particular resonance followed by a single-shot spiral readout. After all excitations for phase encoding along the slab direction are performed the sequence is repeated with the RF shifted to select the next metabolite. (**b**) 2D spectral-spatial profile (bottom) and projection along the slab direction (top) of the RF pulse in the sequence diagram. The frequency offset is such that the lactate resonance is in the passband of the RF pulse. (**c**) Post-gadolinium T1-weighted ^{1}H-MRI from a slice through the brain of a rat with glioma as well as the time-averaged metabolic maps of pyruvate, lactate, and lactate-to-pyruvate ratio. The 3D ^{13}C data were acquired with a FOV of $72 \times 72 \times 50$ mm^3 and a $32 \times 32 \times 10$ matrix (only the central 36 mm are displayed). A constant flip angle of 15° was used for the excitation of both lactate and bicarbonate and 5° for pyruvate. With a TR of 0.125 ms the total acquisition time per metabolite and time point was 1.25 s. The metabolic maps were calculated as the average of the first 8 time points sampled every 3.75 s

3.6 Iterative Reconstruction of Undersampled Data

As already eluded to earlier, one strategy to speed up data acquisition is undersampling, that is, acquiring a fraction of the total nominal *k*-space data. This can be particularly effective if image information is not evenly distributed throughout *k*-space. For example, if 90% of an image's signal is represented by 10% of *k*-space points, an image reconstructed from that fraction of points will probably still be a good representation of the overall image. A clever MR scientist could then collect data from those important points while ignoring regions where little signal is expected. However, undersampling performed this way assumes prior knowledge of what the signal should look like, which is uncertain for unknown samples.

Typical Fourier reconstruction of the undersampled data can lead to aliasing artifacts, especially if there is periodicity in the sampling pattern. A more robust form of undersampling uses a random or pseudo-random sampling pattern, often weighted toward the center of *k*-space where there is usually the most signal.

The pseudo-random pattern causes artifacts to spread out and have the appearance of random noise. These artifacts are then removed by implementing a compressed sensing reconstruction algorithm that iteratively recovers the true data [27–30]. Proper reconstruction requires sparsity, incoherence, and nonlinear reconstruction [31, 32].

3.6.1 Sparsity

Mathematically, data are considered *sparse* if they can be represented by just a few basis functions. It is often the case that data are not sparse in *k*-space or image space themselves, but are sparse in some other algebraic space, and that a transform can be applied to move the data into that sparse space. For example, images are more commonly sparse in the wavelet domain with wavelets as the basis functions. In other cases, data are low-rank, meaning they are sparse in the singular-value domain.

Any appropriate sparse domain can be used for the compressed sensing algorithm, and it is often the case that different transforms need to be tested for a particular type of data to determine what works best. A simple way to test for sparsity is to perform a transformation on the data and plot the component coefficients ordered from largest to smallest. For example, Fig. 3.10 shows that for a typical intensity image, information is present in a large number of pixels (~50%) but only in a small fraction of wavelet components. The image is therefore sparse in the wavelet domain.

Fig. 3.10 (**a**) A typical MR image contains information in a large number of voxels and is therefore not sparse. (**b**) Applying a discreet wavelet transform to the image compresses it into a small number of wavelet coefficients. In the wavelet domain, most coefficients are zero and the data are sparse. (**c**) Plotting the coefficient magnitude for the image and wavelet domains from largest to smallest shows that the wavelet coefficients quickly approach zero

3.6.2 Incoherent Sampling

Due to the Fourier relationship between k-space and image space, the FOV is determined by the k-space step size, whereas resolution is determined by the largest value of k-space sampled. Attempting to undersample by periodically removing k-space lines, for example by eliminating every other PE step, results in a reduced FOV and thus aliasing artifacts from any signal outside the new FOV. However, if k-space points are removed incoherently, such as in a random pattern, the aliasing artifacts also become incoherent. The aliased signals then appear like random noise in the background of the reconstructed image, which results in a recognizable image but with lower SNR. The incoherence of the sampling pattern can be evaluated from its PSF by taking the Fourier transform of the pattern in the appropriate number of dimensions. Fully sampled patterns and those with good incoherence should show most signal in the center of the PSF with few artifacts (Fig. 3.4).

3.6.3 Nonlinear Reconstruction

The nonlinear reconstruction algorithm makes use of sparsity and incoherent sampling to calculate what the image should look like without the overlapping aliasing artifacts. It is an optimization problem that seeks to:

1. Minimize the l_1 norm of the image in the sparse domain, that is, choose the solution that uses the minimum number of basis functions. In the case of matrix data, the l_1 norm is replaced by the nuclear norm. Mathematically, if $\boldsymbol{m}$ represents the reconstructed image and $\boldsymbol{\psi}$ is the transform to the sparse domain, minimize $\|\boldsymbol{\psi}\, \boldsymbol{m}\|_1$.
2. Enforce data consistency. If $\boldsymbol{y}$ is the measured k-space data and $\boldsymbol{F}_s$ is the undersampled Fourier transform, ensure that $\|\boldsymbol{F}_s\, \boldsymbol{m} - \boldsymbol{y}\|_2 < \varepsilon$, where ε controls the desired fidelity.

Numerous iterative algorithms have been developed to solve this problem [28]. An example of one used in the context of dynamic metabolic imaging is shown in Fig. 3.11 [33]. In this case, a set of k-space data was transformed to a set of images M and then to the sparse domain, where thresholding was applied to select only the basis functions containing a minimum amount of signal. The reverse transform was then applied on the remaining basis functions to find a compressed image representation L. The original k-space data were subtracted from the k-space representation of L to find the residual, and the image representation of this residual, representing image artifacts, was subtracted from L to create the next iteration of M. Iterations were continued until the desired level of convergence was achieved. The thresholding level affects the fidelity of reconstruction and can be optimized via simulation by comparing nonlinear reconstruction of full datasets undersampled post-acquisition versus conventional reconstruction. Retaining too many basis functions leads to insufficient removal of aliasing artifacts, while retaining too few causes distortions in the image due to loss of information.

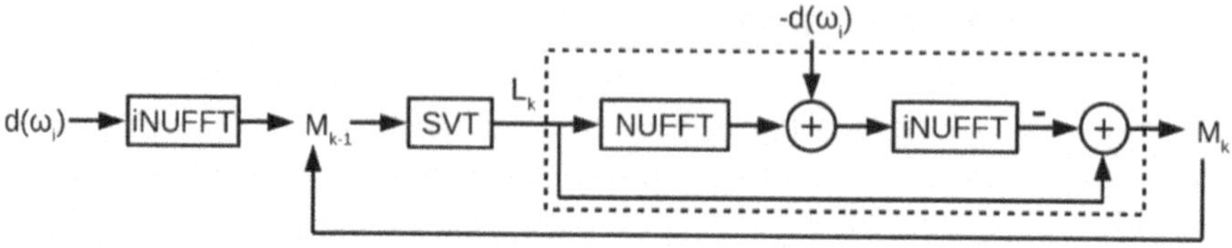

Fig. 3.11 LRMC algorithm. The preprocessed k-space data, $d(\omega_i)$, are converted to image space with an inverse nonuniform fast Fourier transform (iNUFFT). The resulting data M_{k-1} are processed with soft singular value thresholding (SVT), which recreates the dataset L_k using only those singular values $S > \lambda_L S_{max}$, where S_{max} is the maximum singular value and λ_L is a constant. L_k is then converted back to k-space and the original k-space data are subtracted from it. The difference, representing residual aliasing artifacts, is converted to image space and subtracted from L_k to form the reconstructed data M_k. This process is iterated until $|M_k - M_{k-1}| <$ tol, where tol is the desired tolerance for convergence. (Used with permission from S.J. DeVience and D. Mayer, Speeding Up Dynamic Spiral Chemical Shift Imaging with Incoherent Sampling and Low-Rank Matrix Completion. *Magn. Res. Med.* 2017, **77**, 951–960)

3.6.4 Application to Metabolic Imaging

Dynamic metabolic imaging produces additional data dimensions to consider during compressed sensing. Undersampling can also be performed during the spectral acquisition (frequency) and among repetitions (time). This has been demonstrated on HP ^{13}C data with 5D spatial-spectral dynamic imaging using spiral encoding [33]. The sparse spectral dimension is undersampled and reconstructed with a basic unfolding algorithm. Undersampling in space and time is achieved by removing two of the four interleaves pseudo-randomly for each time point. Data are then arranged in matrix form with k_x k_y in one dimension and k_z k_ω in the other. In this case the combined data are low-rank, meaning they are sparse in the singular-value domain. Reconstruction proceeds with the algorithm in Fig. 3.11.

Figure 3.12a shows a series of pyruvate images using full sampling with conventional inverse NUFFT reconstruction, twofold retrospective undersampling with inverse NUFFT reconstruction, and twofold retrospective undersampling with low-rank matrix completion (LRMC) reconstruction. LRMC provides an accurate reconstruction with undersampled data, whereas NUFFT suffers from artifacts. Figure 3.12b compares the results of two scans each with 6-s temporal resolution, one fully sampled with 5-mm resolution and reconstructed with inverse NUFFT (left column) and the other twofold undersampled and reconstructed with LRMC, providing 3.5-mm resolution in the same amount of time (right column). A similar approach uses a series of EPSI sequences in which phase encoding is implemented with a pseudo-random undersampling pattern of blips [34–37]. Different random blips are used for each time point in the series. This simultaneously undersamples in space, frequency, and time. A wavelet-in-time sparsifying transform is used in the nonlinear reconstruction. In multidimensional datasets, there can be a different sparsifying transform in each dimension or group of dimensions. Dynamic data can

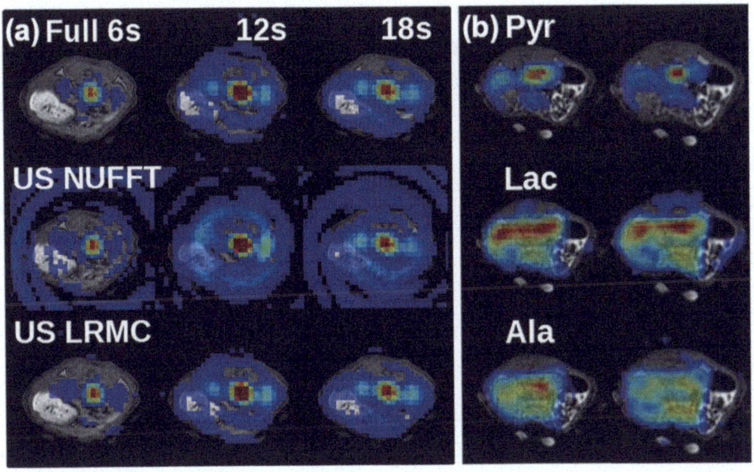

Fig. 3.12 (**a**) A slice through the kidneys of a rat at three time points showing signal from pyruvate. Reconstruction used fully sampled data (top), retrospectively twofold undersampled data reconstructed with inverse NUFFT (middle), and the same twofold undersampled data reconstructed with LRMC (bottom). Inverse NUFFT produces significant spiral artifacts, which are not present in fully sampled or LRMC reconstructions. (**b**) Comparison of fully sampled and twofold prospectively undersampled scans of a rat, each 6 s long. The fully sampled data can achieve 5 mm resolution (left column), while the undersampled data reconstructed with LRMC achieves 3.5-mm resolution in the same amount of time (right column). (Used with permission from S.J. DeVience and D. Mayer, Speeding Up Dynamic Spiral Chemical Shift Imaging with Incoherent Sampling and Low-Rank Matrix Completion. *Magn. Res. Med.* 2017, **77**, 951–960)

often be represented as the sum of a sparse component describing the changing signal and a low-rank static background signal [38, 39]. The algorithm in Fig. 3.11 can be modified to split the image into these two components and perform thresholding on them separately.

This can be implemented for dynamic metabolic measurements with a balanced steady state free precession sequence by performing pseudo-random undersampling on the phase encoding pulses, again using a different pattern for each time point [40]. Using principal component analysis as the sparsifying transform has been found to work best. For the low-rank component, spatial data from various numbers of time points can be analyzed together or each time point can be used separately.

3.7 Summary

This chapter has described the general concepts of spatial encoding in MRI and has summarized commonly used acquisition and processing techniques for metabolic imaging of HP ^{13}C substrates. As each method has advantages and disadvantages, one must make trade-off between various aspects, such as acquisition speed, spectral width, computational complexity, and speed of reconstruction when choosing a method for a specific application. Given the limited print space, the chapter has only surveyed key methods. Several references that dive into more detail or describe techniques that have not been covered, for example, parallel imaging methods leveraging the use of multiple RF receive coils, are listed under Further Studies for the interested reader.

Acknowledgments We acknowledge funding support from NIH (EB009070, DK106395, AA018681, CA202694, CA176836, CA213020, and NS096575), DOD (PC150408), and the Maryland Department of Health's Cigarette Restitution Fund Program.

Problems

1. (a) For your hyperpolarized ^{13}C experiments you are using a 3 T MRI scanner with a maximum gradient strength of 5 G/cm and an RF amplifier that can provide a maximum B_1 of 1 G. You have designed a slice-selective excitation RF pulse that has the following nominal properties: at a duration τ_{nom} of 1.8 ms the pulse has a bandwidth [i.e., full width at half maximum (FWHM) of 2.2 kHz] and requires a $B_{1,nom}$ of 0.5 G for a 90° excitation. The pulse sequence allows using the pulse at a variable duration. What is the duration of the pulse that minimizes the chemical shift displacement artifact on this scanner when exciting a 1-cm slice? The bandwidth of an RF pulse is inversely proportional to its duration and the slice thickness is defined as the FWHM of the slice profile. What is the corresponding amplitude of the slice-section gradient and what is the offset in slice location between [1-^{13}C]pyruvate and [1-^{13}C]lactate? The chemical shift difference between the two resonances is 12.2 ppm and $\gamma_{13C} = 1.071$ kHz/G.
2. (a) For the same scanner you are designing the gradient waveform $G_x(t)$ for a fly-back ^{13}C EPSI pulse sequence with the following parameters: SW of 1 kHz, FOV of 8 cm, nominal in-plane resolution Δx of 0.5 cm. Calculate the amplitude G_{read} of the readout gradient as well as the readout bandwidth for the case the refocusing lobe of the fly-back waveform uses the maximum gradient strength and no gaps in the waveform. You can assume an infinite gradient slew rate, that is, the gradient amplitudes can be changed instantaneously.

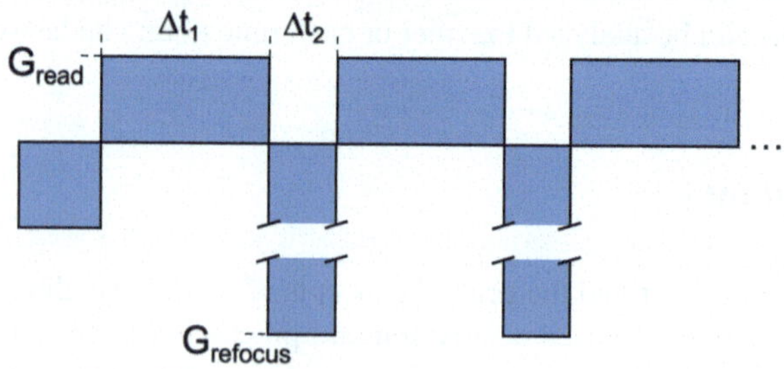

(b) Recalculate the durations and amplitudes for the readout and refocusing gradients considering that both gradient amplitudes can only be updated and data points sampled at an interval Δt_s that is an integer multiple of 4 μs.

References

1. Twieg, D.B.: The k-trajectory formulation of the NMR imaging process with applications in analysis and synthesis of imaging methods. Med. Phys. **10**(5), 610–621 (1983)
2. Mezrich, R.: A perspective on k-space. Radiology. **195**(2), 297–315 (1995). https://doi.org/10.1148/radiology.195.2.7724743
3. Brown, T.R., Kincaid, B.M., Ugurbil, K.: NMR chemical shift imaging in three dimensions. Proc. Natl. Acad. Sci. U. S. A. **79**(11), 3523–3526 (1982)
4. Maudsley, A.A., Hilal, S.K., Perman, W.H., Simon, H.E.: Spatially resolved high resolution spectroscopy by "four-dimensional" NMR. J. Magn. Reson. **51**(1), 147–152 (1983). https://doi.org/10.1016/0022-2364(83)90113-0
5. Maudsley, A.A., Matson, G.B., Hugg, J.W., Weiner, M.W.: Reduced phase encoding in spectroscopic imaging. Magn. Reson. Med. **31**(6), 645–651 (1994). https://doi.org/10.1002/mrm.1910310610
6. Wang, J.-X., Merritt, M.E., Sherry, A.D., Malloy, C.R.: Accelerated chemical shift imaging of hyperpolarized 13C metabolites. Magn. Reson. Med. **76**(4), 1033–1038 (2016). https://doi.org/10.1002/mrm.26286
7. Zhao, L., Mulkern, R., Tseng, C., Williamson, D., Patz, S., Kraft, R., Walsworth, R., Jolesz, F., Albert, M.: Gradient-Echo imaging considerations for hyperpolarized 129Xe MR. J. Magn. Reson. B. **113**(2), 179–183 (1996)
8. Kohler, S.J., Yen, Y., Wolber, J., Chen, A.P., Albers, M.J., Bok, R., Zhang, V., Tropp, J., Nelson, S., Vigneron, D.B., Kurhanewicz, J., Hurd, R.E.: In vivo 13 carbon metabolic imaging at 3T with hyperpolarized 13C-1-pyruvate. Magn. Reson. Med. **58**(1), 65–69 (2007). https://doi.org/10.1002/mrm.21253
9. Mayer, D., Yen, Y.-F., Levin, Y.S., Tropp, J., Pfefferbaum, A., Hurd, R.E., Spielman, D.M.: In vivo application of sub-second spiral chemical shift imaging (CSI) to hyperpolarized 13C metabolic imaging: comparison with phase-encoded CSI. J. Magn. Reson. **204**(2), 340–345 (2010). https://doi.org/10.1016/j.jmr.2010.03.005
10. Posse, S., Tedeschi, G., Risinger, R., Ogg, R., Le Bihan, D.: High speed 1H spectroscopic imaging in human brain by echo planar spatial-spectral encoding. Magn. Reson. Med. **33**(1), 34–40 (1995)
11. Mansfield, P.: Spatial mapping of the chemical shift in NMR. Magn. Reson. Med. **1**(3), 370–386 (1984). https://doi.org/10.1002/mrm.1910010308
12. Hennel, F., Nédélec, J.-F.: Interleaved asymmetric echo-planar imaging. Magn. Reson. Med. **34**(4), 520–524 (1995). https://doi.org/10.1002/mrm.1910340406
13. Cunningham, C.H., Vigneron, D.B., Chen, A.P., Xu, D., Nelson, S.J., Hurd, R.E., Kelley, D.A., Pauly, J.M.: Design of flyback echo-planar readout gradients for magnetic resonance spectroscopic imaging. Magn. Reson. Med. **54**(5), 1286–1289 (2005). https://doi.org/10.1002/mrm.20663
14. Macovski, A.: Noise in MRI. Magn. Reson. Med. **36**(3), 494–497 (1996). https://doi.org/10.1002/mrm.1910360327
15. Adalsteinsson, E., Irarrazabal, P., Topp, S., Meyer, C., Macovski, A., Spielman, D.M.: Volumetric spectroscopic imaging with spiral-based k-space trajectories. Magn. Reson. Med. **39**(6), 889–898 (1998)
16. Jackson, J., Meyer, C., Nishimura, D., Macovski: Selection of a convolution function for Fourier inversion using gridding [computerised tomography application]. IEEE Trans. Med. Imaging. **10**(3), 473–478 (1991)

17. Fessler, J.: Nonuniform fast Fourier transforms using min-max interpolation. IEEE Trans. Signal Process. **51**(2), 560–574 (2003)
18. Spielman, D.M., Pauly, J.M.: Spiral imaging on a small-bore system at 4.7T. Magn. Reson. Med. **34**(4), 580–585 (1995). https://doi.org/10.1002/mrm.1910340414
19. Fessler, J.: Model-based image reconstruction for MRI. IEEE Signal Process. Mag. **27**(4), 81–89 (2010). https://doi.org/10.1109/MSP.2010.936726
20. Levin, Y.S., Mayer, D., Yen, Y.-F., Hurd, R.E., Spielman, D.M.: Optimization of fast spiral chemical shift imaging using least squares reconstruction: application for hyperpolarized (13) C metabolic imaging. Magn. Reson. Med. **58**(2), 245–252 (2007). https://doi.org/10.1002/mrm.21327
21. Reeder, S.B., Brittain, J.H., Grist, T.M., Yen, Y.-F.: Least-squares chemical shift separation for (13)C metabolic imaging. J. Magn. Reson. Imag. **26**(4), 1145–1152 (2007). https://doi.org/10.1002/jmri.21089
22. Wiesinger, F., Weidl, E., Menzel, M.I., Janich, M.A., Khegai, O., Glaser, S.J., Haase, A., Schwaiger, M., Schulte, R.F.: IDEAL spiral CSI for dynamic metabolic MR imaging of hyperpolarized [1-13C]pyruvate. Magn. Reson. Med. **68**(1), 8–16 (2012). https://doi.org/10.1002/mrm.23212
23. Gordon, J.W., Niles, D.J., Fain, S.B., Johnson, K.M.: Joint spatial-spectral reconstruction and k-t spirals for accelerated 2D spatial/1D spectral imaging of (13) C dynamics. Magn. Reson. Med. **71**(4), 1435–1445 (2014). https://doi.org/10.1002/mrm.24796
24. Reeder, S.B., Wen, Z., Yu, H., Pineda, A.R., Gold, G.E., Markl, M., Pelc, N.J.: Multicoil Dixon chemical species separation with an iterative least-squares estimation method. Magn. Reson. Med. **51**(1), 35–45 (2004). https://doi.org/10.1002/mrm.10675
25. Lau, A.Z., Chen, A.P., Hurd, R.E., Cunningham, C.H.: Spectral-spatial excitation for rapid imaging of DNP compounds. NMR Biomed. **24**(8), 988–996 (2011). https://doi.org/10.1002/nbm.1743
26. Meyer, C.H., Pauly, J.M., Macovski, A., Nishimura, D.G.: Simultaneous spatial and spectral selective excitation. Magn. Reson. Med. **15**(2), 287–304 (1990)
27. Candes, E.J., Romberg, J., Tao, T.: Robust uncertainty principles: exact signal reconstruction from highly incomplete frequency information. IEEE Trans. Inf. Theory. **52**(2), 489–509 (2006). https://doi.org/10.1109/TIT.2005.862083
28. Zhang, Z., Xu, Y., Yang, J., Li, X., Zhang, D.: A survey of sparse representation: algorithms and applications. IEEE Access. **3**, 490–530 (2015). https://doi.org/10.1109/ACCESS.2015.2430359
29. Daubechies, I., Defrise, M., Mol, C.D.: An iterative thresholding algorithm for linear inverse problems with a sparsity constraint. Commun. Pure Appl. Math. **57**(11), 1413–1457 (2004). https://doi.org/10.1002/cpa.20042
30. Berthoumieu, Y., Dossal, C., Pustelnik, N., Ricoux, P., Turcu, F.: An evaluation of the sparsity degree for sparse recovery with deterministic measurement matrices. J. Math. Imag. Vis. **48**(2), 266–278 (2014). https://doi.org/10.1007/s10851-013-0453-4
31. Lustig, M., Donoho, D.L., Santos, J.M., Pauly, J.M.: Compressed sensing MRI. IEEE Signal Process. Mag. **25**(2), 72–82 (2008). https://doi.org/10.1109/MSP.2007.914728
32. Hollingsworth, K.G.: Reducing acquisition time in clinical MRI by data undersampling and compressed sensing reconstruction. Phys. Med. Biol. **60**(21), R297–R322 (2015). https://doi.org/10.1088/0031-9155/60/21/R297
33. DeVience, S.J., Mayer, D.: Speeding up dynamic spiral chemical shift imaging with incoherent sampling and low-rank matrix completion. Magn. Reson. Med. **77**(3), 951–960 (2017). https://doi.org/10.1002/mrm.26170
34. Larson, P.E.Z., Hu, S., Lustig, M., Kerr, A.B., Nelson, S.J., Kurhanewicz, J., Pauly, J.M., Vigneron, D.B.: Fast dynamic 3D MR spectroscopic imaging with compressed sensing and multiband excitation pulses for hyperpolarized 13C studies. Magn. Reson. Med. **65**(3), 610–619 (2011). https://doi.org/10.1002/mrm.22650
35. Hu, S., Lustig, M., Balakrishnan, A., Larson, P.E.Z., Bok, R., Kurhanewicz, J., Nelson, S.J., Goga, A., Pauly, J.M., Vigneron, D.B.: 3D compressed sensing for highly accelerated hyper-

polarized (13)C MRSI with in vivo applications to transgenic mouse models of cancer. Magn. Reson. Med. **63**(2), 312–321 (2010). https://doi.org/10.1002/mrm.22233
36. Cao, P., Shin, P.J., Park, I., Najac, C., Marco-Rius, I., Vigneron, D.B., Nelson, S.J., Ronen, S.M., Larson, P.E.Z.: Accelerated high-bandwidth MR spectroscopic imaging using compressed sensing. Magn. Reson. Med. **76**(2), 369–379 (2016). https://doi.org/10.1002/mrm.26272
37. Bhattacharya, I., Jacob, M.: Denoising and deinterleaving of EPSI data using structured low-rank matrix recovery. 2018 IEEE 15th Int. Symp. Biomed. Imag. (ISBI 2018). **2018**, 679–682 (2018)
38. Otazo, R., Candès, E., Sodickson, D.K.: Low-rank plus sparse matrix decomposition for accelerated dynamic MRI with separation of background and dynamic components. Magn. Reson. Med. **73**, 1125 (2014). https://doi.org/10.1002/mrm.25240
39. Ong, F., Lustig, M.: Beyond low rank + sparse: multi-scale low rank matrix decomposition. IEEE J. Select. Topic. Signal. Process. **10**(4), 672–687 (2016). https://doi.org/10.1109/JSTSP.2016.2545518
40. Milshteyn, E., von Morze, C., Reed, G.D., Shang, H., Shin, P.J., Larson, P.E.Z., Vigneron, D.B.: Using a local low rank plus sparse reconstruction to accelerate dynamic hyperpolarized 13C imaging using the bSSFP sequence. J. Magn. Reson. **290**, 46–59 (2018). https://doi.org/10.1016/j.jmr.2018.03.006

Further Reading

De Graaf, R.A.: In vivo NMR spectroscopy. Wiley, Hoboken, NJ (2013)

Bernstein, M.A., King, K.F., Zhou, X.J.: Handbook of MRI pulse sequences. Elsevier, Amsterdam (2004)

Gordon, J.W., Hansen, R.B., Shin, P.J., Feng, Y., Vigneron, D.B., Larson, P.E.Z.: 3D hyperpolarized C-13 EPI with calibrationless parallel imaging. J. Magn. Reson. **289**, 92–99 (2018). https://doi.org/10.1016/j.jmr.2018.02.011

Josan, S., Hurd, R., Park, J.M., Yen, Y.-F., Watkins, R., Pfefferbaum, A., Spielman, D., Mayer, D.: Dynamic metabolic imaging of hyperpolarized [2-(13) C]pyruvate using spiral chemical shift imaging with alternating spectral band excitation. Magn. Reson. Med. **71**(6), 2051–2058 (2014). https://doi.org/10.1002/mrm.24871

Müller, C.A., Hundshammer, C., Braeuer, M., Skinner, J.G., Berner, S., Leupold, J., Düwel, S., Nekolla, S.G., Månsson, S., Hansen, A.E., Elverfeldt, D., Ardenkjaer-Larsen, J.H., Schilling, F., Schwaiger, M., Hennig, J., Hövener, J.B.: Dynamic 2D and 3D mapping of hyperpolarized pyruvate to lactate conversion in vivo with efficient multi-echo balanced steady-state free precession at 3 T. NMR Biomed. **33**, e4291 (2020). https://doi.org/10.1002/nbm.4291

Yen, Y.-F., Kohler, S.J., Chen, A.P., Tropp, J., Bok, R., Wolber, J., Albers, M.J., Gram, K.A., Zierhut, M.L., Park, I., Zhang, V., Hu, S., Nelson, S.J., Vigneron, D.B., Kurhanewicz, J., Dirven, H.A.A.M., Hurd, R.E.: Imaging considerations for in vivo 13C metabolic mapping using hyperpolarized 13C-pyruvate. Magn. Reson. Med. **62**(1), 1–10 (2009). https://doi.org/10.1002/mrm.21987

Chapter 4
Sampling Strategies in Dynamic Hyperpolarized NMR

Ralph E. Hurd and Albert P. Chen

4.1 Introduction

With the fantastic gain in signal available from spins hyperpolarized from parts-per-million at thermal equilibrium to the 20% or more now routinely achieved, the advances needed to propel this technology forward as a productive clinical and research tool may be in the multinuclear hardware and pulse sequences on the magnetic resonance (MR) scanners used to collect these data. Most of the pulse sequence elements used in hyperpolarized (HP) carbon-13 metabolic imaging are those already employed in thermal equilibrium proton imaging and spectroscopy. The optimum sampling of hyperpolarized carbon-13 metabolic imaging requires the management of these pulse sequence elements to accommodate, and where possible, take advantage of nonrecoverable magnetization, relatively long relaxation time, bolus dynamics, and rapid metabolic exchange or flux. Carbon gyromagnetic ratio, proton–carbon spin coupling, and a relatively broad and sparsely populated spectral bandwidth are also essential elements of design that differentiate carbon-13 from proton sampling. In hyperpolarized carbon-13 metabolic imaging, the quality of spectroscopic sampling is often a critical factor in deciding between spectrally selective imaging versus maintaining a full or a sparsely sampled spectroscopic domain. Just as with proton methods, a trade-off between spatial resolution and signal-to-noise is also often necessary. The choice of sampling and reconstruction also depends on the dose used and the quantitative requirements. This chapter reviews the features of spatial, spectral, and dynamic sampling elements and how they impact state-of-the-art hyperpolarized carbon-13 metabolic imaging.

R. E. Hurd (✉)
Radiological Imaging Laboratory, Stanford University, Stanford, CA, USA
e-mail: ralph.hurd@stanford.edu

A. P. Chen
General Electric Healthcare, Toronto, ON, Canada

T. Jue, D. Mayer (eds.), *Dynamic Hyperpolarized Nuclear Magnetic Resonance*, Handbook of Modern Biophysics, https://doi.org/10.1007/978-3-030-55043-1_4

4.2 Carbon-13 Spectral Bandwidth

The most prevalent hyperpolarized carbon-13 metabolic imaging agent, [1-^{13}C]pyruvate, produces 5 primary peaks spread over 24 ppm (around 800 Hz at 3 T as illustrated in Fig. 4.1). These peaks correspond to [1-^{13}C]pyruvate, [1-^{13}C]pyruvate hydrate, [1-^{13}C]lactate, [1-^{13}C] alanine, and [1-^{13}C]bicarbonate. It is a definite advantage that the spectral characteristics of the agent and products are well known and that the spectral distribution of carbon-13 peaks is relatively sparse. The well-characterized chemical shifts and separation of pyruvate and its metabolic product signals allow for aggressive under-sampling, which effectively folds in the signal one or more times beyond the Nyquist frequency, yet it still avoids frequencies already in play. An excellent example of this is used to reduce the gradient demands of spiral chemical-shift imaging (CSI) [1]. Random sparse sampling in conjunction with compressed sensing (CS) reconstruction is also a very effective strategy for the rapid sampling of a volume fast enough to capture the dynamic spectral changes in metabolites [2]. The sparse spectral distribution also makes it possible to excite individual metabolites, freeing the frequency domain to encode just space, rather than chemical shift or chemical shift plus space. Rapid spatial encoding of each metabolite, following spectrally selective excitation, takes advantage of the large nonoverlapping spectral separation for the metabolite peaks of interest [3]. The spectral range of [1-^{13}C]pyruvate and its metabolic products also lends itself to the

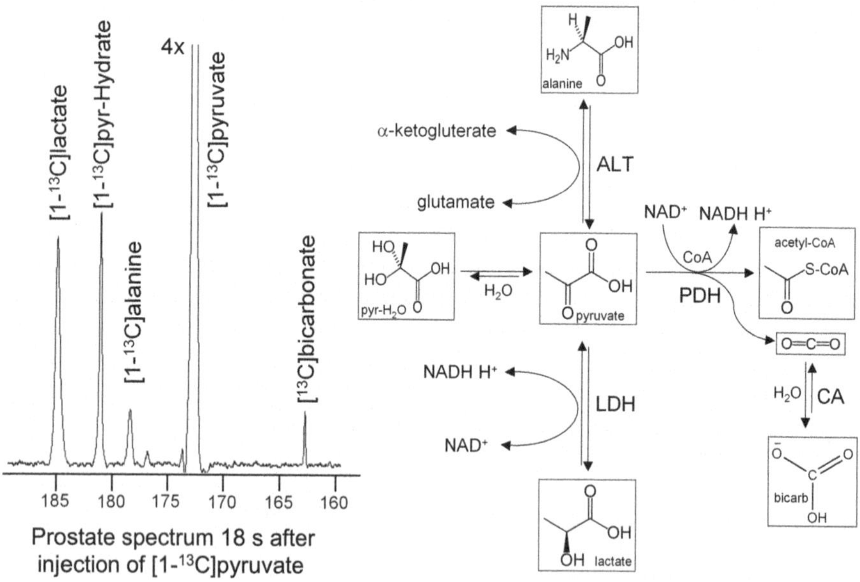

Fig. 4.1 Pulse-and-acquire spectrum taken 18 s after injection of a bolus of hyperpolarized [1-^{13}C]pyruvate. Metabolites observed in carbonyl spectral band include pyruvate, pyruvate-hydrate, alanine, lactate, and bicarbonate

use of **I**terative **D**ecomposition of Water and Fat With **E**cho **A**symmetry and **L**east-Squares Estimation (IDEAL) [4] type encoding, where the spectral domain is sampled using images collected at a few different evolution times. Evolution time spectral encoding was first suggested by Jean Jeener at the Ampere Summer School in 1971 and is the basis for two-dimensional (2D) nuclear magnetic resonance (NMR).

Agents and co-polarization strategies that cover more of the full carbon spectral width include [2-^{13}C]pyruvate and can yield up to a dozen lines over 157 ppm (5000 Hz at 3 T as illustrated in Fig. 4.2 [5]). The full carbon-13 spectral bandwidth is around 230 ppm. Just as with C1-pyruvate, the C2-pyruvate metabolic spectral positions are well known and sparse (with a few resonances closely packed), but the larger spectral bandwidth now restricts the duration of nonlocalized rectangular

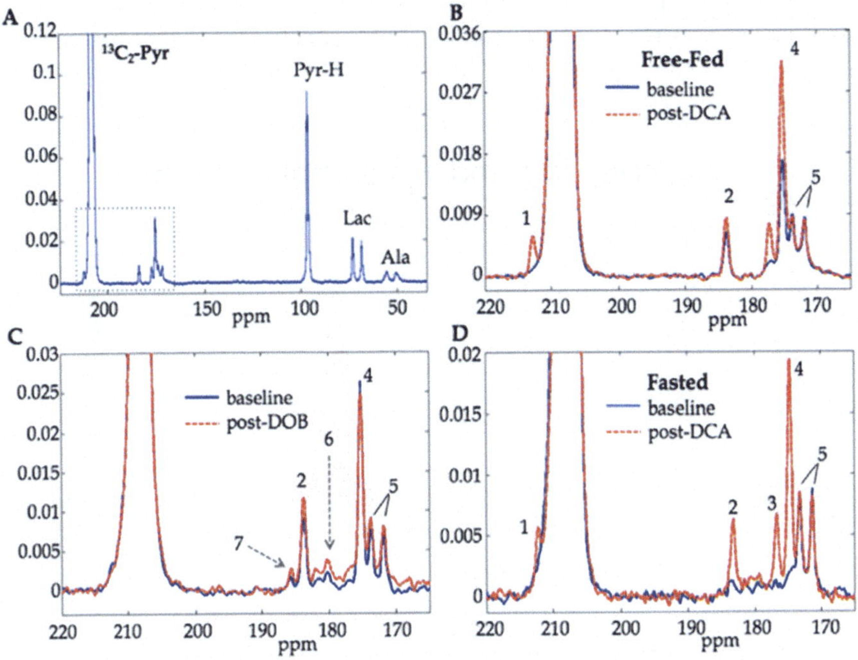

Fig. 4.2 (**a**) Representative ^{13}C MRS spectra from rat heart in vivo showing the peaks observed after hyperpolarized [2-^{13}C]pyruvate bolus injection, normalized to pyruvate signal. Pyr-H represents pyruvate-hydrate. The zoomed-in spectra for the dotted line region are shown in (**b**–**d**). An increase in glutamate (peak 2), acetoacetate (peaks 1 and 3), and acetylcarnitine (4) labeling was observed with increased pyruvate-dehydrogenase (PDH) flux post-dichloroacetate (DCA) for free-fed (**b**) and fasted (**d**) rats. Peak 5 is the natural abundance ^{13}C1 doublet peak from natural abundance [1,2-^{13}C]pyruvate in the injected solution. The spectra in (**c**) show an increase in lactate, alanine, and glutamate after dobutamine (DOB). Peak 6 (180.2 ppm) is tentatively assigned to citrate, but was not reliably observed with sufficient SNR in all animals and could include contributions from one of the peaks of [1,2-^{13}C1]pyruvate-hydrate doublet. Peak 7 is the downfield peak of ^{13}C1-lactate doublet with the upfield peak overlapping with glutamate

excitation pulses. Although the sizeable spectral bandwidth also presents a chemical shift displacement issue for slice-selective elements, the sparse distribution of lines can be an advantage for spectrally selective radiofrequency (RF) designs. One strategy to deal with [2-^{13}C] pyruvate metabolic imaging is to use multiple limited frequency bands to capture spectrally similar groups of signals [6].

4.3 T_2^* and J_{CH}

Since the carbon gamma is only ~1/4th of the proton gamma, the optimum T_2^* sampling interval is correspondingly increased relative to protons. However, two and three bond J_{CH} coupling of 3–4 Hz characterize many of the signals of interest (e.g., [1-^{13}C]lactate and [1-^{13}C]alanine) and often determine the optimum acquisition window. Figure 4.3 illustrates the simulation of J_{CH} coupling in [1-^{13}C]lactate frequency response (top left), time response (bottom), and overlaid on the in vivo pyruvate and lactate response (top right). Note the difference in apparent line-width of lactate and

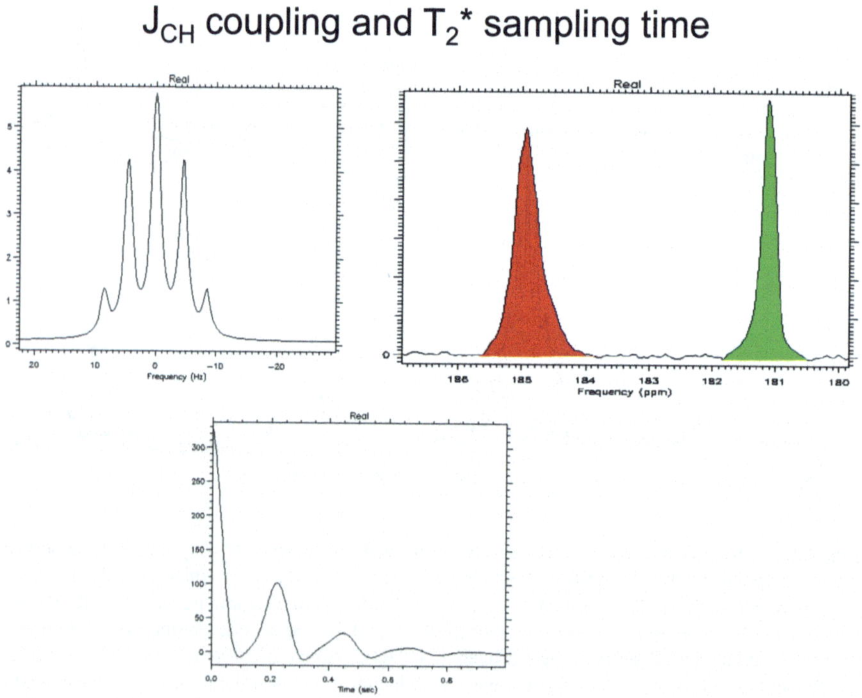

Fig. 4.3 Simulation of J_{CH} coupling in [1-^{13}C]lactate frequency response (top left), time response (bottom), and overlaid on the in vivo pyruvate hydrate and lactate response spectrum from Fig. 4.1. Note the linewidth difference between the [1-^{13}C]lactate (red) and [1-^{13}C]pyruvate-hydrate (green) signals from canine prostate

pyruvate-hydrate in a preclinical canine prostate spectrum. In the graph in the bottom shown in this figure, the sampling of a flame ionization detector (FID) longer than 25 ms will provide no added advantage in C1-lactate signal averaging. Thus, an echo with a T_2^* window over 50 ms would not add to signal-to-noise ratio (SNR).

T_2^* plus J_{CH} decay also lead to losses during the evolution delays in IDEAL type encoding as well as during long spectrally selective pulses. Decoupling can help with the J_{CH} term [7], but the added complications of proton decoupling make this impractical for human studies. Fortunately, the spectral separation of the carbon resonances for many per-polarized substrates and their metabolites is sufficient to limit evolution time losses in IDEAL encoding and allows for relatively short spectral-spatial pulse durations.

4.4 Nonrecoverable Magnetization

Perhaps the most significant difference between thermal equilibrium proton methods and hyperpolarized carbon-13 imaging is dealing with nonrecoverable magnetization. The magnetization of a hyperpolarized carbon-13 agent is dramatically enhanced. Liquid state polarization currently achievable is over 20% (or 200,000 ppm). After dissolution, hyperpolarized magnetization undergoes T_1 relaxation toward thermal equilibrium. This process is similar to that of water protons in the human body after an RF excitation or inversion in a typical magnetic resonance imaging (MRI) scan. The difference is that when protons recover to thermal equilibrium via T_1 relaxation, they can repeatedly be resampled, using RF pulses, to provide additional SNR via averaging. Averaging of T_1 recovered signal is not possible with hyperpolarized magnetization. Hyperpolarized carbon-13 substrate irreversibly returns to thermal equilibrium via T_1 relaxation. Once decayed, the 200,000 ppm HP magnetization is not recoverable and the magnetization of the carbon-13 substrate remains at the thermal equilibrium level, about 2.6 ppm at 3 T.

In addition to the loss of polarization due to T_1 relaxation, RF pulses deplete this nonequilibrium polarization in a nonrecoverable way. Thus, sample delivery and data acquisition of hyperpolarized imaging need to be sufficiently fast to utilize this decaying and nonrecoverable magnetization (for [1-^{13}C]pyruvate, most of the nonequilibrium polarization is lost within 2–3 min postdissolution). Loss of the hyperpolarization by both T_1 and RF excitation are both essential design considerations. T_1 limits the choice of agent and labeling position and favors agents with carbonyl and quaternary carbons. In hyperpolarized carbon-13 metabolic imaging, the label moves from the substrate to the product(s), which makes it essential to know the metabolic rate and the T_1 of the product(s) for pulse sequence design. The objective is often to image the changes in substrate and products dynamically and to measure metabolic rates. Metabolic rate measurement requires the injected bolus and the production of products to be adequately sampled temporally, typically in the range of 3–5 s. Dynamic sampling requires optimizing imaging intervals and careful management of the magnetization throughout the lifetime of the hyperpolarized signals.

For optimum design, preserving the substrate magnetization using small tip angles on the injected agent and larger tip angles for the products is the best practice. Reduced tip angles for the injected agent is especially useful in cases where the substrate levels are significantly more intense than the product. Another best practice is to increase product excitation pulse for the instance of shorter product T_1's. Although relatively long T_1 lifetimes are required for the transfer and physiological delivery of substrate, the product T_1's can be much shorter. For example, in [2-^{13}C]pyruvate, the product [2-^{13}C]lactate has a directly attached proton and consequently a significantly shorter T_1. Optimum tip angle for this product is close to 90° for each dynamic interval. Another beneficial aspect of nonrecoverable magnetization is the lack of a penalty for imaging speed. Without a renewed reservoir of magnetization from the relaxation process, the averaging of the signal from recovered magnetization is not a consideration. Thus, within the limit of sufficient T_2^* sampling, acceleration of the scan is very desirable.

4.5 No Steady State

One consequence of hyperpolarization, inflow, and metabolism is that there is little chance to establish a steady-state condition. In thermal MRI, discarded acquisitions and prep sequences are often used to establish steady-state conditions. Discarded acquisitions can be defaulted on and may need to be disabled for hyperpolarized applications. One strategy to achieve pseudo-steady-state conditions is the use of variable flip angles or progressively larger flip angle during the imaging window. These can improve SNR of a dynamic series, but when used aggressively, it can be challenging to make the magnetization-loss corrections needed for quantification of dynamic rate constants.

4.6 Refocusing of Hyperpolarized Magnetization

Another difference between thermal and hyperpolarized spins is the unwanted impact of RF refocusing. Although perfect inversion of a hyperpolarized spin doesn't spend any potential signal, the transition between full inversion and the unexcited band can excite and destroy signal. The transition band can be defined by the pulse shape of the slice selection or by the B_1 fall-off of the coil itself. Any spins that flow through the transition band from an RF pulse response or the fringe field outside an RF coil will experience excitation and result in a signal loss as illustrated in Fig. 4.4 [8]. As shown in the comparison of an FID versus double spin echo imaging sequence as shown in Fig. 4.5, moving spins in the vascular space are attenuated relative to signals from tissue. In addition to the signal attenuation that would be expected by moving signal during a spin echo and observed in thermal MRI studies, nonrecoverable signal losses occur as hyperpolarized spins move through the

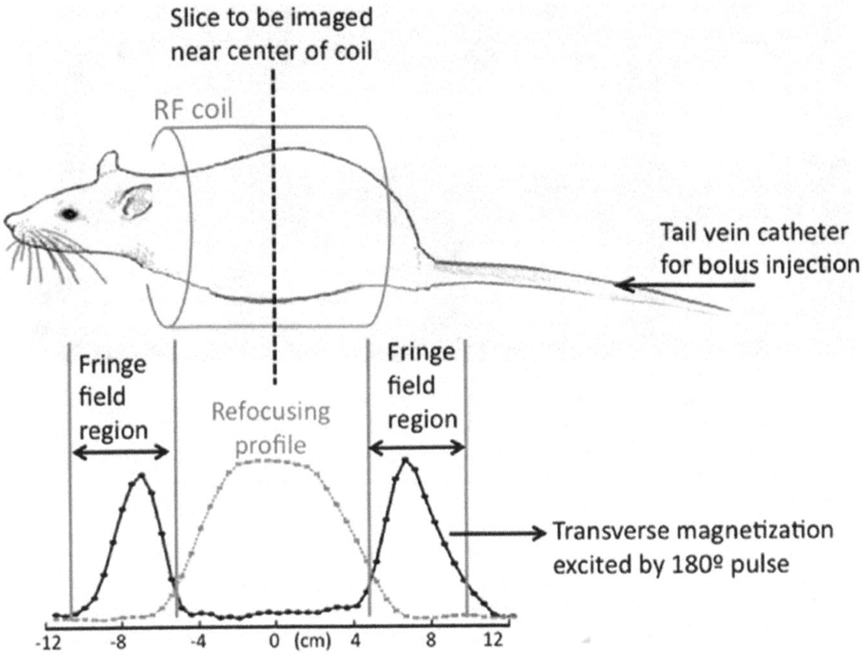

Fig. 4.4 A sketch of the animal position with respect to the RF coil and the fringe-field region at the periphery of the coil. The measured signal profiles plotted here show the region within the coil where the refocusing pulses work well, and also that the refocusing pulses fail in the fringe-field region and attenuate magnetization flowing through that region during acquisition

excitation of the RF transition bands. Designing inversion pulses with adequate bandwidth also adds demands in the B_1 field for the RF hardware, and may be impractical for human size volume-transmit coils.

4.7 Signal Averaging and Improved SNR

Unlike thermal equilibrium MR, hyperpolarized magnetization is not reestablished via relaxation, and thus conventional signal averaging through an increase in number of excitations (NEX) is not possible. Productive signal averaging over a dynamic series can be valuable. Signal weighted averaging or threshold selection can provide a high SNR integral. Also, because the dynamics can usually be characterized as smooth low frequency events, the SNR of individual time points can be improved by leveraging the full signal integral [9]. Signal averaging is also possible via multi-echo T_2. T_2 is substantially longer than the effective T_2* in these agents in vivo. T_2's ranging from 200 to 1200 ms have been reported for [1-^{13}C]lactate with even longer T_2 reported in a TRAMP tumor [10].

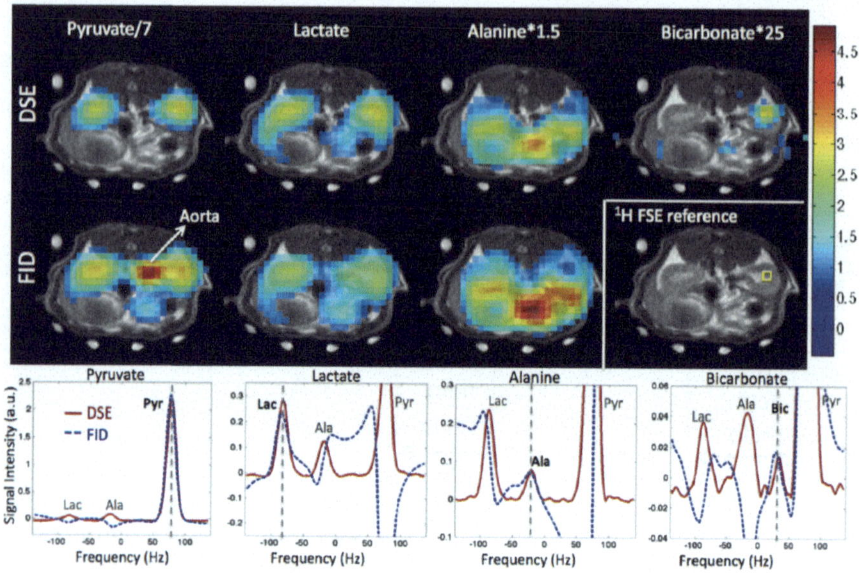

Fig. 4.5 Metabolic images of Pyr, Lac, Ala, and Bic along with spectra from a voxel in the rat right kidney (voxel is shown in ^{1}H anatomical image). DSE-spCSI magnitude mode spectra yield similar linewidth as FID-spCSI absorption mode spectra, eliminating the need for phase correction. Bic image is not shown for FID as it was contaminated by out-of-phase Pyr signal. The DSE sequence also suppresses signal from flowing spins in blood vessels, which is clearly seen in the Pyr map. The spectra for Pyr, Lac, Ala, and Bic are plotted separately due to the independent reconstructions performed for each of them. The vertical dashed lines indicate the frequency of the main (in-focus) spectral components in each of the reconstructed spectra

4.8 Bolus Dynamics

Hyperpolarized carbon-13 metabolic imaging follows the spectral changes that occur as a bolus of the agent makes its way throughout the body. Because of the limit in T_1 lifetimes of the hyperpolarized agents used, this is primarily a first pass process and occurs entirely within the first 90 s following injection for [1-^{13}C]-pyruvate. Figure 4.6 shows a typical bolus response for [1-^{13}C]-pyruvate. Even in human studies, the transit time for an injected bolus falls within this window. The shape of the [1-^{13}C]-pyruvate bolus can vary depending on location and factors such as speed of injection. Even a saline push following the primary injection can affect the bolus. To adequately acquire dynamic images, chemical shift, and the three spatial dimensions require sampling every 3–5 s. For axial scans, quantitation of the bolus signal must account for the competition between RF signal depletion and in-flow of fresh hyperpolarized spins. Once the agent makes it to the tissue of interest, flux or isotope exchange generates the product signal, which is characterized by a metabolic build-up rate convolved with loss due to T_1 and movement of the spins out of the volume of interest. A 3–5 s dynamic resolution is usually sufficient to measure these processes. If quantitation of the dynamics is not necessary, a

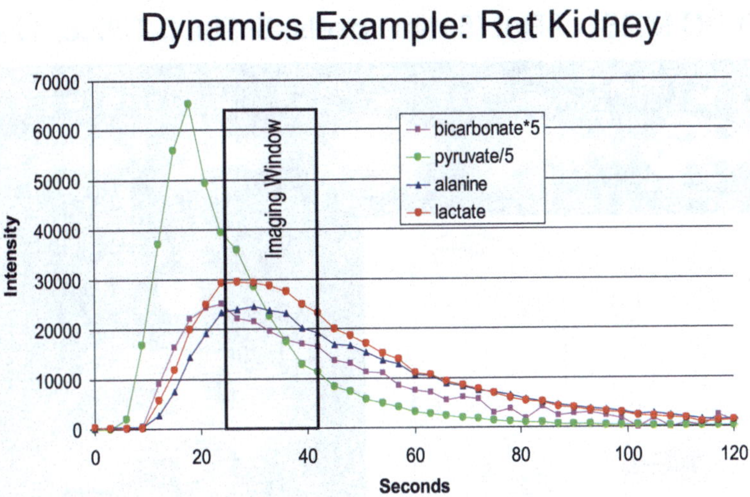

Fig. 4.6 This slide illustrates a typical metabolic time course for hyperpolarized ^{13}C pyruvate. A 5° excitation is applied every 3 s to sample metabolism

relatively fast CSI acquisition can be used to sample with optimal spectral quality. A 20–30 s plateau of product signal is sufficient to achieve a 2D CSI image with a simple matrix of 16 × 16 with a TR of 75 ms [11] (Fig. 4.7). This strategy is very SNR efficient, but lacks resolution of dynamic changes and lacks a quantitative measure of the agent injected. Imaging is started after a delay in order to allow for the build-up of product signal, and thus the signal from the injected bolus is already in decline. For some applications this is not an issue, since the ratio of products is the desired measurement (e.g., lactate/bicarbonate). For applications that need to sample full dynamic evolution of both agent and products faster spatial encoding is needed. Nonlocalized dynamic studies are prevalent for new agents or applications where the spectral quality is of high importance.

4.9 Single Voxel Alternative

Instead of sampling the full field of view (FOV) and slices to obtain spatially, spectrally and temporally resolved data from a specific tissue within the imaging volume, it may also be possible to acquire localized and dynamic ^{13}C spectroscopic data from a single voxel (SV) within the tissue of interest. While a double spin-echo pulse sequence that is similar to PRESS [12] has been used in hyperpolarized ^{13}C applications, data were only localized within a slab through the subject (using the small tip, slice-selective excitation pulse) while the adiabatic refocusing pulses used in these studies were nonselective [8, 13]. The nonselective refocusing scheme was used because the transition bands of any slice selective refocusing pulse contain 90°

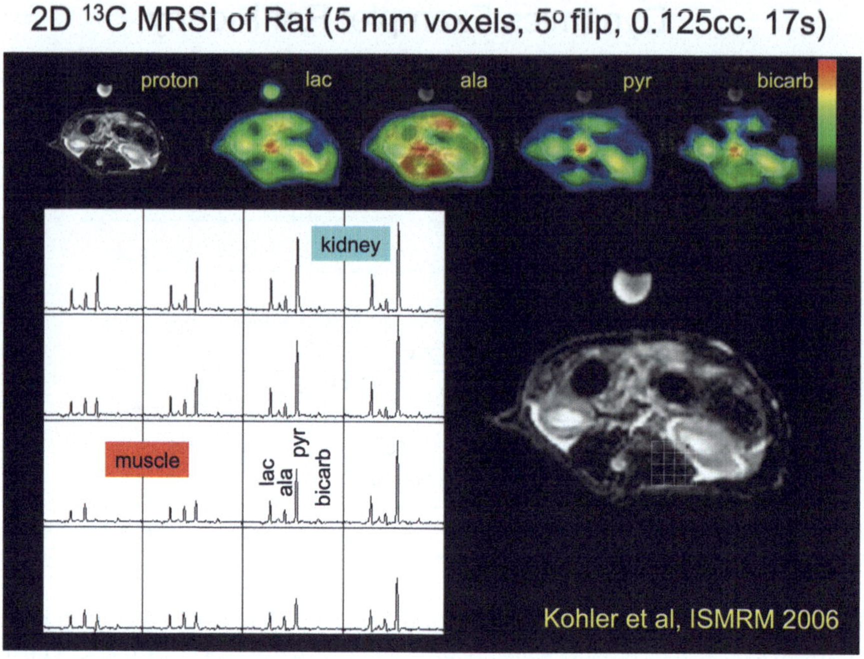

Fig. 4.7 2D ^{13}C MRSI of an axial slice through a normal rat at the level of the kidneys. Metabolite image displays as well as spectral grid insert display is illustrated. Voxel resolution was 5 mm isotropic

tip regions that may saturate the prepolarized substrate spins moving into or out of the selected volume as described above. However, it may be possible to achieve three-dimensional (3D) localization with point-resolved spectroscopy (PRESS) if the refocusing pulses include a "notch," so they do not perturb the hyperpolarized substrate resonance but refocus all other metabolites of interest. A PRESS pulse sequence incorporating spectral-spatial refocusing pulses that have a stop-band positioned at the frequency of the ^{13}C substrate has been demonstrated [14]. From a voxel centered on the rat brains, significantly higher metabolite SNRs were observed in experiments using the spectral-spatial refocusing pulses as compared to the conventional refocusing pulses. The pyruvate signals seen when using the traditional refocusing pulses were quite variable possibly due to the different degree of saturation of the substrate magnetization in these studies. This technique enables acquisition of dynamic ^{13}C magnetic resonance spectroscopy (MRS) data from within the tissue/organ of interest without sacrificing SNR or altering the apparent metabolic kinetics due to saturation of the substrate polarization.

There are some limitations with the aforementioned technique, however, for single voxel localization for hyperpolarized ^{13}C MRS. Relatively high B_1 and long pulse duration are required for notch refocusing pulses that have good spatial and spectral profiles, and hyperpolarized substrates may still be saturated if there are large B_0 inhomogeneities within the coil volume. These issues may be overcome if the localization method requires only a small tip, broadband RF pulse. Spectroscopic Imaging Acquisition Mode (SIAM) was proposed previously to eliminate out of slice artifact for SV MRS acquisitions [15]. In this method, spectroscopic data can be localized to the center slice or voxel of the FOV if *k*-space samples are acquired in the typical pattern for CSI and summed, recovering the single voxel at the center of the FOV. By using a 2D spiral readout gradient trajectory and confining the net duration of each 2D trajectory to a single dwell period of the spectroscopic readout (the digital filter averages the raw data sampled during each trajectory or each dwell time) the output data represent only the center voxel of the image with no *k*-space data sorting and reconstruction required. This proposed sequence can be used practically in the same way as a standard pulse-acquire MRS acquisition for HP ^{13}C experiments, but the spectrum will be localized to a single 3D voxel. It should be feasible that dynamic ^{13}C MRS data can be confined to a single volume from just one pulse-acquire acquisition using this approach Fig. 4.8.

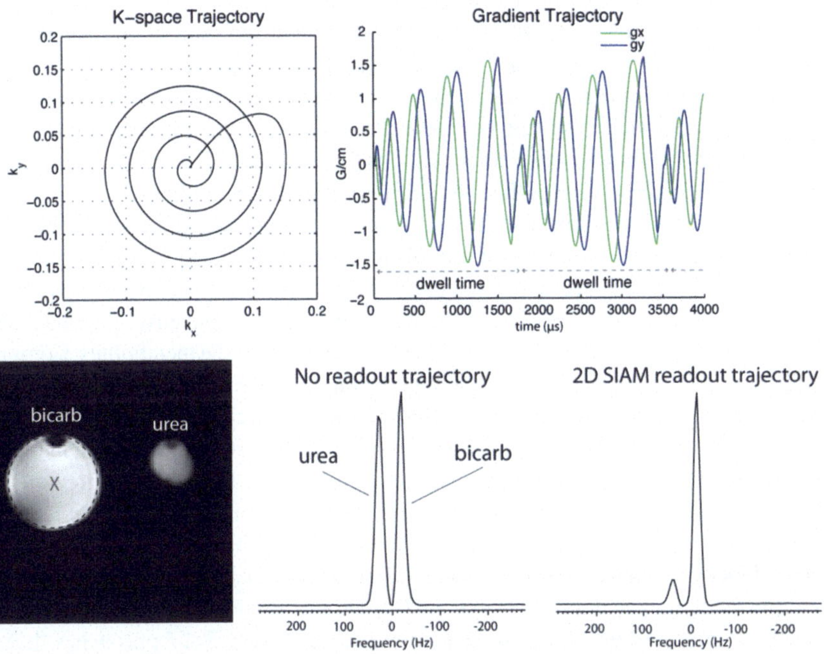

Fig. 4.8 *K*-space (top, left) and gradient trajectories for 2D SIAM readout (top, right) designed to acquire hyperpolarized ^{13}C MRS signal from a single volume without refocusing pulses. This design was incorporated into the readout of a pulse-acquire acquisition and tested with ^{13}C phantoms (bottom, left). With the SIAM readout, signal from outside the volume of interest was largely excluded (bottom, left and right)

4.10 Variable Flip Angles and Multiband Spectral–Spatial Excitation

As mentioned previously, the prepolarized, nonequilibrium magnetization of the ^{13}C nuclei is not recoverable once it decays via T_1 relaxation. All of the hyperpolarized signal can be consumed by just a single 90° RF pulse. However, since the primary interest and utility of hyperpolarized ^{13}C imaging are to investigate dynamic transport and metabolic processes, management of the longitudinal magnetization over its T_1 lifetime becomes very important. The goal would be to not use up the magnetization too quickly before the observation of the substrate uptake and metabolism in the tissue of interest, but also to use an adequately large RF tip-angle, so the data obtained have sufficient SNR. Shortly following administering the prepolarized ^{13}C substrate, the substrate signal is at its maximum, and often only a small flip angle is needed to acquire spectra or images of sufficient SNR. As the substrate signal decreases with T_1 decay, outflow, and metabolic conversion to products, smaller substrate plus the metabolite signals may lack adequate SNR with the same small flip angle. One strategy is to use progressively large tip-angle over the imaging window [16]. This strategy can be helpful when signals from multiple transients (excitations) are used to encode *k*-space, as signal modulations over time would be reduced by the varying flip-angle scheme and lead to more moderate imaging artifacts. This approach may not be optimal for fitting temporally resolved data to derive metabolic fluxes or exchange rates, as flip-angle correction would be required and true flip angle in vivo in the tissue of interest is often not known.

Another strategy to manage the decaying ^{13}C magnetization, as well as the large dynamic range between the maximum substrate signal and the metabolite signals, is to use tailored RF pulses that have spectrally varying flip angle [17]. Using smaller flip angle for the injected ^{13}C substrate can help preserve the nonrecoverable magnetization while using larger flip angle can deliver higher SNR for the metabolites. Combined with fast, dynamic spectroscopic imaging techniques, the use of metabolite specific excitation flip angles may provide the SNR needed to study kinetics in vivo.

For fast volumetric coverage, most of the "speed-up" strategies known in thermal proton imaging and spectroscopy come into play. As touched on in the previous sections, the most potent aspect of hyperpolarized ^{13}C imaging is the potential to observe and quantify metabolic kinetics in real time in vivo. Real-time in vivo metabolic kinetics requires spectral, spatial, and temporal resolution. To adequately sample the dynamic change of the signal due to metabolic conversion, spatially and spectrally resolved data have to be acquired at least every 5 s. Approaches that can accelerate the acquisitions are very much needed.

4.11 EPI and Spiral Readout Trajectories

Echo-planar and spiral readout trajectories are common strategies used in ^{1}H imaging when data need to be acquired in a short time for various reasons. With time-varying gradient trajectories during data acquisition, both techniques are capable of delivering *k*-space information in two dimensions from a single excitation. In many of the applications to hyperpolarized ^{13}C imaging, these approaches are used in conjunction with spectral–spatial excitations for each spectral line or multiecho acquisition (IDEAL) to obtain images from each spectral component [3, 18, 19]. However, both types of trajectories can also be used to acquire spectroscopic data as well and will be discussed in the sections below.

4.12 Compressed Sensing

Another way to increase acquisition speed is to reduce the number of *k*-space points, filling those unsampled regions with zeros. However, under-sampled *k*-space leads to blurring and other artifacts. It has been demonstrated that compressed sensing, a technique based on random under-sampling and nonlinear reconstruction, works well for MR imaging when the signals exhibit sparsity and adequate SNR [20]. The ^{13}C spectrum derived from injection of most hyperpolarized substrates is sparse, due to the choice of enriching the longer T_1 carbonyl with ^{13}C. This spectral sparsity provides an excellent opportunity to use this technique to accelerate the acquisition significantly. It has indeed been applied to hyperpolarized ^{13}C imaging in spectroscopic imaging to achieve higher spatial resolution within the same acquisition window or to increase acquisition speed to enable acquisition of 3D, time-resolved MR spectroscopic data [21, 22], with a close to eightfold reduction in acquisition speed. However, the acceleration level may be limited in human applications as SNR may be lower as compared to that achieved in preclinical studies. It has also been applied to echo-planar imaging (EPI) acquisition to accelerate 3D, time-resolved imaging of ^{13}C substrate and metabolites [23], but since application in nonspectroscopic imaging relays on just the sparsity in the spatial distribution of the signals, lesser acceleration was achieved.

4.13 Array Coil Methods

Parallel imaging is another technique that is commonly applied in ^{1}H MR imaging to allow omitting the acquisition of some *k*-space points but use the spatially varying signals from the receive coil array to encode the missing spatial information [24–26]. This acceleration method requires multichannel receive array coils. Parallel imaging has not been used routinely yet in the setting of hyperpolarized ^{13}C imaging, as array coils at the ^{13}C frequency, are not commonly used for preclinical studies. However, as hyperpolarized ^{13}C imaging moving toward human translation, receive coil arrays will inevitably be used if only to balance SNR (use coil elements that aren't too large) and spatial coverage. Indeed, as hardware for human hyperpolarized imaging has become more common, parallel imaging applications have been demonstrated. With a three channel receive array, twofold under-sampling was shown for MR spectroscopic data acquisition using SENSE reconstruction [27], twofold acceleration was also shown recently with the simultaneous multislice acquisition using hardware suitable for human cardiac ^{13}C imaging [28]. While these two applications of parallel imaging methods require estimates of the coil sensitivity, an auto-calibrated variant of the technique was demonstrated in hyperpolarized ^{13}C imaging of abdomen with modified EPI trajectories [29].

4.14 Spectral Domain: EPSI and Spiral CSI

Similar to EPI and spiral gradient trajectories for fast imaging that was discussed above, echo-planar spectroscopic imaging and spiral chemical shift imaging techniques employ time-varying gradient waveforms during data acquisition. However, instead of encoding just *k*-space (spatial) information, these methods also encode the spectral information. Both methods have been used extensively in ^{1}H MR spectroscopy, but have also been applied in hyperpolarized ^{13}C imaging since the early days of development of this modality [13, 30]. One significant difference between the two approaches is that echo-planar spectroscopic imaging (EPSI) trajectory typically encodes one spatial dimension, while the spiral CSI method encodes two spatial dimensions by default. This difference leads to the trade-off that during a given readout duration and the same gradient strength and slew rate, it is possible to obtain data with larger spectral bandwidth with EPSI, but with less *k*-space coverage as compared to spiral CSI. Although in some cases, it is possible to design the spiral CSI trajectories that lead to favorable spectral aliasing so all detectable resonance can be measured without overlapping [30]. Another interesting trade-off is that while it is possible to perform random under-sampling using spiral CSI for compressed sensing [31], EPSI trajectory designs seem more amenable for nonuniform *k*-space sampling that leads to greater acceleration [21]. Last but not least the EPSI data reconstruction is somewhat more straightforward if the points on the gradient ramps are not used; and the CSI spiral reconstruction is more involved as *k*-space regridding is required.

4.15 Limited Sets of Spectral–Spatial Excitations

For a ^{13}C spectrum with large chemical shifts between different observable resonances, a simple and elegant way to encode the spectral information is to use spectrally selective RF excitation. Following RF excitation of just a single resonance, only spatial information needs to be encoded during data acquisition. Thus, standard imaging readout can be used, and fast imaging readout like EPI and spiral can be readily applied to speed up data acquisition. Data from a ^{13}C substrate and different metabolites can be acquired sequentially following separate RF excitations on each resonance. One advantage of this approach is that different flip angle can be applied to different species. Thus, small flip angle can be used for the injected hyperpolarized substrate, and larger flip angle can be used for the metabolic products (similar to the multiband pulse approach). This method was first demonstrated for imaging of hyperpolarized ^{13}C lactate following injection of hyperpolarized ^{13}C pyruvate, using spectral–spatial excitation following by a single shot 2D EPI readout [3]; with phase encoding in the third spatial dimension, 3D volumetric coverage was achieved with a desirable temporal resolution. Imaging of the injected substrate or other metabolites from separate excitations can easily be added to acquire complete dynamic hyperpolarized ^{13}C imaging data from all five dimensions (3 spatial dimensions, spectral and temporal dimensions, Fig. 4.9). The inherent flexibility of this method was also demonstrated for cardiac gated, multislice ^{13}C imaging of the heart in a preclinical model as well as humans [18, 32]. Improvements of the readout trajectory designs based on this approach have continued [23, 29, 33–35] and making it one of the more attractive options for different hyperpolarized ^{13}C applications in clinical research. One trade-off of this approach is that to have good spectral selectivity (narrow spectral passband), the RF pulses used are relatively long (~10–20 ms), thus making the minimum echo time longer than conventional RF excitation pulses and reduction in ultimate SNR. Also, caution must be taken to set up the scan before injection of the hyperpolarized ^{13}C substrate so that the RF excitation frequencies are at or very near the expected resonances, as off-resonance can lead to a reduction of SNR or failed experiments (Table 4.1).

4.16 IDEAL

Similar to separating fat and water resonances in ^{1}H imaging, ^{13}C spectrum with well separated and less complex spectral patterns may also be imaged using the DIXON/IDEAL type of approach [4, 36]. This method encodes the spectral domain by shift echo time between a set of excitations. The amount of shifting in echo time is determined by the chemical shift separation between the metabolites, and the number of echoes requires is determined by the number of species encoded (usually $n + 1$). Since the spectral encoding is done not by RF pulses or by the readout, conventional RF pulses, as well as standard imaging data acquisition, can be used, making this approach relatively straightforward to implement. However, one disadvantage is that it is not as compatible with the various schemes that use different RF pulses to manage the nonrecoverable magnetization. Moreover, since it requires many excitations to encode the spectral information, a small flip angle must be used, and the temporal resolution may be limited. Even with these constraints, multislice, dynamic hyperpolarized ^{13}C imaging has been demonstrated using this approach [19]. Moreover, this method has also been incorporated in the selective excitation approach to improving the off-resonance robustness [34]. It should be noted that neither the selective excitation approach nor the IDEAL approach is suitable for hyperpolarized imaging when the spectrum is complex and has a large number of resonances, as in the cases of the doubly labeled substrate or co-polarizations of multiple substrates [37, 38].

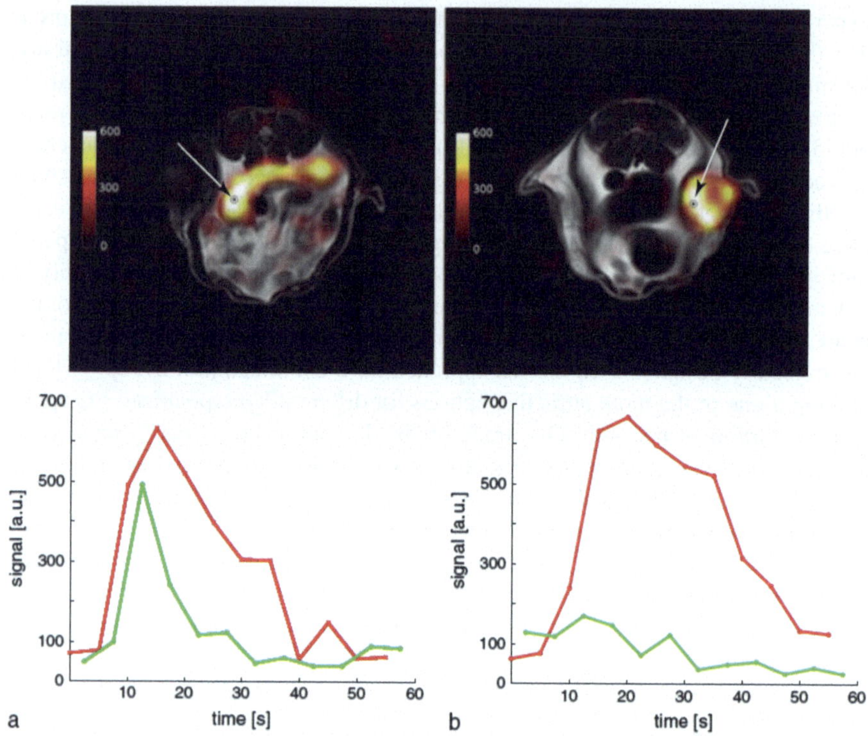

Fig. 4.9 By selective RF excitation of individual spectral components, and using an EPI readout, metabolic imaging that encodes five dimensions (spectral, temporal, and three spatial dimensions) is feasible. ^{13}C lactate images (from the temporal maximum) were overlaid on ^{1}H anatomical images (top row). The signals from individual voxels (left: kidney; right: tumor) and individual metabolites (pyruvate: green; lactate: red) are also plotted as a function of time in this example from a rat model of U87 tumor. Note that larger flip angle (10°) was used on lactate resonance than the pyruvate resonance (3°) and the signal intensities were not corrected for the different flip angle

4.17 Stimulated Echoes and Diffusion Weighting

In almost all hyperpolarized ^{13}C MR data acquired in vivo, the data from each voxel contain a mixture of tissue and vasculature signals. If the signal contributions from these different spaces can be distinguished, it would allow separation of the metabolites that were converted locally in the tissue within the voxel and those that were created by other organs/tissue and carried into the voxel by blood. Separating the substrate signals in the tissue and the blood pools can also potentially allow better estimation of metabolic fluxes in the tissue as the vascular pool often contributes a large substrate signal component that is never taken up by the tissue and used by the cells.

Diffusion-weighted proton MRI has been used extensively to study the effect of fluid motion and tissue structure on image contrast [39]. In this application, dephasing occurs for moving spins under a paired of pulsed gradients [40]. Since the spins

Table 4.1 Comparison of sampling methods

Sequence	Option	Spectral quality	Spatial quality	Dynamic quality	Advantages
Dynamic only	Nonselective	+++	–	+++	High SNR; needed for new agents and to set timing for CSI
	Slice-selective	+++	–	+++	
	Single-voxel	+++	+	+++	
CSI	Single slice	+++	++	–	Easy and robust; added SNR from in flow in axial scans
	Multislice	+++	++	–	
	DSE	+++	++	–	
EPSI	Symmetric	++	++	++	Good spectral quality with some dynamic capability
	Flyback	++	++	++	
spCSI		+	++	+++	Usually needs added gradient performance to fully sample
IDEAL		–	++	–	Poor spectral resolution
Sequential EPI		+	+++	+++	Best option for full dynamics
Sequential spiral		+	+++	+++	Alternative to EPI
STE		++	++	+++	Added specificity in dynamics

in the vascular pool, in general, have greater freedom for motion than the spins in the cells or extracellular space, the vascular component can be suppressed with diffusion-weighted pulse sequences. Depending on the size of the gradient pulses (*b* value), it may also be possible to distinguish components that are intracellular versus extracellular. Diffusion-weighted spectroscopy has been applied in the context to measure apparent diffusion coefficients (ADC) of hyperpolarized metabolites in vitro and in vivo [41–43]. Not surprisingly, the data from these studies suggest that ^{13}C metabolites inside cells, in extracellular space and vascular space have different ADCs. In vivo mapping of hyperpolarized ^{13}C metabolites has also been demonstrated in tumor models using a spin-echo EPI pulse sequence; higher [1-^{13}C]-lactate ADC was observed in brain tumor and lower [1-^{13}C]-lactate ADC was observed in a model prostate tumor [44]. While these results were again not surprising (they correspond to the ^{1}H DWI data from the same models), the potential to distinguish intracellular and extracellular components of the [1-^{13}C]-lactate converted from the injected hyperpolarized [1-^{13}C]-pyruvate may differentiate tumor phenotype and follow disease progression. Besides these studies using spin-echo diffusion-weighted spectroscopy or imaging sequences, a few studies have been done using bipolar gradients in gradient echo imaging sequence for suppressing flowing spins [45–47]. In the context of imaging hyperpolarized ^{13}C pyruvate and its metabolic products, suppressing the flowing spins in the vascular space may improve the accuracy of the metabolic conversion rate measurement. When applying to cardiac perfusion imaging following injection of hyperpolarized ^{13}C urea, the bipolar gradients were shown to be effective in suppressing the signal in the cardiac chambers and allow depiction of first pass perfusion in the myocardium.

In addition to using the pulsed gradient pair with spin-echo based pulse sequences, diffusion weighting can also be created using stimulated echo (STE) pulse sequences since STE is inherently sensitive to motion. In STimulated Echo Acquisition Mode (STEAM) pulse sequence [48], moving spins are suppressed by the motion during dephasing and rephrasing after the initial 90° RF pulse and after the STE readout RF pulse. The diffusion weighting can also be enhanced by the mixing time, thus allowing a higher degree of molecular motion encoding. However, because of the nonrecoverable nature of the hyperpolarized magnetization created by dynamic nuclear polarization (DNP), the STE preparation which utilizes a pair of 90° RF pulses is typically only used once and followed by a series of small tip angle readout pulses to encoding space and temporal information [49]. The primary consequence of performing an STE experiment in this fashion is that the signal acquired in the later readouts would experience longer mixing time thus greater degree of diffusion weighting and can cause imaging blurring or more than expected signal decay [50]. A super-stimulated preparation (super-STEP) described by Larson et al. use a train of RF pulses instead of the traditional 90° pulse pair to generate a square-wave encoding in the magnetization [51]. The benefit of such an approach is that after the preparation spins are encoded along Mz and weighting to motion is frozen before the successive readouts, so the *k*-space data or time-resolved data will all experience the same diffusion weighting created during the super-STE preparation. In models of liver and prostate tumors, the super-STEP was shown to provide greater contrast for [1-^{13}C]lactate converted from the injected hyperpolarized [1-^{13}C]pyruvate in the tumor. A potential disadvantage of the super-STEP is that it requires relatively high B_1 that may not be feasible for larger human size coils. Also, the errors accumulated during the long RF pulse trains may reduce its advantages over the conventional STEAM approach in some situations.

4.18 Agent-Specific Spectral–Spatial Excitation (Substrate Spin-Tagging)

Another way to differentiate ^{13}C metabolic products created in the intracellular space from those arriving via the blood pool is to use sinusoidal spatial tagging of the longitudinal magnetization. Substrate spins that reside in the intravascular space when spin tagging is performed will be strongly suppressed in the data acquisition if a sufficient time interval is allowed between the tagging and the readout, as flow would displace these molecules. Also, if the spin tagging is performed exclusively on the ^{13}C spins of the injected substrate, then the ^{13}C signal from spatially tagged metabolic products in the subsequent data acquisitions can be differentiated from any inflowing products since these will not exhibit the tagging pattern. A spin-tagging pulse sequence based on the STEAM pulse sequence has been demonstrated [52]. In this design, the sinusoidal spatial tagging is created by a gradient pulse following the initial RF excitation (RF1-90°), the prepared magnetization is then stored in the longitudinal plane by another RF pulse (RF2-90°). Similar to the diffusion-weighted hyperpolarized ^{13}C experiment mentioned above using STE, the spin tagging is performed only once, then the tagged longitudinal magnetization is then accessed by consecutive small tip angle RF pulses (RF3). Each small-tip readout RF pulse, followed by a gradient-pulse matching the size as the tagging gradient pulse, is followed by a data acquisition window. Both a conventional slice selective pulse and a spectral–spatial pulse that only excites [1-^{13}C]pyruvate can be used for RF1, resulting in tagging that is achieved either on all resonances in the spectrum or only the injected substrate. In vivo experiments in rat heart and kidney performed using spectral–spatial excitation pulse during the tagging preparation showed that ^{13}C signal from metabolic products originated from the tagged spins can be observed.

4.19 Toward a Routine [1-^{13}C]Pyruvate Metabolic Imaging Protocol

While multinuclear spectroscopy and imaging packages are available from vendors of MR scanners that includes hardware and software that permit non-^{1}H experiments to be performed, the body transmit RF coil (if available), and almost all of the receive coils available are designed for operation only at the ^{1}H resonance frequency. Therefore, many of the ^{13}C studies to date were performed using custom-built coils. These coils are either transmit-receive (TR) volume coils or surface coils, or receive coils with a dedicated ^{13}C transmit coil [11, 27, 53, 54]. While many preliminary preclinical hyperpolarized ^{13}C MRS studies were performed using a simple ^{13}C TR surface coil over the tissue of interest [55, 56], a ^{13}C volume TR coil or a ^{13}C volume transmit-only coil paired with ^{13}C receiver coils are desirable for more advanced hyperpolarized imaging studies in preclinical models and studies in humans [57]. Volume transmit-coils potentially provide uniform B_1 field over the tissue of interest and would allow better management or manipulation of the nonrecoverable prepolarized magnetization through various sampling schemes described in this chapter. Dual-tuned birdcage volume TR coil is a popular choice for small animal hyperpolarized ^{13}C studies because it provides both ^{1}H and ^{13}C coils within the same coil housing and can provide easy subject handling [58, 59]. Having both ^{1}H and ^{13}C coils in the same mechanical support also provides the opportunities to acquire high-resolution anatomical ^{1}H images or other ^{1}H MR data and then perform the hyperpolarized ^{13}C metabolic imaging without disturbing the subject during the study.

The disadvantage of a TR volume coil is the lower SNR as compared to local receive coils. Thus, for the current state of the art ^{1}H MR imaging, studies are performed using the system body ^{1}H coil for RF transmission and array ^{1}H receive coils for signal reception. Similarly, in the first hyperpolarized ^{13}C studies in human subjects, ^{13}C data were acquired from prostates using a clamshell volume transmit coil and a dual-tuned ^{1}H–^{13}C endorectal receive coil [57]. Volume-coil transmit–array receive has also been developed for hyperpolarized ^{13}C cardiac imaging [54]. It is reasonable to assume that similar concept will also be applied to other applications and organs as hyperpolarized ^{13}C metabolic imaging is translated to clinical settings. It is worth noting that one of the reasons that make the volume transmit–array receive setup so widely adapted for ^{1}H imaging is because almost all clinical MR systems have body transmit coils that are integrated into the scanner; it allows receive-coil-arrays of many sizes and shapes depending on the application and anatomy of interest. However, such integrated transmit coil for ^{13}C frequency is not available on any scanner yet. Therefore, a dedicated volume-transmit-coil tuned for ^{13}C needs to reside inside the scanner bore. This requirement limits the size, coverage, and often the performance of both transmit and the receive coils. It may also impact patient handling and comfort and the size of the subject that can be scanned. Ultimately for routine hyperpolarized ^{13}C imaging, body size ^{13}C transmit coil integrated into the scanner (e.g., a dual-tuned ^{1}H–^{13}C body coil or ^{1}H and ^{13}C body coils in layers) will be a crucial step toward that.

Other than some of the dual-tuned ^{1}H–^{13}C coils developed for small animal imaging and the dual-tuned endorectal coil used in the first-in-human study, most efforts in design and construction of ^{13}C coil hardware are made independently from the ^{1}H hardware. This approach is often necessary to reduce the cost and complexity. Thus, in large animal and many future human studies, only the system ^{1}H body coil is used in the TR mode to acquire the necessary anatomical ^{1}H images for prescribing the ^{13}C imaging experiment and for data overlay, either due to all available bore space taken by ^{13}C hardware, or ^{1}H receive array receiver that would be used is not compatible with the ^{13}C coils. As mentioned previously, local receiver arrays provide better SNR as compared to volume receive coil, so for many hyperpolarized ^{13}C studies, the ^{1}H imaging performance is not optimal, and that limits the spatial resolution and amount of data that can be obtained from the subject at the ^{1}H frequency. Alternatively, ^{1}H imaging can be performed with ^{1}H array receive coils without the ^{13}C hardware, and then the ^{1}H receive coils can be removed and replace with the ^{13}C transmit-receive hardware. This approach would potentially allow acquisitions of both ^{1}H and ^{13}C data using the best currently available hardware for each frequency, but it certainly adds experimental complexity and possible issue for data co-registration if the subject has to be moved. Therefore, besides the integrated body transmit coil for both ^{1}H and ^{13}C frequency, integrated receive arrays that work at both ^{1}H and ^{13}C frequency would be ideal.

4.20 Prescan Operation

On modern MR systems, before running any MR experiment, a series of automatic prescan operations are performed to set up parameters that include (but not limited to) transmit/receive frequency, flip angle (transmit power), B_0 shim coil currents, and receiver gain. Typically, these prescan operations utilize the endogenous ^{1}H signal from the subject and requires very little time (<30 s). However, for ^{13}C studies, the low natural abundance of the ^{13}C nucleus and the relatively low concentrations of carbon atom in general (besides adipose tissue) present a challenge for these operations. One popular way to overcome the low endogenous signal issue is to insert a small phantom that contains a concentrated ^{13}C enriched solution into the field of view to provide the ^{13}C MR signal for prescan purpose. A limitation of such an approach is that the phantom is in most cases outside the subject and some distances from the region of interest, so the frequencies, transmit power, and shim currents determined from this signal source may deviate from the optimal settings required for the tissue being imaged. Depending on the coil geometry and the anatomy being studied, the size and location of the ^{13}C insert phantom can be far from optimal. Fortunately for B_0 shimming, B_0 maps acquired using endogenous ^{1}H signal and the shim currents calculated from these maps can be directly applied for the ^{13}C experiments as B_0 that can be achieved is independent of the resonance frequency used in a given experiment.

Similarly, reference data that are often required for artifact reductions in EPI experiments can be acquired using the ^{1}H signal and applied to the ^{13}C data. The receiver gain calibration is something that is more difficult to address either with an insert ^{13}C enriched phantom or using the endogenous ^{1}H signal as the potential dynamic range of the hyperpolarized ^{13}C signal amplitude can be quite large depending on the polarization that is achieved. Fortunately, on modern MR systems, SNR drops across the range of the receiver gain available is not very large (a few %), so manually setting the receiver gain in the lower part of the range would be helpful to ensure that receiver is not saturated.

Appropriate transmit power determined using the ^{13}C phantom insert for desired flip angle has typically been used in hyperpolarized ^{13}C studies. For a volume transmit coil with relatively uniform B_1 profile, such as a birdcage coil, this approach is reasonable. For transmit coil with a large spatially varying B_1, an offset between the flip angle at the phantom and the flip angle at the volume of interest would have to be taken into account. This offset can be predetermined by simulation or by B_1 map measurement using phantom. Regardless the coil or the approach for flip-angle calibration, relying on an external phantom as the signal source can lead to uncertainty that is potentially substantial. Thus, quantitation or data analysis methods that require precise knowledge of flip angle used during the ^{13}C data acquisition may not be suitable for this application.

4.21 Phantoms and Inserts

Phantom is an integral part of MR, regardless of the application or nuclei being studied. Beyond the routine function and performance verifications, appropriate phantoms are essential for pulse sequence development and testing. Since [1-^{13}C]pyruvate is the most widely used hyperpolarized substrate for ^{13}C metabolic imaging, phantoms that simulate the MR spectrum of hyperpolarized [1-^{13}C]pyruvate and its metabolites in vivo have been shown [13, 18, 19]. Enriched and concentrated solutions (1 M or higher) of [1-^{13}C]lactate, [1-^{13}C]alanine, [1-^{13}C]bicarbonate can represent itself in the phantom but substitute for [1-^{13}C]pyruvate is necessary since pyruvate degrades in solution at room temperature. Glycine and formate have been shown as a substitute for pyruvate in some phantoms. These multifrequency phantoms are especially useful for developing sampling schemes that do not include spectroscopic readouts that fully resolves the spectrum, approaches such as spectral–spatial excitations of individual resonances or IDEAL with fast imaging readout.

As discussed in the previous section, a source of a reliable ^{13}C signal within the FOV is required for prescan operation. ^{13}C enriched urea or lactate at high concentration (>2 M) in small spheres or tubes are often used in this case. These phantoms are often positioned externally on the receive coil or placed within the TR volume coil next to the subject for small animal imaging. It is also possible to put the phantom in the coil housing as in the case of the endorectal coil [57]. These phantoms are often doped with Gd to reduce the T_1 of the ^{13}C nuclei from tens of seconds to ~1 s so the phantom can be used for flip-angle calibration efficiently.

Acknowledgments We acknowledge funding support from NIH (P41 EB015891, AA05965, AA018681, AA13521-INIA, and EB009070), The Lucas Foundation, and GE Healthcare.

Problems

1. At 3 T using an RF pulse with a bandwidth of 1000 Hz to obtain a 5 mm slice, what would the spatial offset be between [2-^{13}C]pyruvate at 208 ppm and [2-^{13}C]lactate at 71 ppm?
2. To measure a shift in oxidative and glycolytic metabolism (e.g., the Warburg effect), a ratio of the products [1-^{13}C]lactate and [^{13}C]bicarbonate or [2-^{13}C]lactate and [5-^{13}C]glutamate can be used. In this case, is there value or disadvantage in measuring signal from the injected [^{13}C]pyruvate?

References

1. Mayer, D., Levin, Y.S., Hurd, R.E., Glover, G.H., Spielman, D.M.: Fast metabolic imaging of systems with sparse spectra: application for hyperpolarized 13C imaging. Magn. Reson. Med. **56**(4), 932–937 (2006). https://doi.org/10.1002/mrm.21025
2. Larson, P.E., Hu, S., Lustig, M., Kerr, A.B., Nelson, S.J., Kurhanewicz, J., Pauly, J.M., Vigneron, D.B.: Fast dynamic 3D MR spectroscopic imaging with compressed sensing and multiband excitation pulses for hyperpolarized 13C studies. Magn. Reson. Med. **65**(3), 610–619 (2011). https://doi.org/10.1002/mrm.22650

3. Cunningham, C.H., Chen, A.P., Lustig, M., Hargreaves, B.A., Lupo, J., Xu, D., Kurhanewicz, J., Hurd, R.E., Pauly, J.M., Nelson, S.J., Vigneron, D.B.: Pulse sequence for dynamic volumetric imaging of hyperpolarized metabolic products. J. Magn. Reson. **193**(1), 139–146 (2008)
4. Reeder, S.B., Pineda, A.R., Wen, Z., Shimakawa, A., Yu, H., Brittain, J.H., Gold, G.E., Beaulieu, C.H., Pelc, N.J.: Iterative decomposition of water and fat with echo asymmetry and least-squares estimation (IDEAL): application with fast spin-echo imaging. Magn. Reson. Med. **54**(3), 636–644 (2005). https://doi.org/10.1002/mrm.20624
5. Josan, S., Park, J.M., Hurd, R., Yen, Y.F., Pfefferbaum, A., Spielman, D., Mayer, D.: In vivo investigation of cardiac metabolism in the rat using MRS of hyperpolarized [1-13C] and [2-13C]pyruvate. NMR Biomed. **26**(12), 1680–1687 (2013). https://doi.org/10.1002/nbm.3003
6. Josan, S., Hurd, R., Park, J.M., Yen, Y.F., Watkins, R., Pfefferbaum, A., Spielman, D., Mayer, D.: Dynamic metabolic imaging of hyperpolarized [2-(13) C]pyruvate using spiral chemical shift imaging with alternating spectral band excitation. Magn. Reson. Med. **71**(6), 2051–2058 (2014). https://doi.org/10.1002/mrm.24871
7. Chen, A.P., Tropp, J., Hurd, R.E., Van Criekinge, M., Carvajal, L.G., Xu, D., Kurhanewicz, J., Vigneron, D.B.: In vivo hyperpolarized 13C MR spectroscopic imaging with 1H decoupling. J. Magn. Reson. **197**(1), 100–106 (2009). https://doi.org/10.1016/j.jmr.2008.12.004
8. Josan, S., Yen, Y.F., Hurd, R., Pfefferbaum, A., Spielman, D., Mayer, D.: Application of double spin echo spiral chemical shift imaging to rapid metabolic mapping of hyperpolarized [1-(1)(3)C]-pyruvate. J. Magn. Reson. **209**(2), 332–336 (2011). https://doi.org/10.1016/j.jmr.2011.01.010
9. Granlund, K., Keshari, K.: Temoporal denoising of hyperpolarized data improves SNR while perserving dynamic information, p. 3074. ISMRM, Paris (2018)
10. Milshteyn, E., von Morze, C., Reed, G.D., Shang, H., Shin, P.J., Zhu, Z., Chen, H.-Y., Bok, R., Goga, A., Kurhanewicz, J., Larson, P.E.Z., Vigneron, D.B.: Development of high resolution 3D hyperpolarized carbon-13 MR molecular imaging techniques. Magn. Reson. Imaging. **38**, 152–162 (2017). https://doi.org/10.1016/j.mri.2017.01.003
11. Kohler, S.J., Yen, Y., Wolber, J., Chen, A.P., Albers, M.J., Bok, R., Zhang, V., Tropp, J., Nelson, S., Vigneron, D.B., Kurhanewicz, J., Hurd, R.E.: In vivo 13 carbon metabolic imaging at 3T with hyperpolarized 13C-1-pyruvate. Magn. Reson. Med. **58**(1), 65–69 (2007)
12. Bottomley, P.A.: Spatial localization in NMR spectroscopy in vivo. Ann. N. Y. Acad. Sci. **508**(333), 333–348 (1987)
13. Cunningham, C.H., Chen, A.P., Albers, M.J., Kurhanewicz, J., Hurd, R.E., Yen, Y., Nelson, S.J., Vigneron, D.B.: Double spin-echo sequence for rapid spectroscopic imaging of hyperpolarized 13C. J. Magn. Reson. **287**(2), 357–362 (2007)
14. Chen, A.P., Cunningham, C.H.: Single voxel localization for dynamic hyperpolarized (13)C MR spectroscopy. J. Magn. Reson. **258**, 81–85 (2015). https://doi.org/10.1016/j.jmr.2015.07.002
15. Hurd, R.E.: Volume spectroscopy having image artifact reduction USA Patent US5804966A
16. Yen, Y.F., Kohler, S.J., Chen, A.P., Tropp, J., Bok, R., Wolber, J., Albers, M.J., Gram, K.A., Zierhut, M.L., Park, I., Zhang, V., Hu, S., Nelson, S.J., Vigneron, D.B., Kurhanewicz, J., Dirven, H.A., Hurd, R.E.: Imaging considerations for in vivo 13C metabolic mapping using hyperpolarized 13C-pyruvate. Magn. Reson. Med. **62**(1), 1–10 (2009). https://doi.org/10.1002/mrm.21987
17. Larson, P.E.Z., Kerr, A.B., Chen, A.P., Lustig, M.S., Zierhut, M.L., Hu, S., Cunningham, C.H., Pauly, J.M., Kurhanewicz, J., Vigneron, D.B.: Multiband excitation pulses for hyperpolarized 13C dynamic chemical-shift imaging. J. Magn. Reson. **194**(1), 121–127 (2008). https://doi.org/10.1016/j.jmr.2008.06.010
18. Lau, A.Z., Chen, A.P., Ghugre, N.R., Ramanan, V., Lam, W.W., Connelly, K.A., Wright, G.A., Cunningham, C.H.: Rapid multislice imaging of hyperpolarized 13C pyruvate and bicarbonate in the heart. Magn. Reson. Med. **64**(5), 1323–1331 (2010). https://doi.org/10.1002/mrm.22525
19. Wiesinger, F., Weidl, E., Menzel, M.I., Janich, M.A., Khegai, O., Glaser, S.J., Haase, A., Schwaiger, M., Schulte, R.F.: IDEAL spiral CSI for dynamic metabolic MR imaging of hyper-

polarized [1-13C]pyruvate. Magn. Reson. Med. **68**(1), 8–16 (2011). https://doi.org/10.1002/mrm.23212

20. Lustig, M., Donoho, D., Pauly, J.M.: Sparse MRI: the application of compressed sensing for rapid MR imaging. Magn. Reson. Med. **58**(6), 1182–1195 (2007). https://doi.org/10.1002/mrm.21391
21. Hu, S., Lustig, M., Chen, A.P., Crane, J., Kerr, A., Kelley, D.A., Hurd, R., Kurhanewicz, J., Nelson, S.J., Pauly, J.M., Vigneron, D.B.: Compressed sensing for resolution enhancement of hyperpolarized 13C flyback 3D-MRSI. J. Magn. Reson. **192**(2), 258–264 (2008). https://doi.org/10.1016/j.jmr.2008.03.003
22. Larson, P.E.Z., Hu, S., Lustig, M., Kerr, A.B., Nelson, S.J., Kurhanewicz, J., Pauly, J.M., Vigneron, D.B.: Fast dynamic 3D MR spectroscopic imaging with compressed sensing and multiband excitation pulses for hyperpolarized 13C studies. Magn. Reson. Med. **65**, 610 (2010). https://doi.org/10.1002/mrm.22650
23. Geraghty, B.J., Lau, J.Y.C., Chen, A.P., Cunningham, C.H.: Accelerated 3D echo-planar imaging with compressed sensing for time-resolved hyperpolarized 13C studies. Magn. Reson. Med. **77**(2), 538–546 (2016). https://doi.org/10.1002/mrm.26125
24. Sodickson, D.K., Manning, W.J.: Simultaneous acquisition of spatial harmonics (SMASH): fast imaging with radiofrequency coil arrays. Magn. Reson. Med. **38**(4), 591–603 (1997)
25. Pruessmann, K.P., Weiger, M., Scheidegger, M.B., Boesiger, P.: SENSE: sensitivity encoding for fast MRI. Magn. Reson. Med. **42**(5), 952–962 (1999)
26. Larkman, D.J., Hajnal, J.V., Herlihy, A.H., Coutts, G.A., Young, I.R., Ehnholm, G.: Use of multicoil arrays for separation of signal from multiple slices simultaneously excited. J. Magn. Reson. Imaging. **13**(2), 313–317 (2001)
27. Tropp, J., Lupo, J.M., Chen, A., Calderon, P., McCune, D., Grafendorfer, T., Ozturk-Isik, E., Larson, P.E.Z., Hu, S., Yen, Y.-F., Robb, F., Bok, R., Schulte, R., Xu, D., Hurd, R., Vigneron, D., Nelson, S.: Multi-channel metabolic imaging, with SENSE reconstruction, of hyperpolarized [1–13C] pyruvate in a live rat at 3.0tesla on a clinical MR scanner. J. Magn. Reson. **208**(1), 171–177 (2011). https://doi.org/10.1016/j.jmr.2010.10.007
28. Lau, A.Z., Lau, J.Y.C., Chen, A.P., Cunningham, C.H.: Simultaneous multislice acquisition without trajectory modification for hyperpolarized 13C experiments. Magn. Reson. Med. **100**, 10158–10157 (2018). https://doi.org/10.1002/mrm.27136
29. Gordon, J.W., Hansen, R.B., Shin, P.J., Feng, Y., Vigneron, D.B., Larson, P.E.Z.: 3D hyperpolarized C-13 EPI with calibrationless parallel imaging. J. Magn. Reson. **289**, 92–99 (2018). https://doi.org/10.1016/j.jmr.2018.02.011
30. Mayer, D., Yen, Y.-F., Tropp, J., Pfefferbaum, A., Hurd, R.E., Spielman, D.M.: Application of subsecond spiral chemical shift imaging to real-time multislice metabolic imaging of the rat in vivo after injection of hyperpolarized13C1-pyruvate. Magn. Reson. Med. **62**(3), 557–564 (2009). https://doi.org/10.1002/mrm.22041
31. DeVience, S.J., Mayer, D.: Speeding up dynamic spiral chemical shift imaging with incoherent sampling and low-rank matrix completion. Magn. Reson. Med. **77**(3), 951–960 (2017). https://doi.org/10.1002/mrm.26170
32. Cunningham, C.H., Lau, J.Y.C., Chen, A.P., Geraghty, B.J., Perks, W.J., Roifman, I., Wright, G.A., Connelly, K.A.: Hyperpolarized 13C metabolic MRI of the human HeartNovelty and significance. Circ. Res. **119**(11), 1177–1182 (2016). https://doi.org/10.1161/CIRCRESAHA.116.309769
33. Gordon, J.W., Vigneron, D.B., Larson, P.E.: Development of a symmetric echo planar imaging framework for clinical translation of rapid dynamic hyperpolarized (13) C imaging. Magn. Reson. Med. **77**(2), 826–832 (2017). https://doi.org/10.1002/mrm.26123
34. Lau, J.Y.C., Geraghty, B.J., Chen, A.P., Cunningham, C.H.: Improved tolerance to off-resonance in spectral-spatial EPI of hyperpolarized [1-(13) C]pyruvate and metabolites. Magn. Reson. Med. **80**(3), 925–934 (2018). https://doi.org/10.1002/mrm.27086
35. Geraghty, B.J., Lau, J.Y.C., Chen, A.P., Cunningham, C.H.: Dual-Echo EPI sequence for integrated distortion correction in 3D time-resolved hyperpolarized (13) C MRI. Magn. Reson. Med. **79**(2), 643–653 (2018). https://doi.org/10.1002/mrm.26698

36. Dixon, W.T.: Simple proton spectroscopic imaging. Radiology. **153**(1), 189–194 (1984). https://doi.org/10.1148/radiology.153.1.6089263
37. Chen, A.P., Hurd, R.E., Schroeder, M.A., Lau, A.Z., Gu, Y.P., Lam, W.W., Barry, J., Tropp, J., Cunningham, C.H.: Simultaneous investigation of cardiac pyruvate dehydrogenase flux, Krebs cycle metabolism and pH, using hyperpolarized [1,2-(13)C2]pyruvate in vivo. NMR Biomed. **25**(2), 305–311 (2012). https://doi.org/10.1002/nbm.1749
38. Wilson, D.M., Keshari, K.R., Larson, P.E.Z., Chen, A.P., Hu, S., Criekinge, M.V., Bok, R., Nelson, S.J., Macdonald, J.M., Vigneron, D.B., Kurhanewicz, J.: Multi-compound polarization by DNP allows simultaneous assessment of multiple enzymatic activities in vivo. J. Magn. Reson. **205**(1), 141–147 (2010). https://doi.org/10.1016/j.jmr.2010.04.012
39. Le Bihan, D., Breton, E., Lallemand, D., Grenier, P., Cabanis, E., Laval-Jeantet, M.: MR imaging of intravoxel incoherent motions: application to diffusion and perfusion in neurologic disorders. Radiology. **161**(2), 401–407 (1986)
40. Stejskal, E.O., Tanner, J.E.: Spin diffusion measurements: spin echoes in the presence of a time-dependent field gradient. J. Chem. Phys. **42**(1), 288–292 (1965). https://doi.org/10.1063/1.1695690
41. Schilling, F., Düwel, S., Köllisch, U., Durst, M., Schulte, R.F., Glaser, S.J., Haase, A., Otto, A.M., Menzel, M.I.: Diffusion of hyperpolarized 13C-metabolites in tumor cell spheroids using real-time NMR spectroscopy. NMR Biomed. **26**(5), 557–568 (2012). https://doi.org/10.1002/nbm.2892
42. Kettunen, M.I., Kennedy, B.W.C., Hu, D.-E., Brindle, K.M.: Spin echo measurements of the extravasation and tumor cell uptake of hyperpolarized [1- 13C]lactate and [1- 13C]pyruvate. Magn. Reson. Med. **70**(5), 1200–1209 (2012). https://doi.org/10.1002/mrm.24591
43. Koelsch, B.L., Keshari, K.R., Peeters, T.H., Larson, P.E.Z., Wilson, D.M., Kurhanewicz, J.: Diffusion MR of hyperpolarized 13C molecules in solution. Analyst. **138**(4), 1011 (2013). https://doi.org/10.1039/c2an36715g
44. Koelsch, B.L., Reed, G.D., Keshari, K.R., Chaumeil, M.M., Bok, R., Ronen, S.M., Vigneron, D.B., Kurhanewicz, J., Larson, P.E.Z.: Rapid in vivo apparent diffusion coefficient mapping of hyperpolarized 13C metabolites. Magn. Reson. Med. **74**(3), 622–633 (2014). https://doi.org/10.1002/mrm.25422
45. Gordon, J.W., Niles, D.J., Adamson, E.B., Johnson, K.M., Fain, S.B.: Application of flow sensitive gradients for improved measures of metabolism using hyperpolarized 13c MRI. Magn. Reson. Med. **75**(3), 1242–1248 (2015). https://doi.org/10.1002/mrm.25584
46. Lau, A.Z., Miller, J.J., Robson, M.D., Tyler, D.J.: Cardiac perfusion imaging using hyperpolarized 13c urea using flow sensitizing gradients. Magn. Reson. Med. **75**(4), 1474–1483 (2015). https://doi.org/10.1002/mrm.25713
47. Lau, A.Z., Miller, J.J., Robson, M.D., Tyler, D.J.: Simultaneous assessment of cardiac metabolism and perfusion using copolarized [1-13 C]pyruvate and 13 C-urea. Magn. Reson. Med. **77**, 151 (2016). https://doi.org/10.1002/mrm.26106
48. Frahm, J., Merboldt, K.D., Hanicke, W., Haase, A.: Stimulated echo imaging. J. Magn. Reson. **64**, 81–93 (1985)
49. Merboldt, K.D., Hänicke, W., Frahm, J.: Diffusion imaging using stimulated echoes. Magn. Reson. Med. **19**(2), 233–239 (1991)
50. Larson, P.E.Z., Hurd, R.E., Kerr, A.B., Pauly, J.M., Bok, R.A., Kurhanewicz, J., Vigneron, D.B.: Perfusion and diffusion sensitive 13C stimulated-echo MRSI for metabolic imaging of cancer. Magn. Reson. Imaging. **31**(5), 635–642 (2013). https://doi.org/10.1016/j.mri.2012.10.020
51. Larson, P.E.Z., Kerr, A.B., Reed, G.D., Hurd, R.E., Kurhanewicz, J., Pauly, J.M., Vigneron, D.B.: Generating super stimulated-echoes in MRI and their application to hyperpolarized C-13 diffusion metabolic imaging. IEEE Trans. Med. Imaging. **31**(2), 265–275 (2012). https://doi.org/10.1109/TMI.2011.2168235
52. Chen, A.P., Hurd, R.E., Cunningham, C.H.: Spin tagging for hyperpolarized (1)(3)C metabolic studies. J. Magn. Reson. **214**(1), 319–323 (2012). https://doi.org/10.1016/j.jmr.2011.10.007

53. Lim, H., Thind, K., Martinez-Santiesteban, F.M., Scholl, T.J.: Construction and evaluation of a switch-tuned (13) C - (1) H birdcage radiofrequency coil for imaging the metabolism of hyperpolarized (13) C-enriched compounds. J. Magn. Reson. Imaging. **40**(5), 1082–1090 (2014). https://doi.org/10.1002/jmri.24458
54. Dominguez-Viqueira, W., Lau, A.Z., Chen, A.P., Cunningham, C.H.: Multichannel receiver coils for improved coverage in cardiac metabolic imaging using prepolarized 13C substrates. Magn. Reson. Med. **70**(1), 295–300 (2012). https://doi.org/10.1002/mrm.24460
55. Schroeder, M.A., Cochlin, L.E., Heather, L.C., Clarke, K., Radda, G.K., Tyler, D.J.: In vivo assessment of pyruvate dehydrogenase flux in the heart using hyperpolarized carbon-13 magnetic resonance. Proc. Natl. Acad. Sci. U. S. A. **105**(33), 12051–12056 (2008). https://doi.org/10.1073/pnas.0805953105
56. Day, S.E., Kettunen, M.I., Gallagher, F.A., Hu, D., Lerche, M., Wolber, J., Golman, K., Ardenkjaer-Larsen, J.H., Brindle, K.M.: Detecting tumor response to treatment using hyperpolarized 13C magnetic resonance imaging and spectroscopy. Nat. Med. **13**(11), 1382–1387 (2007)
57. Nelson, S.J., Kurhanewicz, J., Vigneron, D.B., Larson, P.E.Z., Harzstark, A.L., Ferrone, M., Van Criekinge, M., Chang, J.W., Bok, R., Park, I., Reed, G., Carvajal, L., Small, E.J., Munster, P., Weinberg, V.K., Ardenkjaer-Larsen, J.H., Chen, A.P., Hurd, R.E., Odegardstuen, L.I., Robb, F.J., Tropp, J., Murray, J.A.: Metabolic imaging of patients with prostate cancer using hyperpolarized [1-13C]pyruvate. Sci. Transl. Med. **5**(198), 198ra108 (2013). https://doi.org/10.1126/scitranslmed.3006070
58. Derby, K., Tropp, J., Hawryszko, C.: Design and evaluation of a novel dual-tuned resonator for spectroscopic imaging. J. Magn. Reson. **86**, 645–651 (1990)
59. Chen, A.P., Albers, P., Kohler, S.J., Hurd, R.E., Tropp, J., Pels, P., Zierhut, M.L., Bok, R.A., Nelson, S.J., Kurhanewicz, J., Vigneron, D.B.: High-resolution hyperpolarized C-13 spectroscopic imaging of the TRAMP mouse at 3T, p. 587. Proc., ISMRM, 14th Annual Meeting, Seattle (2006)

Further Reading

With translation of this metabolic imaging method well under way (Kurhanewicz J et al., NEO 2019), selecting and optimizing data acquisition protocol for different applications will require careful consideration and potentially further experimentations. Incorporating different sampling strategy to highlight or suppress certain metabolic contrasts may also help to distinguish between different tissue types or interrogate metabolic changes due to diseases. For further reading see also:

Chen, H.Y., Gordon, J.W., Bok, R.A., Cao, P., von Morze, C., van Criekinge, M., Milshteyn, E., Carvajal, L., Hurd, R.E., Kurhanewicz, J., Vigneron, D.B., Larson, P.E.Z.: Pulse sequence considerations for quantification of pyruvate-to-lactate conversion kPL in hyperpolarized (13) C imaging. NMR Biomed. **32**(3), e4052 (2019). https://doi.org/10.1002/nbm.4052

Hurd, R.E., Yen, Y.F., Chen, A., Ardenkjaer-Larsen, J.H.: Hyperpolarized 13C metabolic imaging using dissolution dynamic nuclear polarization. J Magn Reson Imaging. **36**(6), 1314–1328 (2012). https://doi.org/10.1002/jmri.23753

Chapter 5
Kinetic Modeling of Enzymatic Reactions in Analyzing Hyperpolarized NMR Data

Daniel M. Spielman and Jae Mo Park

5.1 Overview

Exploiting a technique known as dissolution dynamic nuclear polarization (dDNP) [1–3], hyperpolarized (HP) ^{13}C magnetic resonance spectroscopy (MRS) is an emerging imaging technology offering a >10,000-fold increased signal-to-noise ratio (SNR) over conventional MRS methods [4]. This technology, which employs an intravenous bolus injection of an HP ^{13}C agent followed by dynamic imaging, now enables the direct in vivo investigation of key metabolic pathways inaccessible by other imaging technologies [5, 6].

Multiple ^{13}C-labeled substrates have been investigated in preclinical studies, with pyruvate (Pyr), having favorable magnetic resonance (MR) properties and an excellent in vivo safety profile, the first of these agents to reach clinical trials [7]. Pyr, occupying a critical branch point in glucose metabolism, can be reduced to lactate (Lac) via glycolysis, decarboxylated to produce bicarbonate (Bic) and acetyl-co-enzyme A or acetyl CoA (as the first step toward the entry into the Krebs cycle and subsequent oxidative phosphorylation), or transaminated to alanine (Ala). Observing pyruvate metabolism has generated considerable excitement with initial human studies targeting the Warburg effect in solid tumors [5, 8–10] and energy metabolism in the heart [11–13]. Figure 5.1 shows the results of a representative preclinical HP [1-^{13}C]Pyr study highlighting the targeted metabolic pathways, resulting in spectral peaks, and signal temporal dynamics.

D. M. Spielman (✉)
Radiology, Stanford University, Stanford, CA, USA
e-mail: spielman@stanford.edu

J. M. Park
Advanced Imaging Research Center, University of Texas Southwestern Medical Center, Dallas, TX, USA

T. Jue, D. Mayer (eds.), *Dynamic Hyperpolarized Nuclear Magnetic Resonance*, Handbook of Modern Biophysics, https://doi.org/10.1007/978-3-030-55043-1_5

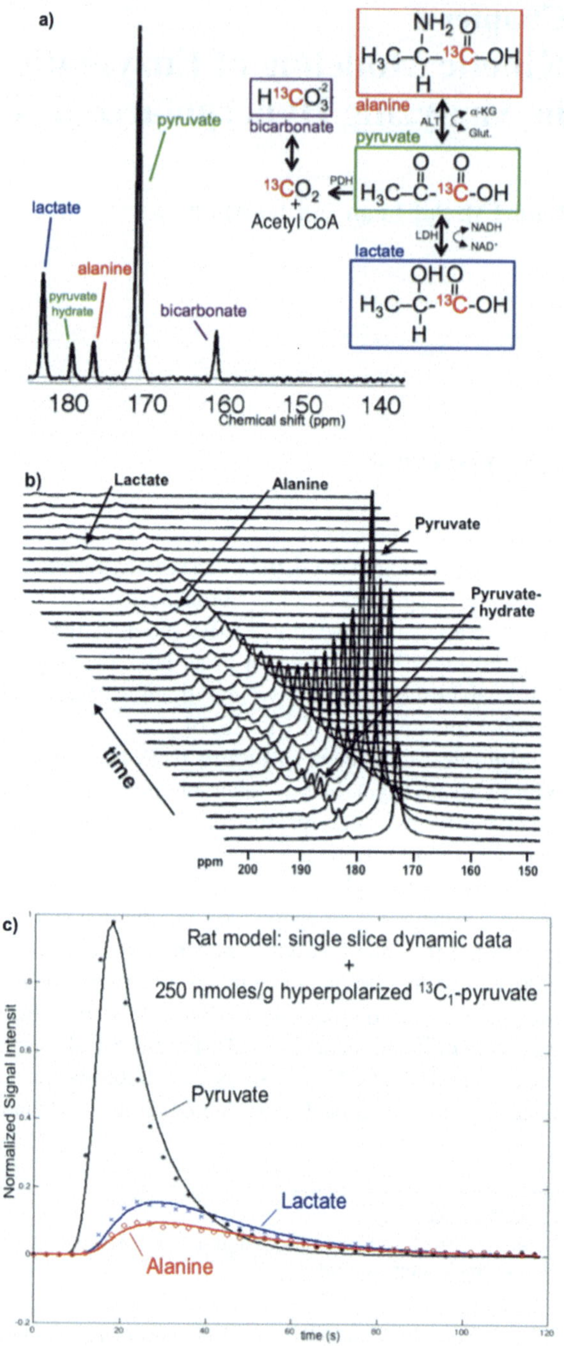

Fig. 5.1 Representative in vivo HP [1-^{13}C]Pyr study in a rat model. (**a**) In vivo spectrum of metabolic products corresponding to different metabolic pathways. (**b**) Stack plot of spectra acquired every 3 s following the bolus injection of [1-^{13}C]Pyr. (**c**) Corresponding temporal curves for Pyr, Lac, and Ala obtained by integrating the corresponding ^{13}C-peaks in each time frame

Specifically, the injection of HP [1-^{13}C]Pyr results in prominent in vivo peaks from [1-^{13}C]Pyr, [1-^{13}C]Lac, [1-^{13}C]Ala, and ^{13}C-Bic. With the conversion of the injected Pyr into downstream products, clearly some metabolism has occurred, and this chapter focuses on how the resulting data can be both quantified and associated with specific biochemical processes.

The dynamic nature of the HP ^{13}C experiment is a critical feature. Substrates are injected as boluses, and, despite the enhanced SNR, injected substrate concentrations are invariably supraphysiological as well as time variant [6]. For example, at the arrival to the organ of interest, in vivo blood concentration of ^{13}C-Pyr is typically in the 2–5 mM range, well above normal physiological levels [14]. This strongly contrasts with the positron emission tomography (PET), in which higher sensitivity (many orders of magnitude higher than MR) permits the use of tracer doses, thereby allowing the interrogation of unperturbed biological processes.

A second key feature of HP studies is that MRS signal intensities, while proportional to substrate concentrations, are strongly dependent on T_1 decay. Upon leaving the polarizer, the HP MR signal decays with the characteristic T_1 relaxation time associated with each particular molecule and the labeling position. For [1-^{13}C]Pyr and its primary products, T_1s range from 60–70 s at 3 T [3]. As a second example, when using [2-^{13}C]Pyr is used as the injected substrate, the T_1 of [2-^{13}C]Pyr is ~60 s but that of subsequently produced [2-^{13}C]Lac is close to 10 s [15]. Thus, in Fig. 5.1b, the loss of HP signal is in large part due to T_1 decay rather than any metabolic processes. Taken together, T_1 decay and bolus injections limit in vivo HP ^{13}C measurements to a very short time scale, with all data typically acquired within 2-min post-injection. One positive feature of this signal decay process is that, despite the high concentrations, boluses (at least in the case of Pyr) appear to have minimal effects on underlying metabolic pool sizes during the restricted measurement time intervals [16].

Given supraphysiological bolus injections and T_1 decay, the common in vivo MRS metric of absolute metabolite concentration is typically not meaningful. For example, the observed concentration of [1-^{13}C]Lac directly depends on the local delivery of [1-^{13}C]Pyr. Although, the ratio of the downstream metabolite signal to that of the injected substrate, for example, the Lac/Pyr, Ala/Pyr, or Bic/Pyr ratios, makes more sense, but caution is needed as, following a bolus injection, such ratios are strongly time variant. For example, the results of the study shown in Fig. 5.1 have a minimum Lac/Pyr ratio at the peak of Pyr bolus and a maximum at ~30 s after injection.

Despite these difficulties, investigators have been quite successful in the development of robust, quantitative, in vivo HP ^{13}C metrics. However, directly relating these metrics to underlying metabolism has been much more challenging. Although HP ^{13}C studies do not directly measure unperturbed steady-state intermediary metabolism, at best providing a "snapshot" [6], we need first provide a better understanding of these experiments followed by a discussion of the various methods proposed for data analysis.

5.2 An In Vivo Enzyme Assay

In its most basic form, many HP ^{13}C experiments closely resemble enzyme assays. For example, consider an in vitro tissue assay for the activity of pyruvate dehydrogenase (PDH), the enzyme catalyzing the conversion of Pyr and co-enzyme A (CoA) to acetyl-CoA and CO_2 (see Fig. 5.2).

For the in vitro assay, tissue or cells are first mixed into a buffer containing CoA, nicotinamide adenine dinucleotide (NAD^+), and necessary co-factors such as thiamine pyrophosphate (TPP) and Mg^{2+}. Excess Pyr is then added, and the rate of nicotinamide adenine dinucleotide + hydrogen (NADH) accumulation, as measured by a spectrophotometer, is proportional to PDH activity.

This is quite similar to the in vivo HP experiment whereby excess substrate is provided to the organ or tissue of interest via an intravenous bolus injection and the accumulation of metabolic products is observed. However, details matter. Consider more carefully the seemingly simple example. Following an HP [1-^{13}C]Pyr bolus injection, and assuming (1) the necessary substrates of Pyr, NAD^+ and CoA are available in abundance, (2) all co-factors are present, and (3) CO_2 is in rapid equilibrium with Bic, strongly favoring Bic, then the observed rate of appearance of the ^{13}C-Bic signal can provide a good measure of PDH activity. For typical in vivo conditions, these assumptions are likely met [17], however, not always, and accounting for effects such as differential metabolite T_1 relaxation times, bolus shape and duration, local blood flow, and saturation effects may be required. Consider the case of a low [1-^{13}C]Pyr dose for which the Pyr concentration may be rate limiting rather than PDH activity. As will be further discussed under Sect. 5.5 Metabolic Flux Versus Isotopic Exchange, interpreting such experimental results can be even more complicated with reversible reactions as, for example, in the case of Pyr to Lac conversion.

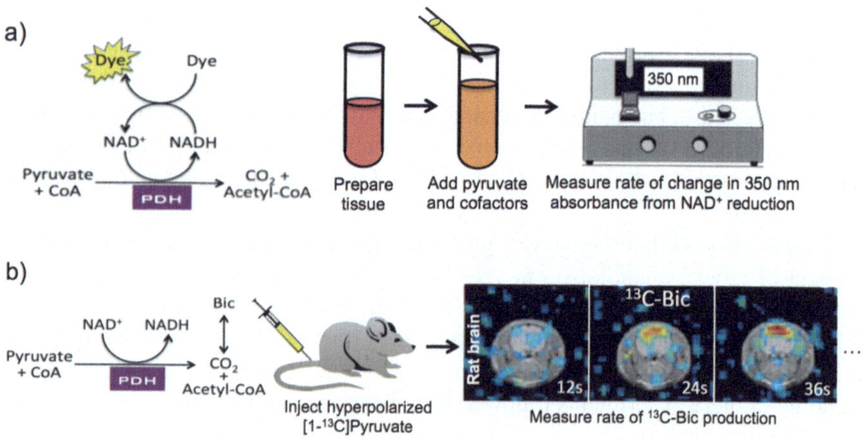

Fig. 5.2 In vitro versus in vivo enzyme assays. (**a**) Schematic for an in vitro pyruvate dehydrogenase (PDH) assay. (**b**) Equivalent in vivo HP ^{13}C experiment using [1-^{13}C]Pyr and its product ^{13}C-Bic

5.3 The Apparent Rate Constant

A major appeal of HP ^{13}C is the potential to quantify in vivo metabolic kinetics. However, given the difficulties controlling in vivo conditions, apparent rate constants rather than the true constants are typically targeted, with first order multisite exchange models the most common approach [6, 18, 19]. An example of two-site exchange model for ^{13}C-Lac labeling following an HP ^{13}C-Pyr injection is shown in Fig. 5.3.

In this simplified model k_{LP} is assumed to be zero, the validity of which depends on the existing unlabeled ^{12}C-Lac pool size. Further the arterial input function, u, is derived from the ^{13}C-Pyr signal itself [20], which in this case is likely valid due to the Pyr concentration being approximately ~20× that of Lac. Here, the decay terms, Pyr T_1 (T_{1P}) and Lac T_1 (T_{1L}), account for all the losses of Pyr and Lac signal including T_1 relaxation, Rf excitations, and conversion to other unspecific metabolic products. For in vivo studies that often suffer from B_1 inhomogeneities, accurately accounting for the loss of magnetization induced by Rf pulses can be problematic [6].

Numerous extensions to this highly simplified model have also been proposed in order to better accommodate the multiple underlying transport and metabolic processes. Schematics for four such representative dynamic models are shown in Fig. 5.4. Model extensions tend to fall into two main categories: (1) the inclusion of additional metabolic products, and (2) adding exchange rate parameters to account for the transport across various tissue compartments and cell membranes. The complexity of the chosen model is typically influenced by the SNR of the data set [21, 22], experimental setup, such as cell cultures [18, 23, 24], ex vivo perfused organs [16, 25, 26], in vivo organs of interest [27], or targeted disease process [28]. Other model extensions include direct measurement of the arterial input function [29], including either ^1H or ^{13}C measures of perfusion [30, 31], and nonlinear kinetics to account for effects such as NADH depletion [32]. Bankson et al. proposed extending the well-known Tofts model for kinetic analysis of dynamic contrast enhanced-MRI [33], using Akaike's information criterion (AIC) to select among candidate

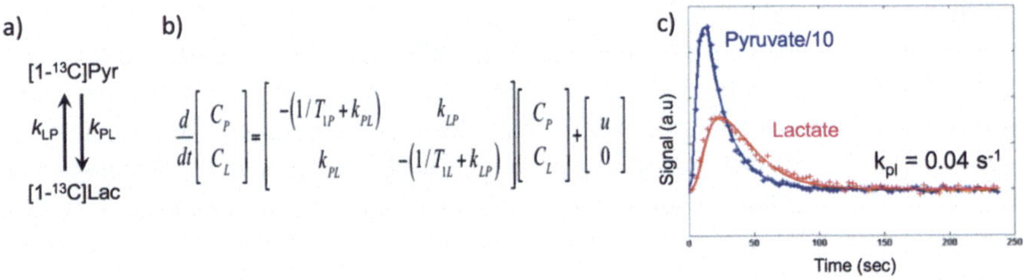

Fig. 5.3 Representative two-site exchange model for [1-^{13}C]Pyr (C_P) to [1-^{13}C]Lac conversion (C_L): (**a**) chemical exchange reaction, (**b**) first-order kinetic model, and (**c**) fit for apparent rate of conversion of ^{13}C-Pyr to ^{13}C-Lac, k_{PL}

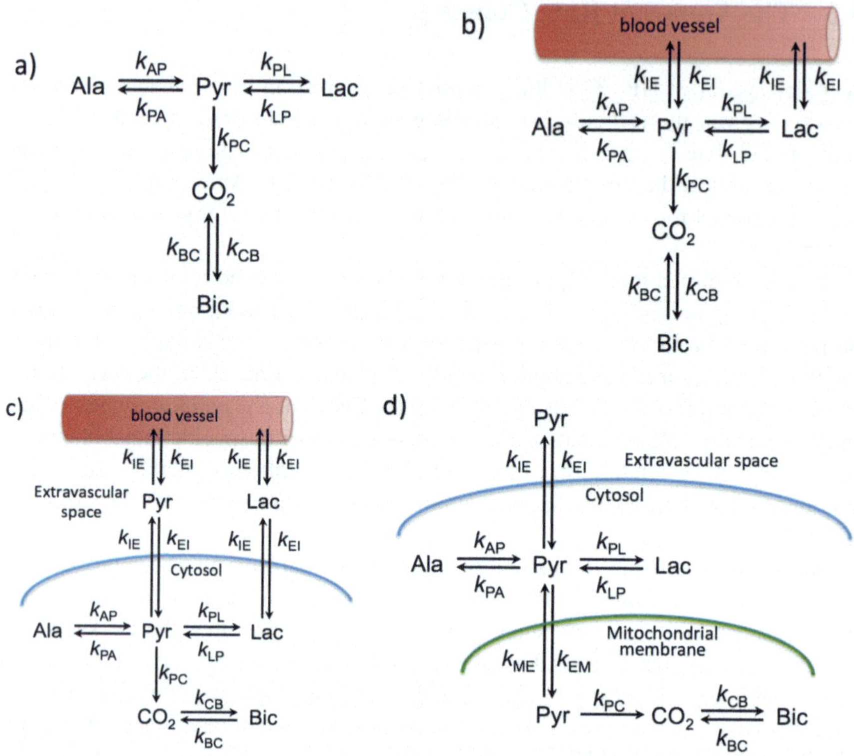

Fig. 5.4 Examples of metabolic exchange models for an HP ^{13}C Pyr experiment. (**a**) Multisite exchange model. (**b–d**) Multisite, multicompartment models incorporating transport across the vascular, cellular, and mitochondrial spaces. Models (**b**, **c**) are adapted from Bankson et al. [20] and (**d**) is adapted from Comment and Merritt [6]

models, and then integrate the pharmacokinetics into a constrained reconstruction framework to allow more efficient data acquisitions and analyses [20].

Models (a) and (b) of Fig. 5.4 seem to be the most popular, with investigators largely focusing on estimated net flux without trying to differentiate many of the multiple transport processes and chemical reactions. Although there is temptation to equate these apparent rate constants to actual metabolic fluxes (and sometimes even to corresponding enzyme activities), considerable caution is advised, and this will be discussed further in the next two sections. Nonetheless, the use of the dynamic models is widespread and quite successful. Changes in apparent rate constants, in particular k_{PL}, have been correlated with prostate cancer aggressiveness and response to therapy [28, 34], lung inflammation [35], breast cancer progression [36], tumor metastatic risk [37], alterations in the cancer Warburg effect [38], and cardiac metabolism [39].

5.4 Saturation Effects

A basic question is whether there is indeed a single well-defined apparent rate constant for a given tissue. For many cases, the answer is no. Foremost, apparent rate constants are dose dependent, with Fig. 5.5 showing a representative in vivo rat experiment in which the production of ^{13}C-Lac is minimally affected despite increasing the HP ^{13}C-Pyr dose by a factor of 2.5.

In terms of apparent rate constants, Zierhut et al., using a two-site kinetic metabolic model to analyze dynamic HP ^{13}C data from a rat model, found that k_{PL} and k_{PA} ranged from 0.06 to 0.01 s^{-1} and from 0.04 to 0.01 s^{-1}, respectively as the Pyr bolus increased from 50 to 725 μmol/kg [40]. This "saturation" effect is best modeled using a Michaelis-Menton like formulation [40].

Namely, consider the conversion of ^{13}C-Pyr to ^{13}C-Lac. Instead of modeling the reaction velocity for the production of ^{13}C-Lac as proportional to the Pyr dose (Dose_{Pyr}), k_{PL} becomes an explicit function of the dose:

$$k_{\text{PL}} = \frac{V_{\text{max}}}{k_{\text{m}} + \text{Dose}_{\text{Pyr}}}$$

where the Michaelis-Menten parameters V_{max} is the maximum reaction velocity and K_M is known as the Michaelis constant defined to be the dose for which the reaction velocity = ½ V_{max}. Although this proposed saturable-kinetics model has a similar mathematical formulation to the original Michaelis–Menten equations, the formulation presented here is conceptually different. Rather than reflecting pure enzyme kinetics, the observed ^{13}C-label dose-dependent exchange rate constants and reaction velocities reflect the collective effects of factors including the organ perfusion rate, substrate transport kinetics, enzyme activities, and the sizes of the unlabeled (^{12}C) intrinsic substrate pools.

Apparent rate constant estimates are also a function of bolus shape. Following an injection of HP [1-^{13}C]Pyr, the Pyr signal reaches a maximum shortly after the end of the injection and then decays as a result of T_1 relaxation, metabolic flux, isotopic exchange (to be discussed below), and dilution into the blood and extracellular space. By exploiting this inherently time-varying Pyr concentration, Xu et al. showed that the Michaelis-Menten parameters V_{max} and K_M can be calculated from a single bolus injection [41] (see Fig. 5.6).

The importance of including Michaelis-Menten corrections depends on dose, and significant saturation effects have clearly been demonstrated in in vitro experiments [6]. Human Pyr studies to date have used a relative low dose (100 μmol/kg) [7], whereas doses for preclinical studies are often much higher [40]. Controlling for these effects is important when comparing results.

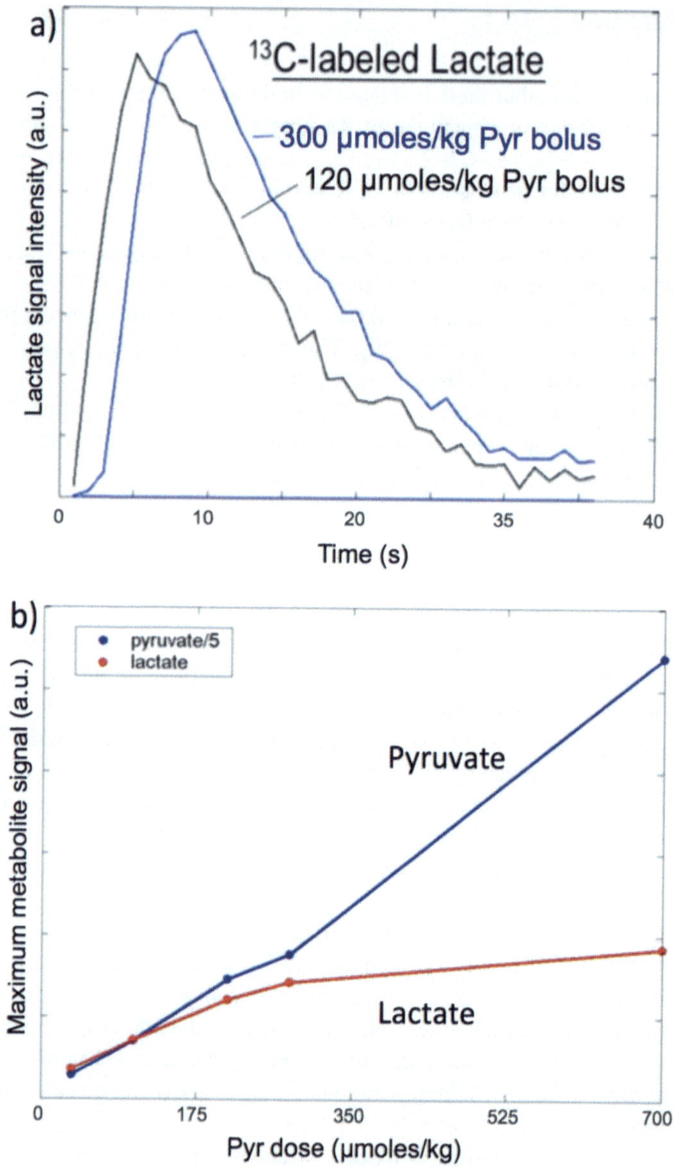

Fig. 5.5 In vivo demonstration of ^{13}C-Pyr to ^{13}C-Lac saturation in the normal rat kidney. (**a**) Despite a 2.5-fold increase in the Pyr, ^{13}C-Lac production is changed by much than that a factor of 2.5. (**b**) Maximum ^{13}C-Pyr and ^{13}C-Lac signals as a function of HP ^{13}C-Pyr dose

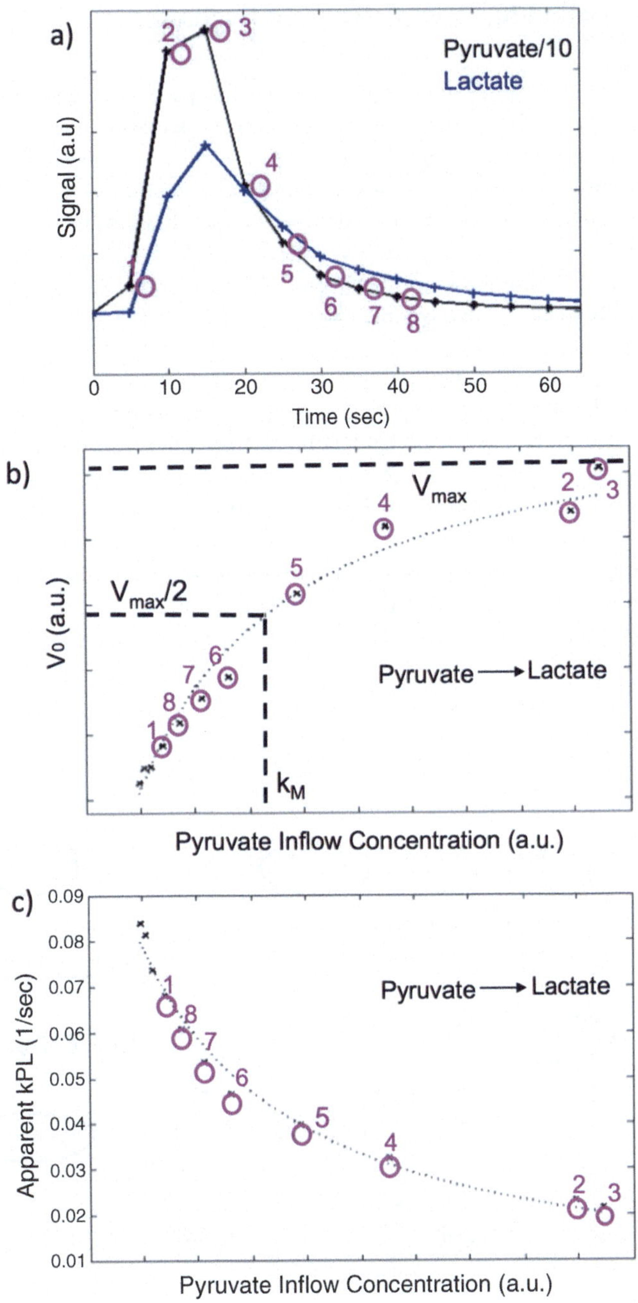

Fig. 5.6 Bolus dynamics following the experimental design of Xu et al. [41]. (**a**) The local production of ^{13}C-Lac during each TR interval can be sequentially measured during a single HP ^{13}C-Pyr bolus injection. By reordering the individual data points, both (**b**) Michaelis-Menten parameters (V_{max} and K_M) and (**c**) k_{PL}(t) can be computed

5.5 Metabolic Flux Versus Isotopic Exchange

Despite obvious difficulties with saturation effects, apparent rate constants are often described incorrectly in the literature as measures of metabolic flux. More accurate terms, such as "flux of label" or "^{13}C flux" are better, but even these obscure potentially important underlying processes. Detectable downstream ^{13}C products result from both irreversible and reversible reactions, or combinations thereof, with examples for both [1-^{13}C]Pyr and [2-^{13}C]Pyr shown in Fig. 5.7. Equilibrium rates are important as are differentiating between true metabolic fluxes from isotopic exchange mechanisms, with the conversion of ^{13}C-Pyr to ^{13}C-Lac serving as a quintessential example.

As depicted in Fig. 5.8, the forward reaction mechanism of lactate dehydrogenase (LDH), the enzyme catalyzing Pyr to Lac conversion, follows an ordered sequence in which the coenzyme NADH binds first followed by Pyr. The two ternary complexes, enzyme-NADH-Pyr and enzyme-NAD$^+$-Lac are in rapid equilibrium. Lac then dissociates prior to NAD$^+$, with the binding of NADH being the overall rate-limiting step. The reverse reaction follows an analogous process [42, 43]. This ordered sequential binding leads to potentially very different kinetics for a net flux, in which Pyr and NADH are converted to Lac and NAD$^+$, versus isotopic exchange, for which no net flux of Pyr or NADH is required. Isotopic exchange

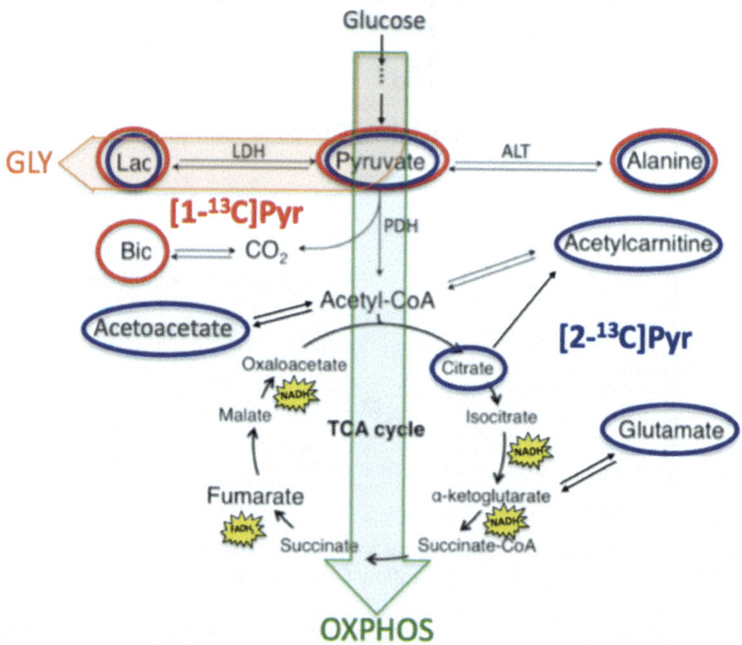

Fig. 5.7 Primary metabolic products detectable following an in vivo bolus injection of either [1-^{13}C]Pyr (red) or [2-^{13}C]Pyr (blue)

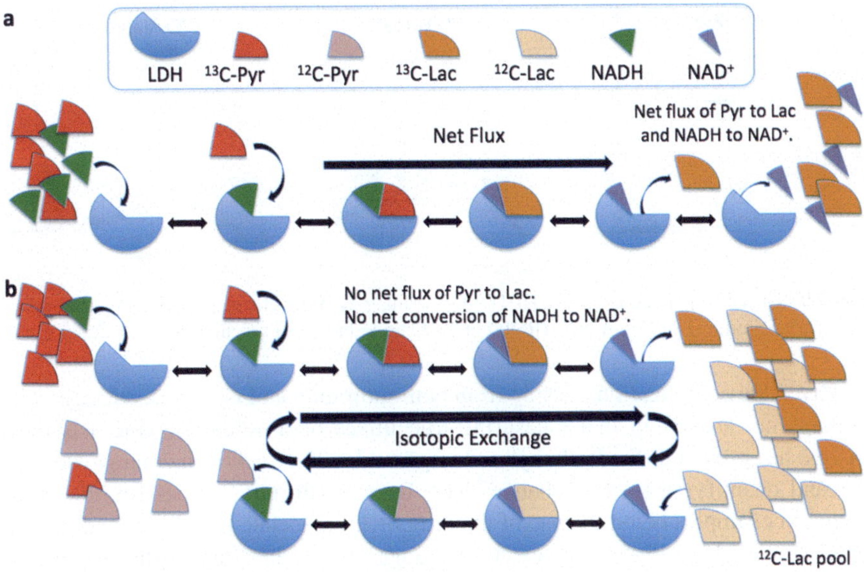

Fig. 5.8 Flux versus isotopic exchange example for models ^{13}C-Pyr to ^{13}C-Lac conversion. (**a**) Model showing a net flux of ^{13}C-Pyr and NADH to ^{13}C-Lac and NAD^{+}. Note, this reaction will result in a change of the cell's NADH/NAD^{+} redox ratio. (**b**) Alternative model in which ^{13}C-Lac signal appears with no net flux of Pyr to Lac, only isotopic exchange, which has the effect of partially labeling the underlying Lac pool. Note, under isotopic exchange there is no net change in cellular redox

simply results in the partial labeling of the existing endogenous Lac pool. That is, for every ^{13}C-Pyr that is converted to ^{13}C-Lac, a ^{12}C-Lac is converted back to ^{12}C-Pyr, and the size of the Lac pool determines the probability that a ^{12}C-Lac rather than a ^{13}C-Lac is involved in the backward exchange. Another important point is that this isotopic exchange process can occur quite quickly with no overall change in cellular redox levels.

Both flux and isotopic exchange occur in vivo, with underlying metabolic pool sizes playing a key role [44]. Given the rapid exchange rate of Pyr-to-Lac and the short time scale of the HP ^{13}C experiment, several studies have shown that in most cases the appearance of ^{13}C-Lac is mostly, if not entirely, due to isotopic exchange [43]. Exploiting this effect, via a technique named exchange-linked dissolution agents (ELDA), was clearly demonstrated in a paper by Hurd et al., in which the co-administration of NMR-invisible ^{12}C-Lac in combination with a conventional HP ^{13}C-Pyr bolus results in significantly increased ^{13}C-Lac production in multiple tissues with an essentially no change in the ^{13}C-Pyr or ^{13}C-Ala signals [45]. Fig. 5.9 shows representative in vivo data. Increasing the intrinsic Lac pool size, with the addition of the ^{12}C-Lac, results in increased isotopic exchange and consequently, greater observed production of ^{13}C-Lac.

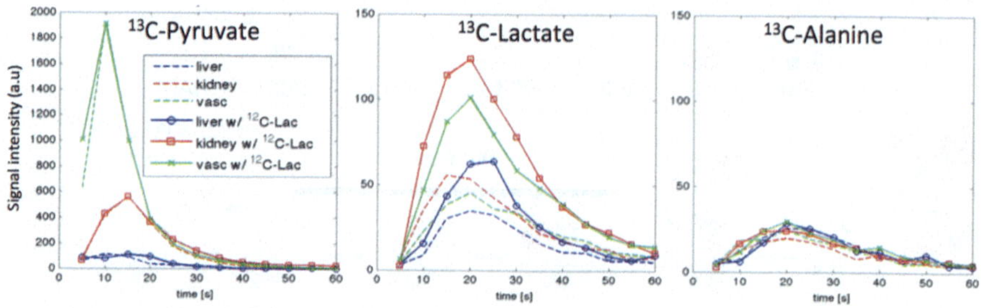

Fig. 5.9 The ELDA effect. Dynamic HP [1-^{13}C]Pyr imaging results from a rat showing metabolite time curves from regions of interest (ROIs) in the liver, kidney, and vasculature

Given the complications arising from both saturation and isotopic exchange, it is perhaps not surprising that measuring true fluxes or absolute enzyme activities in vivo has proven difficult, though not impossible [26, 46, 47]. In contrast, detecting and quantifying relative changes have been much more successful, the most common examples being the use of the drug dichloroacetate (DCA) to modulation PDH activity [16, 48–53] or detecting changes in ^{13}C labeling in the fed versus fasted state [25, 54, 55]. ^{13}C-Lac labeling in some preclinical tumor models has been correlated with LDH activity [28, 56–58], but for other cases, observed differences are unlikely to be due to variations in LDH. Examples being experiments employing the addition unlabeled Lac [45] or inducing changes in tissue redox levels as reflected by the NADH/NAD$^+$ ratio [59].

Importantly, exchange processes are not necessarily problematic for all HP ^{13}C studies, but, in some cases, can be exploitable features. Fast exchange = slow on the NMR time scale ($\gg 10^{-6}$ s), but fast on HP ^{13}C experimental time scale ($\ll 1$ s), allows measurement of important biochemical steady-state ratios. The first example of such was presented by Gallagher et al. in which tumor pH was determined by measuring the ^{13}C Bic/CO_2 ratio [60]. The importance of such ratios has been one of the motivations to examine the model-free approaches to quantification.

5.6 Model-Free Approaches

Given the complexity of in vivo biochemistry and metabolism, in our opinion, model-free approaches should be considered as no less valid, and in some cases more appropriate, than metrics derived from sophisticated kinetic models. An alternative to estimating apparent rate constants is to simply focus on repeatable data quantitation, with, in many cases, not making any strong claims regarding underlying metabolic fluxes or processes. While a conventional MRS peak ratio is problematic due to the time-varying nature of a bolus injection, quantifying ^{13}C peaks at single or fixed time points [61, 62] or, more commonly, by calculating the areas under the metabolic time curves [10, 63] has been widely adopted. Such areas under the curve (AUCs), or more appropriately ratios of the corresponding metabolite AUCs have been favorably compared with kinetic modeling, and have been shown to have the advantage of being robust, easily computed, and independent of the arterial input function and other kinetic modeling assumptions [63]. AUCs are also applicable even for low with the metabolite time curves calculated by only summing over SNR-thresholded time points [52]. Metabolite ratios based on AUCs, however, can be more sensitive than apparent conversion rates to the bolus shape of the injected hyperpolarized substrate [64]. Other metrics that have been investigated include maximum metabolite signals or their ratios [61, 62, 65], fitting with gamma-variate functions [66], and full-width half-maximum measures [66].

Metabolic ratios can also have other advantages. Although the intracellular Lac/Pyr ratio is usually inaccessible during an HP ^{13}C-Pyr study due to a large fraction of the observed ^{13}C-Pyr signal coming from the vascular space, injecting HP [1-^{13}C]Ala, which does indeed generate an intracellular ^{13}C Pyr/Lac ratio, can be used in the measurement of redox in the liver [67]. The brain ^{13}C-Lac/^{13}C-Bic ratio in a rat model was also shown to be invariant to vasodilation effects of anesthesia, whereas Lac/Pyr and Bic/Lac were strongly dependent on isoflurane levels [68]. Interestingly, dehydroascorbic acid (DHA) was first proposed as an HP ^{13}C substrate based on its potential use as a redox imaging biomarker via the simultaneous detection of its redox partner ascorbate (Asc) [69, 70]. However, subsequent studies, at least in a murine prostate cancer model, showed that the ^{13}C-DHA/^{13}C-Asc ratio did not correspond to tissue redox, rather the size of the ^{13}C-Asc peak was most strongly dependent on the reduced glutathione (GSH) concentrations [71].

5.7 Summary

The excitement of HP ^{13}C technology centers on the opportunity to image previously inaccessible in vivo metabolic processes. Among most important goals are the goals to both reliably quantify HP ^{13}C studies and relate those metrics to specific underlying metabolic parameters. The most common approaches are estimating apparent rate constants using a variety of multisite exchange models or using model-free measures such as AUCs. The goal of this chapter is to provide an overview of the underlying physical concepts and modeling considerations for quantifying in vivo HP ^{13}C data sets, with the most appropriate quantitation algorithm often being specific to the metabolic pathway, preclinical model, disease process, or patient population being studied. Care is needed in generating specific hypothesis or interpreting results, with distinct biochemical processes such as changes in LDH activity versus changes in Lac pool sizes often being very hard to distinguish. Perhaps the safest approach is to first establish HP ^{13}C findings as robust imaging biomarkers important for the assessment of clinical disease, with subsequent experiments delving into unraveling underlying biochemical parameters. As with many MRI applications, absolute quantitation remains challenging, and the most fruitful near-term studies will likely focus on measuring the relative changes in response to therapy.

Acknowledgments We acknowledge funding support from NIH (R01EB01901802, R01CA17683603, P41EB0158912, R01NS107409-01A1, and P41EB015908), The Mobility Foundation, The Texas Institute for Brain Injury and Repair, The Welch Foundation (I-2009-20190330), and GE Healthcare.

Problems

Q1. A solution containing hyperpolarized [1-^{13}C]-pyruvate is injected into a brain tumor patient. A significantly higher signal of [1-^{13}C]-lactate is observed in the tumor than in peripheral normal-appearing brain regions. Before concluding that the higher lactate production is due to the increased pyruvate flux to lactate, you remembered that this observation can be also from isotopic exchange with larger intrinsic lactate pool size in the tumor. As you want to compare the metabolic flux in the tumor and exclude the contribution of isotopic exchange, you decide to exploit the two-site exchange model from the Fig. 5.3b to estimate k_{PL}. Is this going to work?

Q2. You co-injected ^{12}C-alanine with hyperpolarized [1-^{13}C]-pyruvate into a rat to test the exchange-linked dissolution agents (ELDA) effect in liver. However, you did not see any increase in the hyperpolarized [1-^{13}C]-alanine signal. How can you explain this?

References

1. Golman, K., Ardenkjaer-Larsen, J.H., Petersson, J.S., Mansson, S., Leunbach, I.: Molecular imaging with endogenous substances. Proc. Natl. Acad. Sci. U. S. A. **100**(18), 10435–10439 (2003). https://doi.org/10.1073/pnas.1733836100
2. Ardenkjaer-Larsen, J.H., Fridlund, B., Gram, A., Hansson, G., Hansson, L., Lerche, M.H., Servin, R., Thaning, M., Golman, K.: Increase in signal-to-noise ratio of > 10,000 times in liquid-state NMR. Proc. Natl. Acad. Sci. U. S. A. **100**(18), 10158–10163 (2003). https://doi.org/10.1073/pnas.1733835100
3. Hurd, R.E., Yen, Y.F., Chen, A., Ardenkjaer-Larsen, J.H.: Hyperpolarized 13C metabolic imaging using dissolution dynamic nuclear polarization. J. Magn. Reson. Imaging. **36**(6), 1314–1328 (2012). https://doi.org/10.1002/jmri.23753
4. Ardenkjaer-Larsen, J.H.: On the present and future of dissolution-DNP. J. Magn. Reson. **264**, 3–12 (2016). https://doi.org/10.1016/j.jmr.2016.01.015
5. Kurhanewicz, J., Vigneron, D.B., Brindle, K., Chekmenev, E.Y., Comment, A., Cunningham, C.H., Deberardinis, R.J., Green, G.G., Leach, M.O., Rajan, S.S., Rizi, R.R., Ross, B.D., Warren, W.S., Malloy, C.R.: Analysis of cancer metabolism by imaging hyperpolarized nuclei: prospects for translation to clinical research. Neoplasia. **13**(2), 81–97 (2011)
6. Comment, A., Merritt, M.E.: Hyperpolarized magnetic resonance as a sensitive detector of metabolic function. Biochemistry. **53**(47), 7333–7357 (2014). https://doi.org/10.1021/bi501225t
7. Nelson, S.J., Kurhanewicz, J., Vigneron, D.B., Larson, P.E., Harzstark, A.L., Ferrone, M., van Criekinge, M., Chang, J.W., Bok, R., Park, I., Reed, G., Carvajal, L., Small, E.J., Munster, P., Weinberg, V.K., Ardenkjaer-Larsen, J.H., Chen, A.P., Hurd, R.E., Odegardstuen, L.I.,

Robb, F.J., Tropp, J., Murray, J.A.: Metabolic imaging of patients with prostate cancer using hyperpolarized [1-(1)(3)C]pyruvate. Sci. Transl. Med. **5**(198), 198ra108 (2013). https://doi.org/10.1126/scitranslmed.3006070

8. Brindle, K.M., Bohndiek, S.E., Gallagher, F.A., Kettunen, M.I.: Tumor imaging using hyperpolarized 13C magnetic resonance spectroscopy. Magn. Reson. Med. **66**(2), 505–519 (2011). https://doi.org/10.1002/mrm.22999
9. Hesketh, R.L., Brindle, K.M.: Magnetic resonance imaging of cancer metabolism with hyperpolarized (13)C-labeled cell metabolites. Curr. Opin. Chem. Biol. **45**, 187–194 (2018). https://doi.org/10.1016/j.cbpa.2018.03.004
10. Chaumeil, M.M., Radoul, M., Najac, C., Eriksson, P., Viswanath, P., Blough, M.D., Chesnelong, C., Luchman, H.A., Cairncross, J.G., Ronen, S.M.: Hyperpolarized (13)C MR imaging detects no lactate production in mutant IDH1 gliomas: implications for diagnosis and response monitoring. Neuroimag. Clin. **12**, 180–189 (2016). https://doi.org/10.1016/j.nicl.2016.06.018
11. Schroeder, M.A., Clarke, K., Neubauer, S., Tyler, D.J.: Hyperpolarized magnetic resonance: a novel technique for the in vivo assessment of cardiovascular disease. Circulation. **124**(14), 1580–1594 (2011). https://doi.org/10.1161/CIRCULATIONAHA.111.024919
12. Cunningham, C.H., Lau, J.Y., Chen, A.P., Geraghty, B.J., Perks, W.J., Roifman, I., Wright, G.A., Connelly, K.A.: Hyperpolarized 13C metabolic MRI of the human heart: initial experience. Circ. Res. **119**(11), 1177–1182 (2016). https://doi.org/10.1161/CIRCRESAHA.116.309769
13. Malloy, C.R., Merritt, M.E., Sherry, A.D.: Could 13C MRI assist clinical decision-making for patients with heart disease? NMR Biomed. **24**(8), 973–979 (2011). https://doi.org/10.1002/nbm.1718
14. Johnson, R.E., Edwards, H.T.: Lactate and pyruvate in blood and urine after exercise. J. Biol. Chem. **118**(2), 427–432 (1937)
15. Taglang, C., Korenchan, D.E., von Morze, C., Yu, J., Najac, C., Wang, S., Blecha, J.E., Subramaniam, S., Bok, R., VanBrocklin, H.F., Vigneron, D.B., Ronen, S.M., Sriram, R., Kurhanewicz, J., Wilson, D.M., Flavell, R.R.: Late-stage deuteration of (13)C-enriched substrates for T1 prolongation in hyperpolarized (13)C MRI. Chem. Commun. (Camb.). **54**(41), 5233–5236 (2018). https://doi.org/10.1039/c8cc02246a
16. Atherton, H.J., Schroeder, M.A., Dodd, M.S., Heather, L.C., Carter, E.E., Cochlin, L.E., Nagel, S., Sibson, N.R., Radda, G.K., Clarke, K., Tyler, D.J.: Validation of the in vivo assessment of pyruvate dehydrogenase activity using hyperpolarised 13C MRS. NMR Biomed. **24**(2), 201–208 (2011). https://doi.org/10.1002/nbm.1573
17. Chen, A.P., Hurd, R.E., Schroeder, M.A., Lau, A.Z., Gu, Y.P., Lam, W.W., Barry, J., Tropp, J., Cunningham, C.H.: Simultaneous investigation of cardiac pyruvate dehydrogenase flux, Krebs cycle metabolism and pH, using hyperpolarized [1,2-(13)C2]pyruvate in vivo. NMR Biomed. **25**(2), 305–311 (2012). https://doi.org/10.1002/nbm.1749
18. Day, S.E., Kettunen, M.I., Gallagher, F.A., Hu, D.E., Lerche, M., Wolber, J., Golman, K., Ardenkjaer-Larsen, J.H., Brindle, K.M.: Detecting tumor response to treatment using hyperpolarized 13C magnetic resonance imaging and spectroscopy. Nat. Med. **13**(11), 1382–1387 (2007). https://doi.org/10.1038/nm1650
19. Bohndiek, S.E., Kettunen, M.I., Hu, D.E., Brindle, K.M.: Hyperpolarized (13)C spectroscopy detects early changes in tumor vasculature and metabolism after VEGF neutralization. Cancer Res. **72**(4), 854–864 (2012). https://doi.org/10.1158/0008-5472.CAN-11-2795
20. Bankson, J.A., Walker, C.M., Ramirez, M.S., Stefan, W., Fuentes, D., Merritt, M.E., Lee, J., Sandulache, V.C., Chen, Y., Phan, L., Chou, P.C., Rao, A., Yeung, S.C., Lee, M.H., Schellingerhout, D., Conrad, C.A., Malloy, C., Sherry, A.D., Lai, S.Y., Hazle, J.D.: Kinetic modeling and constrained reconstruction of hyperpolarized [1-13C]-pyruvate offers improved metabolic imaging of tumors. Cancer Res. **75**(22), 4708–4717 (2015). https://doi.org/10.1158/0008-5472.CAN-15-0171
21. Santarelli, M.F., Positano, V., Giovannetti, G., Frijia, F., Menichetti, L., Ardenkjaer-Larsen, J.H., De Marchi, D., Lionetti, V., Aquaro, G., Lombardi, M., Landini, L.: How the signal-to-noise ratio influences hyperpolarized 13C dynamic MRS data fitting and parameter estimation. NMR Biomed. **25**(7), 925–934 (2012). https://doi.org/10.1002/nbm.1813

22. Sun, C.Y., Walker, C.M., Michel, K.A., Venkatesan, A.M., Lai, S.Y., Bankson, J.A.: Influence of parameter accuracy on pharmacokinetic analysis of hyperpolarized pyruvate. Magn. Reson. Med. **79**(6), 3239–3248 (2018). https://doi.org/10.1002/mrm.26992
23. Harris, T., Eliyahu, G., Frydman, L., Degani, H.: Kinetics of hyperpolarized 13C1-pyruvate transport and metabolism in living human breast cancer cells. Proc. Natl. Acad. Sci. U. S. A. **106**(43), 18131–18136 (2009). https://doi.org/10.1073/pnas.0909049106
24. Harrison, C., Yang, C., Jindal, A., DeBerardinis, R.J., Hooshyar, M.A., Merritt, M., Dean Sherry, A., Malloy, C.R.: Comparison of kinetic models for analysis of pyruvate-to-lactate exchange by hyperpolarized 13 C NMR. NMR Biomed. **25**(11), 1286–1294 (2012). https://doi.org/10.1002/nbm.2801
25. Merritt, M.E., Harrison, C., Sherry, A.D., Malloy, C.R., Burgess, S.C.: Flux through hepatic pyruvate carboxylase and phosphoenolpyruvate carboxykinase detected by hyperpolarized 13C magnetic resonance. Proc. Natl. Acad. Sci. U. S. A. **108**(47), 19084–19089 (2011). https://doi.org/10.1073/pnas.1111247108
26. Merritt, M.E., Harrison, C., Storey, C., Jeffrey, F.M., Sherry, A.D., Malloy, C.R.: Hyperpolarized 13C allows a direct measure of flux through a single enzyme-catalyzed step by NMR. Proc. Natl. Acad. Sci. U. S. A. **104**(50), 19773–19777 (2007). https://doi.org/10.1073/pnas.0706235104
27. Khegai, O., Schulte, R.F., Janich, M.A., Menzel, M.I., Farrell, E., Otto, A.M., Ardenkjaer-Larsen, J.H., Glaser, S.J., Haase, A., Schwaiger, M., Wiesinger, F.: Apparent rate constant mapping using hyperpolarized [1-(13)C]pyruvate. NMR Biomed. **27**(10), 1256–1265 (2014). https://doi.org/10.1002/nbm.3174
28. Chen, H.Y., Larson, P.E.Z., Bok, R.A., von Morze, C., Sriram, R., Delos Santos, R., Delos Santos, J., Gordon, J.W., Bahrami, N., Ferrone, M., Kurhanewicz, J., Vigneron, D.B.: Assessing prostate Cancer aggressiveness with hyperpolarized dual-agent 3D dynamic imaging of metabolism and perfusion. Cancer Res. **77**(12), 3207–3216 (2017). https://doi.org/10.1158/0008-5472.CAN-16-2083
29. Kazan, S.M., Reynolds, S., Kennerley, A., Wholey, E., Bluff, J.E., Berwick, J., Cunningham, V.J., Paley, M.N., Tozer, G.M.: Kinetic modeling of hyperpolarized (13)C pyruvate metabolism in tumors using a measured arterial input function. Magn. Reson. Med. **70**(4), 943–953 (2013). https://doi.org/10.1002/mrm.24546
30. Smith, M.R., Peterson, E.T., Gordon, J.W., Niles, D.J., Rowland, I.J., Kurpad, K.N., Fain, S.B.: In vivo imaging and spectroscopy of dynamic metabolism using simultaneous 13C and 1H MRI. IEEE Trans. Biomed. Eng. **59**(1), 45–49 (2012). https://doi.org/10.1109/TBME.2011.2161988
31. von Morze, C., Bok, R.A., Reed, G.D., Ardenkjaer-Larsen, J.H., Kurhanewicz, J., Vigneron, D.B.: Simultaneous multiagent hyperpolarized (13)C perfusion imaging. Magn. Reson. Med. **72**(6), 1599–1609 (2014). https://doi.org/10.1002/mrm.25071
32. Mariotti, E., Orton, M.R., Eerbeek, O., Ashruf, J.F., Zuurbier, C.J., Southworth, R., Eykyn, T.R.: Modeling non-linear kinetics of hyperpolarized [1-(13)C] pyruvate in the crystalloid-perfused rat heart. NMR Biomed. **29**(4), 377–386 (2016). https://doi.org/10.1002/nbm.3464
33. Tofts, P.S., Brix, G., Buckley, D.L., Evelhoch, J.L., Henderson, E., Knopp, M.V., Larsson, H.B., Lee, T.Y., Mayr, N.A., Parker, G.J., Port, R.E., Taylor, J., Weisskoff, R.M.: Estimating kinetic parameters from dynamic contrast-enhanced T(1)-weighted MRI of a diffusable tracer: standardized quantities and symbols. J. Magn. Reson. Imaging. **10**(3), 223–232 (1999)
34. Chen, H.Y., Larson, P.E.Z., Gordon, J.W., Bok, R.A., Ferrone, M., van Criekinge, M., Carvajal, L., Cao, P., Pauly, J.M., Kerr, A.B., Park, I., Slater, J.B., Nelson, S.J., Munster, P.N., Aggarwal, R., Kurhanewicz, J., Vigneron, D.B.: Technique development of 3D dynamic CS-EPSI for hyperpolarized (13) C pyruvate MR molecular imaging of human prostate cancer. Magn. Reson. Med. **80**(5), 2062–2072 (2018). https://doi.org/10.1002/mrm.27179
35. Xu, H.N., Kadlececk, S., Shaghaghi, H., Zhao, H., Profka, H., Pourfathi, M., Rizi, R., Li, L.Z.: Differentiating inflamed and normal lungs by the apparent reaction rate constants of lactate dehydrogenase probed by hyperpolarized (13)C labeled pyruvate. Quant. Imaging Med. Surg. **6**(1), 57–66 (2016). https://doi.org/10.3978/j.issn.2223-4292.2016.02.04

36. Asghar Butt, S., Sogaard, L.V., Ardenkjaer-Larsen, J.H., Lauritzen, M.H., Engelholm, L.H., Paulson, O.B., Mirza, O., Holck, S., Magnusson, P., Akeson, P.: Monitoring mammary tumor progression and effect of tamoxifen treatment in MMTV-PymT using MRI and magnetic resonance spectroscopy with hyperpolarized [1-13C]pyruvate. Magn. Reson. Med. **73**(1), 51–58 (2015). https://doi.org/10.1002/mrm.25095
37. Xu, H.N., Kadlececk, S., Profka, H., Glickson, J.D., Rizi, R., Li, L.Z.: Is higher lactate an indicator of tumor metastatic risk? A pilot MRS study using hyperpolarized (13)C-pyruvate. Acad. Radiol. **21**(2), 223–231 (2014). https://doi.org/10.1016/j.acra.2013.11.014
38. Lin, G., Andrejeva, G., Wong Te Fong, A.C., Hill, D.K., Orton, M.R., Parkes, H.G., Koh, D.M., Robinson, S.P., Leach, M.O., Eykyn, T.R., Chung, Y.L.: Reduced Warburg effect in cancer cells undergoing autophagy: steady- state 1H-MRS and real-time hyperpolarized 13C-MRS studies. PLoS One. **9**(3), e92645 (2014). https://doi.org/10.1371/journal.pone.0092645
39. Flori, A., Liserani, M., Frijia, F., Giovannetti, G., Lionetti, V., Casieri, V., Positano, V., Aquaro, G.D., Recchia, F.A., Santarelli, M.F., Landini, L., Ardenkjaer-Larsen, J.H., Menichetti, L.: Real-time cardiac metabolism assessed with hyperpolarized [1-(13) C]acetate in a large-animal model. Contrast Media Mol. Imaging. **10**(3), 194–202 (2015). https://doi.org/10.1002/cmmi.1618
40. Zierhut, M.L., Yen, Y.F., Chen, A.P., Bok, R., Albers, M.J., Zhang, V., Tropp, J., Park, I., Vigneron, D.B., Kurhanewicz, J., Hurd, R.E., Nelson, S.J.: Kinetic modeling of hyperpolarized 13C1-pyruvate metabolism in normal rats and TRAMP mice. J. Magn. Reson. **202**(1), 85–92 (2010). https://doi.org/10.1016/j.jmr.2009.10.003
41. Xu, T., Mayer, D., Gu, M., Yen, Y.F., Josan, S., Tropp, J., Pfefferbaum, A., Hurd, R., Spielman, D.: Quantification of in vivo metabolic kinetics of hyperpolarized pyruvate in rat kidneys using dynamic 13C MRSI. NMR Biomed. **24**(8), 997–1005 (2011). https://doi.org/10.1002/nbm.1719
42. Shoemark, D.K., Cliff, M.J., Sessions, R.B., Clarke, A.R.: Enzymatic properties of the lactate dehydrogenase enzyme from Plasmodium falciparum. FEBS J. **274**(11), 2738–2748 (2007). https://doi.org/10.1111/j.1742-4658.2007.05808.x
43. Brindle, K.: Watching tumours gasp and die with MRI: the promise of hyperpolarised 13C MR spectroscopic imaging. Br. J. Radiol. **85**(1014), 697–708 (2012). https://doi.org/10.1259/bjr/81120511
44. Kettunen, M.I., Hu, D.E., Witney, T.H., McLaughlin, R., Gallagher, F.A., Bohndiek, S.E., Day, S.E., Brindle, K.M.: Magnetization transfer measurements of exchange between hyperpolarized [1-13C]pyruvate and [1-13C]lactate in a murine lymphoma. Magn. Reson. Med. **63**(4), 872–880 (2010). https://doi.org/10.1002/mrm.22276
45. Hurd, R.E., Spielman, D., Josan, S., Yen, Y.F., Pfefferbaum, A., Mayer, D.: Exchange-linked dissolution agents in dissolution-DNP (13)C metabolic imaging. Magn. Reson. Med. **70**(4), 936–942 (2013). https://doi.org/10.1002/mrm.24544
46. Witney, T.H., Kettunen, M.I., Brindle, K.M.: Kinetic modeling of hyperpolarized 13C label exchange between pyruvate and lactate in tumor cells. J. Biol. Chem. **286**(28), 24572–24580 (2011). https://doi.org/10.1074/jbc.M111.237727
47. Josan, S., Xu, T., Yen, Y.F., Hurd, R., Ferreira, J., Chen, C.H., Mochly-Rosen, D., Pfefferbaum, A., Mayer, D., Spielman, D.: In vivo measurement of aldehyde dehydrogenase-2 activity in rat liver ethanol model using dynamic MRSI of hyperpolarized [1-(13) C]pyruvate. NMR Biomed. **26**(6), 607–612 (2013). https://doi.org/10.1002/nbm.2897
48. Serrao, E.M., Kettunen, M.I., Rodrigues, T.B., Lewis, D.Y., Gallagher, F.A., Hu, D.E., Brindle, K.M.: Analysis of (13) C and (14) C labeling in pyruvate and lactate in tumor and blood of lymphoma-bearing mice injected with (13) C- and (14) C-labeled pyruvate. NMR Biomed. **31**(5), e3901 (2018). https://doi.org/10.1002/nbm.3901
49. Hu, S., Yoshihara, H.A., Bok, R., Zhou, J., Zhu, M., Kurhanewicz, J., Vigneron, D.B.: Use of hyperpolarized [1-13C]pyruvate and [2-13C]pyruvate to probe the effects of the anticancer agent dichloroacetate on mitochondrial metabolism in vivo in the normal rat. Magn. Reson. Imaging. **30**(10), 1367–1372 (2012). https://doi.org/10.1016/j.mri.2012.05.012

50. Park, J.M., Josan, S., Jang, T., Merchant, M., Watkins, R., Hurd, R.E., Recht, L.D., Mayer, D., Spielman, D.M.: Volumetric spiral chemical shift imaging of hyperpolarized [2-(13) c]pyruvate in a rat c6 glioma model. Magn. Reson. Med. **75**(3), 973–984 (2016). https://doi.org/10.1002/mrm.25766
51. Dodd, M.S., Ball, V., Bray, R., Ashrafian, H., Watkins, H., Clarke, K., Tyler, D.J.: In vivo mouse cardiac hyperpolarized magnetic resonance spectroscopy. J. Cardiovasc. Magn. Reson. **15**, 19 (2013). https://doi.org/10.1186/1532-429X-15-19
52. Josan, S., Park, J.M., Hurd, R., Yen, Y.F., Pfefferbaum, A., Spielman, D., Mayer, D.: In vivo investigation of cardiac metabolism in the rat using MRS of hyperpolarized [1-13C] and [2-13C]pyruvate. NMR Biomed. **26**(12), 1680–1687 (2013). https://doi.org/10.1002/nbm.3003
53. Lauritzen, M.H., Laustsen, C., Butt, S.A., Magnusson, P., Sogaard, L.V., Ardenkjaer-Larsen, J.H., Akeson, P.: Enhancing the [13C]bicarbonate signal in cardiac hyperpolarized [1-13C]pyruvate MRS studies by infusion of glucose, insulin and potassium. NMR Biomed. **26**(11), 1496–1500 (2013). https://doi.org/10.1002/nbm.2982
54. Hu, S., Chen, A.P., Zierhut, M.L., Bok, R., Yen, Y.F., Schroeder, M.A., Hurd, R.E., Nelson, S.J., Kurhanewicz, J., Vigneron, D.B.: In vivo carbon-13 dynamic MRS and MRSI of normal and fasted rat liver with hyperpolarized 13C-pyruvate. Mol. Imaging Biol. **11**(6), 399–407 (2009). https://doi.org/10.1007/s11307-009-0218-z
55. Moreno, K.X., Satapati, S., DeBerardinis, R.J., Burgess, S.C., Malloy, C.R., Merritt, M.E.: Real-time detection of hepatic gluconeogenic and glycogenolytic states using hyperpolarized [2-13C]dihydroxyacetone. J. Biol. Chem. **289**(52), 35859–35867 (2014). https://doi.org/10.1074/jbc.M114.613265
56. Keshari, K.R., Sriram, R., Van Criekinge, M., Wilson, D.M., Wang, Z.J., Vigneron, D.B., Peehl, D.M., Kurhanewicz, J.: Metabolic reprogramming and validation of hyperpolarized 13C lactate as a prostate cancer biomarker using a human prostate tissue slice culture bioreactor. Prostate. **73**(11), 1171–1181 (2013). https://doi.org/10.1002/pros.22665
57. Saito, K., Matsumoto, S., Takakusagi, Y., Matsuo, M., Morris, H.D., Lizak, M.J., Munasinghe, J.P., Devasahayam, N., Subramanian, S., Mitchell, J.B., Krishna, M.C.: 13C-MR spectroscopic imaging with hyperpolarized [1-13C]pyruvate detects early response to radiotherapy in SCC tumors and HT-29 tumors. Clin. Cancer Res. **21**(22), 5073–5081 (2015). https://doi.org/10.1158/1078-0432.CCR-14-1717
58. Scroggins, B.T., Matsuo, M., White, A.O., Saito, K., Munasinghe, J.P., Sourbier, C., Yamamoto, K., Diaz, V., Takakusagi, Y., Ichikawa, K., Mitchell, J.B., Krishna, M.C., Citrin, D.E.: Hyperpolarized [1-(13)C]-pyruvate magnetic resonance spectroscopic imaging of prostate Cancer in vivo predicts efficacy of targeting the Warburg effect. Clin. Cancer Res. **24**(13), 3137–3148 (2018). https://doi.org/10.1158/1078-0432.CCR-17-1957
59. Spielman, D.M., Mayer, D., Yen, Y.F., Tropp, J., Hurd, R.E., Pfefferbaum, A.: In vivo measurement of ethanol metabolism in the rat liver using magnetic resonance spectroscopy of hyperpolarized [1-13C]pyruvate. Magn. Reson. Med. **62**(2), 307–313 (2009). https://doi.org/10.1002/mrm.21998
60. Gallagher, F.A., Kettunen, M.I., Brindle, K.M.: Imaging pH with hyperpolarized 13C. NMR Biomed. **24**(8), 1006–1015 (2011). https://doi.org/10.1002/nbm.1742
61. Kohler, S.J., Yen, Y., Wolber, J., Chen, A.P., Albers, M.J., Bok, R., Zhang, V., Tropp, J., Nelson, S., Vigneron, D.B., Kurhanewicz, J., Hurd, R.E.: In vivo 13 carbon metabolic imaging at 3T with hyperpolarized 13C-1-pyruvate. Magn. Reson. Med. **58**(1), 65–69 (2007). https://doi.org/10.1002/mrm.21253
62. Chen, A.P., Albers, M.J., Cunningham, C.H., Kohler, S.J., Yen, Y.F., Hurd, R.E., Tropp, J., Bok, R., Pauly, J.M., Nelson, S.J., Kurhanewicz, J., Vigneron, D.B.: Hyperpolarized C-13 spectroscopic imaging of the TRAMP mouse at 3T-initial experience. Magn. Reson. Med. **58**(6), 1099–1106 (2007). https://doi.org/10.1002/mrm.21256
63. Hill, D.K., Orton, M.R., Mariotti, E., Boult, J.K., Panek, R., Jafar, M., Parkes, H.G., Jamin, Y., Miniotis, M.F., Al-Saffar, N.M., Beloueche-Babari, M., Robinson, S.P., Leach, M.O., Chung, Y.L., Eykyn, T.R.: Model free approach to kinetic analysis of real-time hyperpolarized 13C magnetic resonance spectroscopy data. PLoS One. **8**(9), e71996 (2013). https://doi.org/10.1371/journal.pone.0071996

64. Park, J.M., Josan, S., Jang, T., Merchant, M., Yen, Y.F., Hurd, R.E., Recht, L., Spielman, D.M., Mayer, D.: Metabolite kinetics in C6 rat glioma model using magnetic resonance spectroscopic imaging of hyperpolarized [1-(13)C]pyruvate. Magn. Reson. Med. **68**(6), 1886–1893 (2012). https://doi.org/10.1002/mrm.24181
65. Hansen, A.E., Gutte, H., Holst, P., Johannesen, H.H., Rahbek, S., Clemmensen, A.E., Larsen, M.M.E., Schoier, C., Ardenkjaer-Larsen, J., Klausen, T.L., Kristensen, A.T., Kjaer, A.: Combined hyperpolarized (13)C-pyruvate MRS and (18)F-FDG PET (hyperPET) estimates of glycolysis in canine cancer patients. Eur. J. Radiol. **103**, 6–12 (2018). https://doi.org/10.1016/j.ejrad.2018.02.028
66. Lupo, J.M., Chen, A.P., Zierhut, M.L., Bok, R.A., Cunningham, C.H., Kurhanewicz, J., Vigneron, D.B., Nelson, S.J.: Analysis of hyperpolarized dynamic 13C lactate imaging in a transgenic mouse model of prostate cancer. Magn. Reson. Imaging. **28**(2), 153–162 (2010). https://doi.org/10.1016/j.mri.2009.07.007
67. Park, J.M., Khemtong, C., Liu, S.C., Hurd, R.E., Spielman, D.M.: In vivo assessment of intracellular redox state in rat liver using hyperpolarized [1-(13) C]alanine. Magn. Reson. Med. **77**(5), 1741–1748 (2017). https://doi.org/10.1002/mrm.26662
68. Josan, S., Hurd, R., Billingsley, K., Senadheera, L., Park, J.M., Yen, Y.F., Pfefferbaum, A., Spielman, D., Mayer, D.: Effects of isoflurane anesthesia on hyperpolarized (13)C metabolic measurements in rat brain. Magn. Reson. Med. **70**(4), 1117–1124 (2013). https://doi.org/10.1002/mrm.24532
69. Bohndiek, S.E., Kettunen, M.I., Hu, D.E., Kennedy, B.W., Boren, J., Gallagher, F.A., Brindle, K.M.: Hyperpolarized [1-13C]-ascorbic and dehydroascorbic acid: vitamin C as a probe for imaging redox status in vivo. J. Am. Chem. Soc. **133**(30), 11795–11801 (2011). https://doi.org/10.1021/ja2045925
70. Keshari, K.R., Kurhanewicz, J., Bok, R., Larson, P.E., Vigneron, D.B., Wilson, D.M.: Hyperpolarized 13C dehydroascorbate as an endogenous redox sensor for in vivo metabolic imaging. Proc. Natl. Acad. Sci. U. S. A. **108**(46), 18606–18611 (2011). https://doi.org/10.1073/pnas.1106920108
71. Keshari, K.R., Sai, V., Wang, Z.J., Vanbrocklin, H.F., Kurhanewicz, J., Wilson, D.M.: Hyperpolarized [1-13C]dehydroascorbate MR spectroscopy in a murine model of prostate cancer: comparison with 18F-FDG PET. J. Nucl. Med. **54**(6), 922–928 (2013). https://doi.org/10.2967/jnumed.112.115402

Further Reading

Harrison, C., Yang, C., Jindal, A., DeBerardinis, R.J., Hooshyar, M.A., Merritt, M., Dean Sherry, A., Malloy, C.R.: Comparison of kinetic models for analysis of pyruvate-to-lactate exchange by hyperpolarized 13 C NMR. NMR Biomed. **25**(11), 1286–1294 (2012). https://doi.org/10.1002/nbm.2801

Hurd, R.E., Spielman, D., Josan, S., Yen, Y.F., Pfefferbaum, A., Mayer, D.: Exchange-linked dissolution agents in dissolution-DNP (13)C metabolic imaging. Magn Reson Med. **70**(4), 936–942 (2013). https://doi.org/10.1002/mrm.24544

Mariotti, E., Orton, M.R., Eerbeek, O., Ashruf, J.F., Zuurbier, C.J., Southworth, R., Eykyn, T.R.: Modeling non-linear kinetics of hyperpolarized [1-(13)C] pyruvate in the crystalloid-perfused rat heart. NMR Biomed. **29**(4), 377–386 (2016). https://doi.org/10.1002/nbm.3464

Zierhut, M.L., Yen, Y.F., Chen, A.P., Bok, R., Albers, M.J., Zhang, V., Tropp, J., Park, I., Vigneron, D.B., Kurhanewicz, J., Hurd, R.E., Nelson, S.J.: Kinetic modeling of hyperpolarized 13C1-pyruvate metabolism in normal rats and TRAMP mice. J Magn Reson. **202**(1), 85–92 (2010). https://doi.org/10.1016/j.jmr.2009.10.003

Chapter 6
Using Hyperpolarized NMR to Understand Biochemistry from Cells to Humans

Renuka Sriram, Celine Baligand, and John Kurhanewicz

Abbreviations

ATP	Adenosine triose phosphate
DNP	Dynamic nuclear polarization
FDA	Food and Drug Administration
FDG	Fluoro deoxy glucose
GMP	Good manufacturing practice
HP	Hyperpolarization
HR-MAS	High resolution-magic angle spinning
IND	Investigational new drug
LDH	Lactate dehydrogenase
NCI	National Cancer Institute
PBS	Phosphate buffered saline
PDH	Pyruvate dehydrogenase
PET	Positron emission tomography
RF	Radio frequency
SNR	Signal to noise ratio
TR	Repetition time
TRAMP	Transgenic adenocarcinoma of the murine prostate

6.1 Introduction

Imaging of hyperpolarized ^{13}C-labeled biomolecules (probes) provides significant new insights into previously inaccessible aspects of the biology of human diseases. Many of the biomolecules crucial for understanding and monitoring metabolism

R. Sriram · C. Baligand · J. Kurhanewicz (✉)
Radiology and Biomedical Imaging, University of California San Francisco,
San Francisco, CA, USA
e-mail: John.kurhaewicz@ucsf.edu

T. Jue, D. Mayer (eds.), *Dynamic Hyperpolarized Nuclear Magnetic Resonance*,
Handbook of Modern Biophysics, https://doi.org/10.1007/978-3-030-55043-1_6

(glycolysis, tricarboxylic acid cycle, fatty acid/lipid metabolism, redox, and acid/base chemistry) are often present in low concentrations that are often beyond the detection threshold of traditional magnetic resonance spectroscopy and metabolic imaging approaches. One solution is to improve sensitivity by a factor of 10,000 or more by temporarily redistributing the populations of nuclear spins in a magnetic field, using a process termed "hyperpolarization." Hyperpolarization (HP) using the dissolution dynamic nuclear polarization (DNP) approach is based on polarizing nuclear spins in the amorphous solid state at ~0.8–1.2 K in 3.35–5 T magnetic fields through coupling of the nuclear spins with unpaired electrons that are added to the sample via an organic free radical (Chap. 1). Critical to the in vivo use of this method is the HP probe dissolution method introduced by Ardenkjaer-Larsen et al. [1]. The dissolution step rapidly transforms frozen polarized samples into solutions at physiologic temperature and pH, while preserving nuclear polarization. An important milestone for HP ^{13}C MR was the demonstration that hyperpolarized ^{13}C-labeled molecules (probes) can be delivered to living systems (cells, tissues, preclinical animal models, and humans) where the molecular probe is metabolized and the HP probe and its metabolic products can be imaged [2]. Therefore, HP ^{13}C MR lifts the primary constraint on magnetic resonance metabolic imaging, poor sensitivity, while preserving the advantage of biochemical information.

Due both to its excellent HP MR properties and being at a metabolic crossroad, HP [1-^{13}C]pyruvate has been extensively studied preclinically, and is the first HP probe used in patients [3]. Recent studies have also shown the clinical potential of a number of other HP probes including [2-^{13}C]fructose, [U-$^{13}C_6$, $^{2}H_7$]glucose, [1-^{13}C] alpha-ketoglutarate, and [5-^{13}C]glutamine, ^{13}C urea, ^{13}C bicarbonate, and [1-^{13}C] dehydroascorbate to provide information about metabolism, perfusion, interstitial pH, and the redox status of human disease. Additionally, techniques to simultaneous polarize multiple ^{13}C-enriched probes as well as fast dynamic ^{13}C MR sequences are being developed to provide information concerning several enzymatic pathways and other physiologic properties in a single imaging exam [4].

Arguably, the most significant advantages of HP ^{13}C MR are its ability to uniquely image metabolic fluxes in patients and its ease of integration into clinical ^{1}H MRI exams. Although the first NCI-sponsored white paper describing the potential of this new molecular imaging technique was published only 8 years ago [5], over 80 ^{13}C-labeled biomolecules have been hyperpolarized and tested in preclinical studies. These HP probes can be used to study basic biochemical reactions mediated by dehydrogenases, transaminases, decarboxylases, peptidases, acetyltransferases, acylases, kinases, and hydratases, to name a few. Moreover, a phase 1 clinical trial of hyperpolarized [1-^{13}C]pyruvate in prostate cancer patients [3] and subsequent patient studies [6–8] have demonstrated that this powerful technology can be translated to the clinic. This chapter will focus on the development and preclinical testing of hyperpolarized ^{13}C-labeled probes and the required steps necessary for their translation to patient studies, with an emphasis on the platforms and techniques necessary to achieve this goal.

6.2 Identifying Metabolic Targets for Hyperpolarized ^{13}C Probe Development: HR-MAS and Isotope Labeling Studies

Metabolic dysfunction can be implicated in almost all pathologies. Specifically, in cancer, it is now considered a necessary and coordinated change that accompanies genomic alterations that bestow the uncontrolled proliferation of this pathology. Steady-state proton magnetic resonance spectroscopy has long been used to understand the metabolic alterations that occur in many pathologies using tissues and biopsies from patients in an ex situ condition. These often involve destructive extraction of metabolites from cells and tissues that precludes correlative biology on the same sample, which would confound the metabolic landscape in heterogenous diseases. Consequently, in the case of precious patient specimens, we often have to employ techniques such as high-resolution magic angle spinning (HR-MAS) [9] which would preserve the intact tissue for concurrent clinical routine diagnostics on the same tissue. However, implementing the same technology in the clinical setting becomes complicated due to (a) motion of the patient (b) lack of more efficient RF coils and sequences, thereby limiting sensitivity as well as (c) overlapping signals from nontargeted tissue regions. Furthermore, due to metabolic homeostasis, the variation in concentration of metabolites inside cells in response to changes in conditions is generally relatively small, thereby necessitating the determination of isotopomer fluxes. However, using ^{13}C-enriched labeling analysis in the clinic is an even more complicated endeavor which has now been bolstered with the use of DNP technology owing to its assessment of metabolic fluxes within a minute. Thus, in order to choose the optimal probe for hyperpolarization using DNP for its ultimate application in the clinic as a biomarker, it is imperative to first characterize the metabolic dysfunctions in the pathology of interest using the above mentioned tools of (1) proton steady-state estimates of the pool sizes and (2) isotopic labeling measurements to shed light on the pathways that are altered. Figure 6.1 illustrates the comprehensive development of lactate as a biomarker of prostate malignancy using HR-MAS [10] as well as 2D NMR techniques in patient derived tissues and biopsies [11].

6.3 In Vitro Testing of Hyperpolarized ^{13}C Probes

In developing and optimizing a new hyperpolarized probe for a given disease, the general approach consists of the following steps. Initial in vitro chemistry studies are performed in order to maximize HP probe concentration, polarization, and T_1 relaxation, while minimizing contaminants. Usually as part of these chemistry studies both the probe polarization and dissolution processes are also optimized and the magnetic field dependence of the T_1 relaxation of the HP probe and low field probe transfer effects (quadrupolar and paramagnetic relaxation effects) are determined

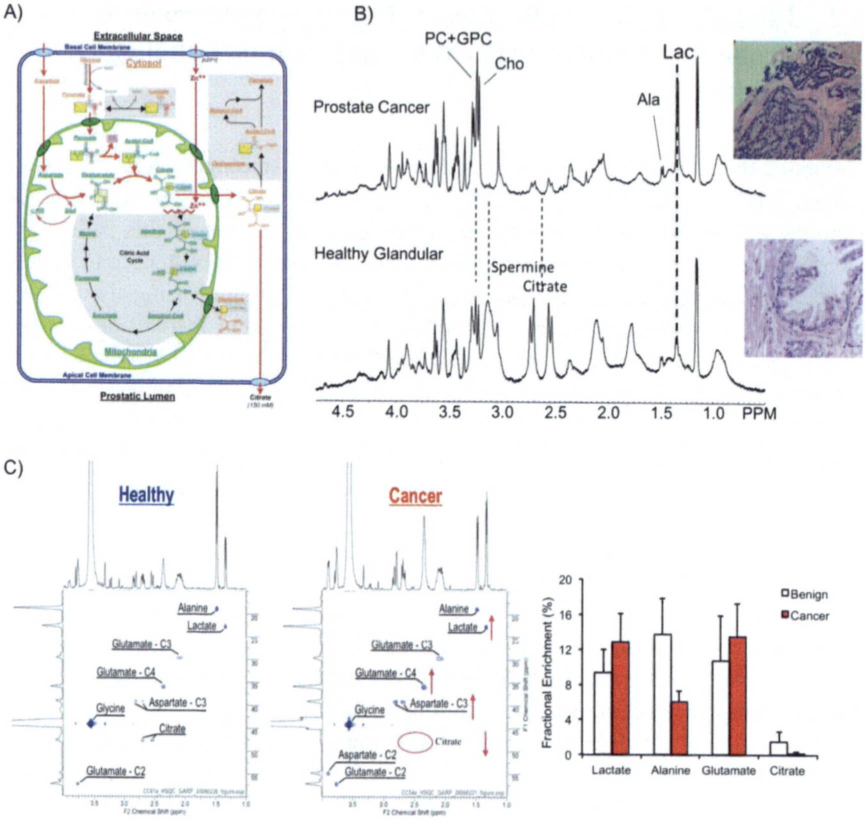

Fig. 6.1 Identifying metabolic targets for hyperpolarized MR. This illustration shows the various tools that were used to identify lactate as a biomarker of prostate malignancy. (**a**) Pictorial representation of the underlying metabolic alterations with the onset of disease in prostate epithelial cell [63]. The red lines represent the pathways that are active in healthy secretory prostatic cells and the black lines along with the gray background highlight the metabolic adaptations proposed to occur with prostate cancer. (**b**) The HR-MAS spectra show the differences in observed metabolites in the cancerous (top) and normal healthy tissue (bottom). There is increased lactate and choline components in the cancer tissue while the abundance of citrate and spermine observed in the healthy tissues is lacking. The inset of the hematoxylin and eosin immunohistochemical stain is the gold standard of pathology assessment and confirms the presence of 50% cancerous cells of Gleason grade 3 + 4 in the malignant tissue and 40% glandular tissue in the healthy specimen. (**c**) The 2D heteronuclear single quantum coherence plot shows the cross-peaks arising from the protons coupled to ^{13}C nuclei in the metabolite in tissues (obtained from prostatectomy patients) after labeling with [1-^{13}C] pyruvate, adapted from Keshari et al. [11]

(Chap. 3). Following which, the probe is tested in enzyme systems to assess the detection and rate of metabolic byproducts formation. The subsequent step often involves the injection of the HP probe in living cells or tissues to demonstrate feasibility of HP probe uptake and metabolism under controlled conditions. Many sites have used cell slurry's in order to accomplish this step, but some sites have also employed NMR compatible 3D cell and tissue culture perfusion systems, often called bioreactor. The next step usually involves in vivo studies in murine models to demonstrate biodistribution, probe uptake and metabolism, and overall imaging feasibility and clinical significance. The potential clinical utility of HP probe can be established in these studies and used in the FDA IND application for its eventual use in patients. The clinical translation of a HP probe also requires the further optimization of the probe in terms of cost, determination of HP probe contaminants shelf-life, optimization of clinical polarization, dissolution steps (considered to be sterile compounding by FDA) delivery, establishing protocol and quality assurance procedures, and working with stable isotope companies to produce lower cost GMP HP probes.

6.3.1 *In Vitro Characterization of the Hyperpolarized Probe*

The development of new HP probes is based both on the MR properties of the molecule, as well as transport, endogenous pool size, and metabolism considerations, and therefore the optimization of the HP probes will differ according to the specific disease and probe being investigated. In general, HP probes require a high concentration and long T_1 relaxation to facilitate efficient spin diffusion during the process of dynamic nuclear polarization (DNP) and to maintain the polarization until the time of in vivo imaging (Chap. 3). Specifically, for effective in vivo imaging, the probe needs to be soluble in aqueous solution, and have a sufficiently long T_1 lifetime and/or degree of spin polarization to permit transport, administration, biodistribution, and/or metabolism in vivo, and the acquisition of images and/or spectra with high spatial and/or temporal resolution.

6.3.1.1 Probe Concentration

If the neat carbon-13-enriched substrate does not glass readily, it is added to an appropriate glassing agent before it can be polarized (see Chap. 3 for more details). This addition of the glassing agent needs to be minimized as it directly impacts the achievable polarization as well as the net concentration of the substrate upon dissolution.

6.3.1.2 Solution State T_1

T_1 is the characteristic decay time of the hyperpolarized signal which follows an exponential function: $T_1 = a \times \exp(-t/T_1)$, where a is the constant and t is time in seconds. This is measured by simply adding the hyperpolarized sample that is ejected out of the polarizer to an NMR tube and rapidly inserting it into a magnet. It is important to note that the T_1 is different at different field strengths. This becomes a very important factor in determining the transfer path taken to move the sample from the polarizer to the magnet. For example, if the polarized sample passes through a zero-field point during the transfer of the dissolution buffer, all the signals will be lost (see Chaps. 1–3). Repetitive small angle (1°–5°) RF pulse is then applied with a TR of 2–5 s. The ensuing peak is then integrated and the dynamic data are fit to the above equation to determine the solution state T_1. A short T_1 of the order of 10 s precludes the successful use of the probe for in vivo metabolic studies. For a more realistic estimate of in vivo T_1, the dissolved hyperpolarized substrate is added to serum to account for any alterations due to protein binding.

6.3.1.3 Polarization Measurement

Liquid state polarization refers to the increase in signal gained by the DNP process, compared against the signal measured using the same sample at thermal equilibrium. This is calculated by using the following equation [12, 13]:

$$P_{\mathrm{hp}}(\%) = P_{\mathrm{th}}\left(A_{\mathrm{hp}} / A_{\mathrm{th}}\right)\left(\sin\theta_{\mathrm{th}} / \sin\theta_{\mathrm{hp}}\right),$$

where the subscripts "th" and "hp" refer to thermal and hyperpolarized, respectively. "A" is the integral peak area and θ is the RF flip angle used for acquisition, which is typically 90° for the thermal spectrum acquisition. In order to estimate both the T_1 as well as the polarization simultaneously, the first spectrum in the dynamic acquisition (used to estimate T_1) is used for the polarization measurement and hence would be a small flip angle of 1–5°. By carefully monitoring the time taken for the transfer of solution from the polarizer to the magnet (t), the polarization upon dissolution can be back-calculated by incorporating the signal loss due to T_1 by modifying the above equation to

$$P_{\mathrm{hp}}(\%) = P_{\mathrm{th}}\left(A_{\mathrm{hp}} / A_{\mathrm{th}}\right)\left(\sin\theta_{\mathrm{th}} / \sin\theta_{\mathrm{hp}}\right)\exp\left(t / T_1\right).$$

6.3.2 *Naïve Enzyme in a Tube*

This is the simplest way of testing a new hyperpolarized probe and can be used to study the initial rate kinetics at physiological conditions. It simply entails the addition of isolated enzymes at an appropriate concentration in a physiologic buffer and temperature in an MR compatible receptacle with the appropriate cofactors and ligands. For example, in order to measure the phosphorylation of fructose, one can use the enzyme hexokinase, with supplementary ATP and $MgCl_2$ in a TRIS or PBS buffer at pH 7.4. It is necessary to add ATP and $MgCl_2$ as the enzyme hexokinase facilitates the reaction

$$\mathrm{Hexose-CH_2OH + MgATP^{2-} \rightarrow Hexose-CH_2O-PO_{2\text{-}3} + MgADP^{-} + H^{+}}$$

The magnesium ion is required as the reactive form of ATP is the complex with magnesium (II) ion. Selection of appropriate isoform of the enzyme is also critical in testing hyperpolarized substrates. In the previous example, hexokinase isoform I–III is capable of converting fructose to fructose-1-phospahte but not isoform IV. Also, the enzyme concentration needs to be high pursuant to the concentration of the substrate used. If 80 μmol of hyperpolarized [2-^{13}C]-fructose is used, it needs to be reacted with at least 80 units of the enzyme hexokinase (1 unit catalyzes the conversion of 1 μmol of substrate/min) with ATP and Mg salts in excess. Hyperpolarized substrates are then added to the appropriately formulated enzyme mixture, and the reaction is quickly measured usually using a simple dynamic pulse and acquire sequence (see Chap. 4). This enables the study of a single enzymatic conversion and can help elucidate the mechanism of binding (i.e., molecular interactions [14] and detect otherwise invisible low-populated intermediate species [14, 15]). An inherent drawback of using this model system for initial testing prior to an in vivo application is the lack of intact cell membrane, which obfuscates any transport concerns. In the case of testing multiple enzymes in a test tube, assessment of further downstream metabolism needs careful consideration of substrate/product inhibition and cofactor concentration. That being said, this powerful technique has unequivocal ability to shed insights into protein–ligand interactions and protein folding [16].

6.3.3 Cell Lysates

In case of commercially unavailable enzyme, cell lysate is a popular alternative. For a standard 5 mm NMR tube experiment, 5×10^7 cells are required. Once harvested from adherent plates or suspension, the cells need to be lysed with detergents in a buffer containing protease and phosphatase cocktail mixture to prevent inhibition of enzymes. Lysates are then further homogenized using mechanical grinding to ensure maximal disruption of intracellular organelle membrane when studying noncytosolic enzymes, all while keeping the cell material ice-cold [16, 17]. The cell homogenate is then spun down, and the supernatant containing the soluble proteins/enzyme mixture is used for the hyperpolarized study. This set-up also requires addition of cofactors specific to the enzyme of interest. The cofactors should be in two- to three-fold excess to the hyperpolarized substrate concentration to account for nonavailability due to nonspecific binding to other enzymes present in the lysate.

6.3.4 Cell Slurries

Cell slurry, which simply denotes cells in a tube, is also known as "cells in suspension." A good majority of published hyperpolarized ^{13}C MR studies uses this simple system of an aggregate of cells. This requires cells of the order of 10^8 [18] harvested from adherent plates or from suspension as for any biochemical assay immediately prior to the addition of hyperpolarized experiment. This needs to be timed diligently to minimize the latent period between cell preparation and hyperpolarized MR measurements to minimize alteration of the metabolic state of the cells and their viability. Typically, once the substrate is polarized (anywhere from 1 to 3 h), 15 min before the dissolution is commenced, the cells will be prepared. There are two ways of acquiring data using this system—(1) add cells in solution via a catheter to the hyperpolarized substrate already in the magnet as this provides vital information regarding the initial metabolic period [18, 19]. Alternatively, the cells are added to the hyperpolarized substrate to ensure thorough mixing, and then the sample is inserted into the magnet for acquisition. Major drawbacks of this cell suspension method are: (1) This method requires a large number of cells (a minimum of $4e^7$ cells/mL of suspension buffer) to do one hyperpolarized injection. A fresh batch of cells is required for each replicate as the cells are subjected to warm ischemia and can also result in huge acidification of the extracellular media in case of hyperpolarized lactate that quickly effluxes out of the cell. (2) The metabolic state of the cells can be highly variable from one experiment to another as the cell transition from being on ice (a common practice to slow down metabolic processes) to biologically relevant temperatures all under the auspices of nonhomogenous oxygen availability. (3) Longitudinal studies are not feasible in this model system as the viability of the cells is greatly reduced over time. (4) Improper mixing of the cells with the hyperpolarized substrate and settling of cells can sometimes yield unsuccessful results due to decreased sensitivity (as cells might be displaced away from the sensitive region of the coil and poor inhomogeneity due to improper mixing). Nonetheless, this is the easiest in vitro model that can be used to study metabolism of hyperpolarized substrates as it incorporates all the basic elements of an intact system with the exception of the microenvironment and have been successfully employed to test almost all new probes using immortalized cancer cells. Usually, cell count and viability are performed after the hyperpolarized experiment as a confirmation of intact cell metabolism. Although no other correlative biochemical measures are done on this simple system, it has nonetheless been used for extensive kinetic models [19, 20] (see Chap. 6), to assess the rate of transport [20, 21], as well as to interrogate simultaneous, alternate pathways of metabolism [21, 22].

6.3.5 Bioreactors

Bioreactors are MR compatible, engineered perfusion systems that offer a sterile environment, physiologically pertinent culture conditions (e.g., temperature, CO_2 level, nutrient supply, waste drain), and growth factors that enable MR measurements in a sterile, well-controlled manner over long periods (up to 72 h have been published [10, 11]) to ensure normal cell metabolism. Such a system provides the ability to serially monitor the uptake and metabolism of injected HP probes using ^{13}C spectroscopy while establishing the viability of the cells using ^{31}P spectroscopy. The schematic shown in Fig. 6.2 clearly details the multiple components of the cell incubator adapted for hyperpolarized NMR studies. The continuous nutrient supply is established via circulating media, maintained at 37 °C with optimal dissolved oxygen content. The cells are immobilized by either growing them on microcarrier beads, or various encapsulation process in alginate or hollow fibers [22, 23] to spatially restrict the cells in the NMR sensitive region of the tube. The choice of cell immobilization depends on the cell under study; usually, cells that are strongly adherent are studied on microcarrier beads, such as macrophages [23, 24] or primary cells [24, 25], while most immortalized cells do well in alginate encapsulation. One major disadvantage of microcarrier beads is the limited cell density (translating to decreased NMR sensitivity) that can be achieved as the beads occupy nearly 70% of the volume.

This platform overcomes all the shortcomings of a cell lysate technique mentioned in the prior section. However, this is a sophisticated complex system that is developed and implemented in individual academic labs and involves elaborate setup of plastic tubes, usually 2–6 m long, for the circulating media, effective sterilization of the bioreactor parts in contact with cells for prolonged experiments and ensuring bubble free environment in the RF coil sensitive region for good shim homogeneity.

This powerful platform has several unique advantages: (1) simultaneous multinuclear detection such as ^{31}P for bioenergetics monitoring, ^{13}C for combined hyperpolarized and thermal isotopic flux monitoring; (2) ability to carefully control multiple physiologic parameters such as pH, perfusion, oxygen, and temperature, and observe its consequence on cellular metabolism; (3) perform longitudinal studies on the same cells to assess therapeutic response and proliferative status; (4) minimized sample requirements; and (5) adaptable for concurrent PET applications. The same system was expanded to have four reaction chambers perfused from a single-media reservoir which could directly be injected with the PET probe. Each sample chamber contained cells treated with different concentrations or cell concentrations to assess simultaneously the drug efficacy or PET probe sensitivity [26].

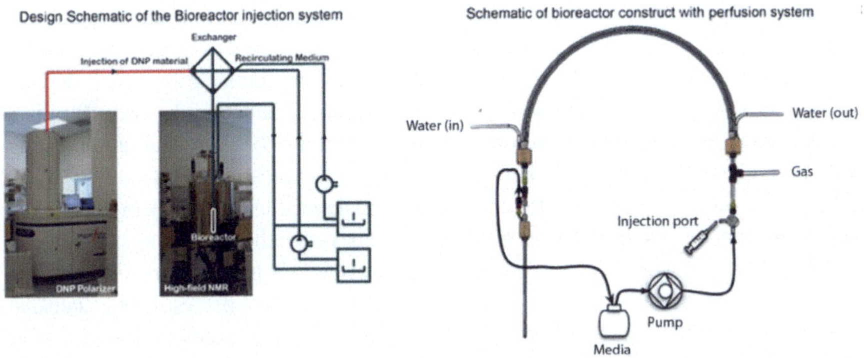

Fig. 6.2 Bioreactor schematic. Adapted from Keshari et al. [26], this figure shows the schematic of the bioreactor. It comprises of a fluid path containing the circulating media heated by a heated water jacket. The media also has a separate gas port for high-efficiency oxygen dissolution into the media. The injection port is provided to infuse hyperpolarized substrate

6.4 Ex Vivo Applications of Hyperpolarized NMR to Understand Biochemistry

The ex vivo model systems that have been employed to study biochemistry using hyperpolarized NMR predominantly involve perfused organs, the bread and butter of physiology, as well as intact tissue derived from patients without long-term culture. These systems present a simplified alternative to whole organism studies especially in the absence of pertinent in vivo animal models. Table 6.1 lists the various hyperpolarized probes that have been successfully used to interrogate enzymatic reactions in situ.

6.4.1 Intact Tissue

As an extension to cells in the bioreactor, that platform has been used to study the metabolism of living patient derived tissues [10, 11, 26, 27] of both renal and prostate cancer. This provides an exclusive method to study certain diseases where appropriate model systems are sorely lacking such as benign renal tumors. It has an added advantage of low tissue sample requirements (<100 mg) and can be used to do correlative biochemical and immunohistochemical assays on the same sample as has been demonstrated. A rather large infrastructure as well as specialized tools is required to execute this preclinical yet highly valuable platform to study clinically relevant disease processes. Some of the essential elements are (1) robust tissue acquisition immediately after surgical removal of tumors, subsequent tissue coring and slicing using specialized apparatus followed by sterile culture of the tissues in specially formulated and tested media and ultimately, (2) careful loading of the tissues into specifically designed cartridge-tissue holders for bioreactor chambers, and (3) optimized oxygen and media flow conditions prior to hyperpolarized MR acquisitions. The nutrient and oxygen requirements vary depending on the tissue under study. For example, prostate tissue has low metabolism and requires slower perfusion compared to renal tissue, which is metabolically very active. This also would necessitate more mass of tissue to be used for the study. An outstanding advantage of this platform is all correlative biochemical and immunohistochemical assays can be done on the same tissue. We have used this system to study renal carcinoma from patient-derived tissues and showed that in addition to increased lactate production, lactate efflux is a distinguishing factor of malignant renal cancer compared to benign renal tumors along with correlative pathological confirmation, enzyme expression, and activity [27]. Figure 6.3 shows the apparent differences in hyperpolarized lactate levels in benign and malignant renal tissues in comparison to normal kidneys and the underlying biochemical mechanisms. This work exemplified the fact that even in benign tumors, the Warburg effect is very much in effect and it is the compartmentalization of the resultant lactate that is the distinguishing factor between malignant and benign renal tumors. Based on this observation, in situ further work was carried out using orthotopic murine models and advanced diffusion-encoded pulse sequences using hyperpolarized pyruvate to measure the lactate compartmentation in real time [28].

Table 6.1 List of enzymes that have been successfully probed using hyperpolarized MR

	^{13}C-labeled molecules	Models systems used
Lactate dehydrogenase	[1-^{13}C]pyruvate, [2-^{13}C]pyruvate, [1-^{13}C] lactate, [1-^{13}C]alanine, ^{13}C glycerate, [2-^{13}C]dihydroxyacetone, [U-$^{13}C_6$, $^2H_7$] glucose	Enzymes, cells, perfused tissues, perfused organs, and in vivo
Alanine aminotransferase	[1-^{13}C]pyruvate, [2-^{13}C]pyruvate, [1-^{13}C] lactate, [1-^{13}C]alanine, ^{13}C glycerate	Cells, perfused tissues, perfused organs, and in vivo
Pyruvate dehydrogenase	[1-^{13}C]pyruvate, [2-^{13}C]pyruvate, [1-^{13}C] lactate, [1-^{13}C]alanine	Cells, perfused organs, and in vivo
Carnitine acetyl transferase	[2-^{13}C]pyruvate, [1-^{13}C]acetate, [1-^{13}C] butyrate, [1-^{13}C]propionate	Enzymes, cells, and in vivo
Carbonic anhydrase	^{13}C bicarbonate, ^{13}C-enriched ethyl acetyl carbonate	Enzymes and in vivo
Fumarase	[1,4-$^{13}C_2$]fumarate	Cells and in vivo
Glutaminase	[5-^{13}C]glutamine	Enzymes, cells, and in vivo
Acetyl-CoA-synthase	[1-^{13}C]acetate, [1-^{13}C]butyrate, [1-^{13}C] propionate	Enzymes, cells, perfused organs, and in vivo
Hexokinase	[U-$^{13}C_6$, $^2H_7$]glucose, [2-^{13}C]-fructose	Enzymes and in vivo
Branched chain amino acid transferase	[1-^{13}C]-α-ketoisocaproate, [1-^{13}C]-α-ketoglutarate, [1-^{13}C]-glutamate	Enzymes, cells, and in vivo
Acetoacetyl-CoA thiolase	^{13}C glycerate	Fumarase
3-Oxoacid CoA transferase	[1-^{13}C]butyrate	In vivo
β-Hydroxybutyrate dehydrogenase	[1-^{13}C]butyrate	In vivo
Pyruvate carboxylase	[1-^{13}C]pyruvate, [2-^{13}C]dihydroxyacetone	Perfused organs and in vivo
Enolase	[2-^{13}C]dihydroxyacetone	Perfused organs and in vivo
Glucose 6 phosphatase	[2-^{13}C]dihydroxyacetone	Perfused organs and in vivo
Phosphoglycerate kinase	[2-^{13}C]dihydroxyacetone	Perfused organs and in vivo
Carboxyl esterase	[1,3-13C2]ethyl acetoacetate	Enzymes, cells and in vivo
Aminoacylase-1	[1-^{13}C]*N*-acetyl-L-methionine	Enzymes and cells
Glyoxalase	[2-^{13}C]methylglyoxal	Cells and in vivo
Esterase	^{13}C-enriched ethyl acetyl carbonate	Enzymes, cells/tissue homogenates, and in vivo

6.4.2 Perfused Organs

While perfused organs have been used for a long time to study intermediate metabolism in intact organs, only the perfused heart and liver has been employed for hyperpolarized NMR. This requires a very specialized skill set, and is an order of magnitude more complex than cell studies in bioreactors. While the media and perfusion conditions are well worked out, this requires expertise in leak proof cannula-

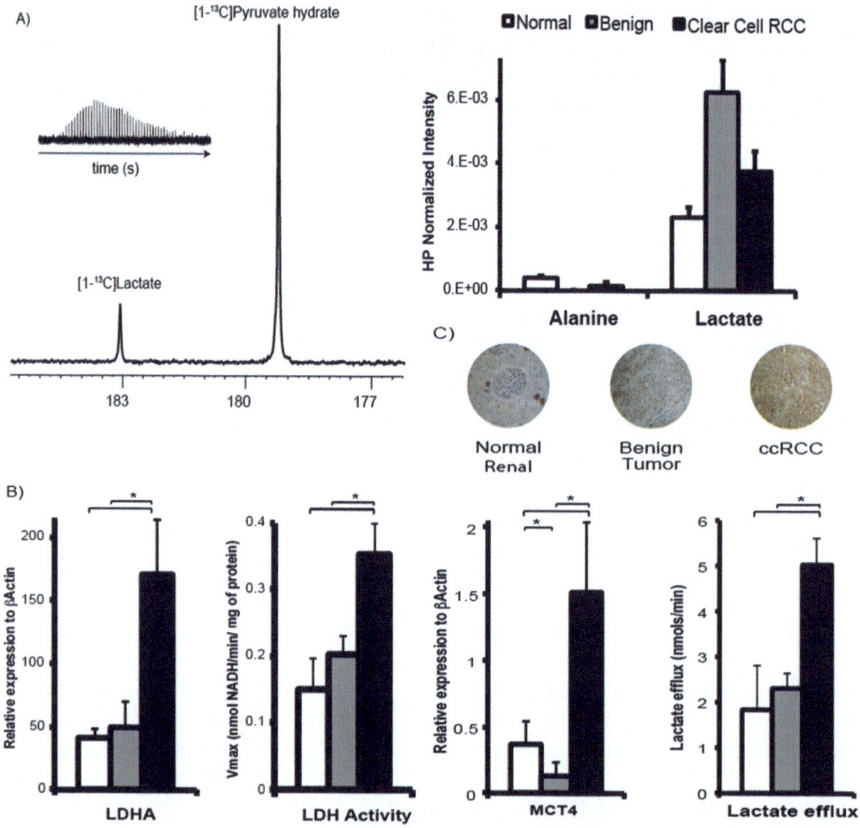

Fig. 6.3 Hyperpolarized lactate dynamics in patient-derived living renal tissue. Adapted from Sriram et al. [26, 27], this figure shows the compendium of correlative biochemical assays that is feasible on the same tissue to understand the underlying mechanisms of the hyperpolarized lactate signal. (**a**) Representative example of lactate dynamics in clear cell subtype of renal cell carcinoma and summary findings of apparent increase of lactate signal in both benign and malignant renal cancer over normal tissue. (**b**) Increased lactate dehydrogenase activity as well as expression is confirmed to be highest in the malignant tissue using RT-qPCR and calorimetric assay. (**c**) The cause of apparent decrease in observed hyperpolarized lactate signal is confirmed to be due to quick washout of produced lactate via the increased expression of the monocarboxylate transporter (isoform 4) by mRNA expression, immunohistochemical staining as well as stable isotope measurement of rate of lactate export from the cell

tion of the aorta or portal vein (for heart and liver, respectively). Additional requirements similar to the bioreactor are controlled media flow, temperature, and oxygenation. For a detailed protocol on liver and heart surgical preparation, refer to Lumata et al. [17, 18]. Although all of the concerns relating to NMR measurements of perfused organs regarding shim and proper position are still valid, extra caution needs to be taken with respect to the injection of hyperpolarized substrates to ensure nonturbulent, well-mixed solutions to minimize any disturbances in osmolarity and flow. This has a distinct advantage over in vivo measures of these organs due to the increased sensitivity from inherent localization and high filling factor enabling the study of metabolic intermediates otherwise inaccessible for a better understanding of the biochemical pathways. Furthermore, since these are ex vivo organs that are time intensive to set up, constant monitoring of tissue health via ^{31}P or functional monitoring probes such as respiratory and oxygen sensors are required. When combined with steady-state tracer methods, hyperpolarized studies on these model organs can be used to generate a comprehensive picture of the metabolic function as shown in the heart [22, 29]. These works showed that flux of hyperpolarized pyruvate to bicarbonate reflect pyruvate dehydrogenase (PDH) flux and its production could be quantified using rigorous models in cell cultures.

6.5 Preclinical Animal Models Used in Hyperpolarized NMR

Animal models represent the penultimate step before clinical translation of any new diagnostic or therapeutic technology. While the use of large animals, especially nonhuman primates, is closer to human biology such as organ morphology and physiology, logistics dictates the extensive use of smaller animal models like rats and mice. While the biological relevance of the choice of animal model transcends all applications, we highlight some of the considerations that are specific to its use for hyperpolarized MR studies to study biochemical processes.

6.5.1 Animal Preparation for MR Imaging

Planning of hyperpolarized experiments on animals requires some specific physiological and technical considerations.

6.5.1.1 Vascular Access

Unlike most other MR or other imaging modality needs, the hyperpolarized studies require patent vascular catheterization over prolonged time periods (1–4 h) for injection of hyperpolarized substrates. Tail vein is by far the most often used route as it provides for easy repetitive cannulation. Other options are jugular and femoral veins but are more complicated and have a lower throughput. However, depending on the murine model used, they can be purchased precannulated which remain patent for 10 days. To minimize trauma and to enable repeated cannulation, it is recommended to use catheters that are devoid of needle and instead employ a 1F outer diameter polyurethane tubing. This also minimizes any artifacts from the needle. Furthermore, due to the flexible nature of the tubing, secure taping is required and allows for easy handling of the animal while placing in the animal holder and insertion into the magnet for imaging. Additionally, periodic flushing of the cannula with heparinized saline is required to prevent blood clot, which will preclude the delivery of the hyperpolarized agent. One important factor to consider is to minimize the catheter length to avoid injecting large fluid volumes into the animal during the course of a study. In our experience, the saline flush following the injection of hyperpolarized agent (to avoid increased probe injection) when accounting for the dead volume of the catheter results in a second peak of the signal, which often complicates the kinetic modeling.

6.5.1.2 Physiological Monitoring

Similar to most other imaging studies, rigorous control of functional parameters is required:

1. Heart and respiration rate—for those applications that require gating to prevent motion-related artifacts in addition to maintaining normal physiology.
2. Body temperature—it is absolutely essential to maintain the body temperature as the hyperpolarized study is a metabolic readout of enzymatic activity and is subject to the normal animal physiology.
3. Level of anesthesia—similar to other functional imaging applications, hyperpolarized signals are sensitive to the depth of anesthesia. Josan et al. [29, 30] showed that in rats, the pyruvate signal observed in the brain was dependent on the anesthetic depth (see Fig. 6.4). In conjunction, it has been recently shown that type of anesthetic also determines the rate of conversion [31].
4. Nutritional state—hyperpolarized signals are acutely sensitive to the fed/fasted state of the animal. With this powerful technology, multiple pathways of metabolism can be measured. For example, in the liver, pyruvate can be metabolized via multiple pathways depending on the state of the animal. It has been shown in rat livers that the pyruvate is metabolized via the PDH (pyruvate dehydrogenase) when fed, while in the fasted condition, pyruvate is catabolized via pyruvate carboxylase [28]. Hence, in order to measure hyperpolarized bicarbonate signal, it is necessary to ensure that the animal is not fasted. Similarly, a reduction in hyperpolarized alanine signal has been observed in rat livers [30–32]. Careful thought also needs to be given to the hyperpolarized agent being used to interrogate a particular organ. As illustrated by Bastiaansen et al. [32, 33], in the rodent heart, hyperpolarized butyrate and pyruvate gives a measure of fatty acid oxidation and carbohydrate metabolism, respectively. While hyperpolarized butyrate transport remained unchanged in either fed or fasted state, the pyruvate

metabolism was diminished. Hence, to decrease any confounding factors and variability in the results between animals, it is recommended to carefully control for the fed/fasted state of the animal and the time of day the experiments are conducted.

6.5.1.3 Hyperpolarized Substrate Dosing Considerations

Unlike PET tracers, the injected doses of hyperpolarized substrates often are much higher than the equilibrium concentrations found at normal physiology. While it has been shown that the lactate and alanine signal is not saturated when using pyruvate concentrations of up to 0.7 mmol/kg in the fed state in mouse and rat kidney, the dose-dependent signals of the downstream metabolites are variable in the different organs for different metabolites. For instance, in the rodent cardiac metabolism, pyruvate concentrations greater than 0.4 mmol/kg cause saturation of bicarbonate and lactate signal while there is a linear correspondence with alanine signal [34], whereas hyperpolarized alanine and bicarbonate signals plateau above 0.2 mmol/kg in the rodent liver [34–36]. Furthermore, additional considerations of the metabolic state of the organ under study for consecutive hyperpolarized studies should be included when determining the dose to be used. Furthermore, physical impediments of certain organs like the blood–brain barrier highly influences the signal in brain studies and is dependent on the hyperpolarized probe's permeability.

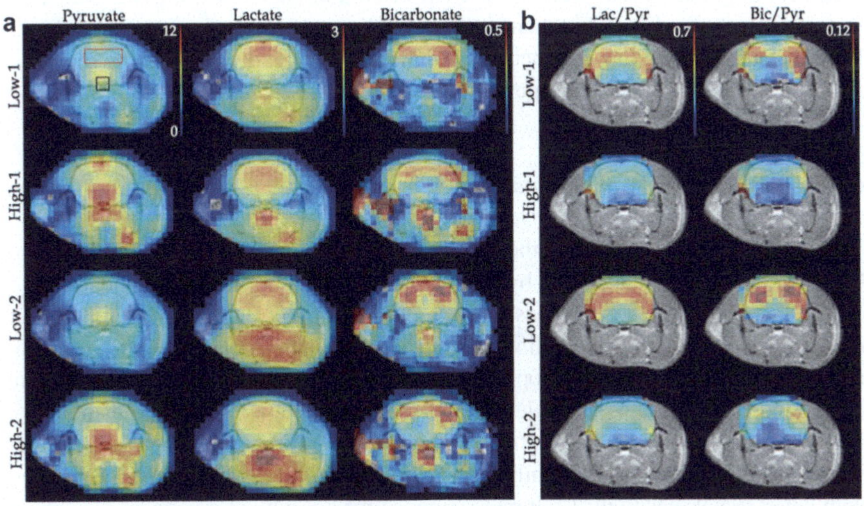

Fig. 6.4 Effect of anesthesia on hyperpolarized signals. Adapted from Josan et al. [29, 30], this figure shows the different levels of anesthesia

6.5.1.4 Imaging Considerations Dictated by Animal Physiology

Some of the key elements in designing the hyperpolarized imaging protocol are restricted by the animal parameters such as perfusion and respiration rate. The rate of blood flow, and hence the delivery of the hyperpolarized substrate to the organ of choice dictates the timing as well as the RF flip angles of the imaging sequence. For instance, in our lab, we use 35 s from the time of injection to be the optimal time to image the murine tumors for maximal lactate signal and significant pyruvate signal. This was arrived at by initial slab dynamic studies (see Chap. 4) which employs dynamic small flip angle RF pulsing to obtain the time-dependent signal changes in the organ of interest. The optimal time point is considered to be the one where the metabolite signal is plateauing (before it starts decreasing due to T_1) while the precursor still has sufficient SNR. A more comprehensive image acquisition usually involves dynamic serial sampling (see Chap. 4) the rate of perfusion/recirculation in the tissue of interest dictates the sampling interval as well as the RF flip angle choice. The respiration/heart rate is also a factor in determining the sequence repetition time for dynamic acquisitions especially when gating is used. Furthermore, when studying organs such as kidney and liver, the recirculating lactate (that is produced in the other parts of the body) needs to be accounted for via modeling or using suppression techniques.

6.5.2 Animal Models Used for Hyperpolarized Studies

To date, cancer is the main application of ^{13}C hyperpolarized studies. The second dominant application has been in the field of cardiac metabolism. Liver and renal metabolism have been biochemically investigated using hyperpolarized ^{13}C MR. This technology has also been used in other malignancies including diabetes, stroke as well as to assess therapeutic interventions for biomarker development.

6.5.2.1 Oncologic Applications

Tumor metabolism is considered the "killer app" for hyperpolarized ^{13}C NMR. This has been predominantly championed by the Warburg effect (upregulation of oxidative glycolysis) in tumors and the robustness of hyperpolarized pyruvate. Many different cancer models have been studied, including genetically modified models (TRAMP), subcutaneous tumor models, and orthotopic models. Orthotopic models (over xenografts) are considered the most clinically relevant and better suited to treatment evaluation [36, 37]. However, they present with some challenges pertaining to ^{13}C-hyperpolarized studies: (1) the depth in the body entails the use of volume coils, which is often associated with lower signal sensitivity, (2) tumor sizing is time consuming, as opposed to subcutaneous models that can be visually inspected. This is an important consideration as a minimal tumor volume of 0.1 cm^3 is required owing to the lower spatial resolution achievable in ^{13}C hyperpolarized imaging (see Chaps. 4 and 5). On the other hand, a key consideration to a successful hyperpolarized study in the context of implanted tumors is the development of adequate vascularization and sufficient perfusion for the substrate to reach tumor cells. This is a challenging factor in subcutaneous tumors to obtain substantial tumor growth while still maintaining tumor vascularization without forming necrotic cores, which is often the case. Patient-derived xenografts are a highly relevant model; however, they have limited success in hyperpolarized applications as only the highly aggressive tumors grow into sizeable tumors and serial passaging of the tissue, to enable repeated measures and model propagation, leads to natural selection of only the aggressive subcolony of cells.

While hyperpolarized glucose would be the ideal substrate to study the upregulated glycolysis in cancer [37, 38], the short T_1 limits the effective use of the probe. Hyperpolarized pyruvate, however, is a powerful alternative and its conversion to lactate has shown to correlate to tumor grade [39], lactate dehydrogenase (LDH) activity and expression in multiple cancers, brain, kidney, prostate, lung, breast to name a few. Alternative biochemical pathway activation can also be monitored using pyruvate. For example, with the use of dichloroacetate (promotes pyruvate oxidation), reduced pyruvate to lactate conversion and subsequent increase in hyperpolarized bicarbonate have been observed [39, 40]. Furthermore, hyperpolarized lactate production from pyruvate has shown to be an early predictor of therapeutic efficacy be it a small molecule inhibitor, radiation [41–43], or chemotherapy [43, 44]. One key aspect of therapeutic studies is the decreased perfusion, which is a confounding factor as no metabolism can be observed if the probe cannot reach the tissue. However, this can be overcome with the copolarization with a hyperpolarized perfusion probe such as urea [44, 45].

6.5.2.2 Cardiology and Other Applications

Changes in substrate utilization in cardiac diseases are now clearly seen as a causative factor rather than a consequence. While fatty acid is the major substrate utilized by the heart muscle, it has limited translational application due to the low polarization and shorter T_1 than hyperpolarized pyruvate. However, since pyruvate sits at the key regulating point of multiple pathways, it has proven to be very useful in detecting many cardiac pathologies such as ischemic-reperfusion, hypertrophy, acidosis, and cardiomyopathy [46]. However, the most challenging aspect of in vivo cardiac hyperpolarized imaging is the choice of optimal pulse sequence and RF hardware (see Chaps. 9 and 10).

Skeletal muscle has yet been a challenging tissue to study with hyperpolarized NMR owing to its relative lower perfusion. However, significant real-time metabolism reflection muscle physiology has been observed with the use of hyperpolarized pyruvate [46, 47], lactate [47, 48], and acetate [49]. Metabolic dysregulation in diabetes has been interrogated in rats and mice using multiple probes such as ethylacetoacetate, dihydroxyacetone. This is one application in which the fed/fasted state of the animal when the hyperpolarized exam is conducted needs to be well controlled (see Chaps. 11).

6.6 Clinical Translation of Hyperpolarized NMR

The first-in-human study showed that both the injection of HP [1-^{13}C]pyruvate and imaging hyperpolarized [1-^{13}C]pyruvate metabolism was safe and feasible, respectively. Establishing safety was important, given the difference in pharmacologic dose between ^{13}C HP MR and other metabolic imaging technologies such as PET (the difference between HP [1-^{13}C] pyruvate and FDG doses is of the order of 10^7). Based on a large body of pharmacology and toxicology studies, and the findings from preclinical studies and pyruvate dose escalation studies in healthy volunteers, an FDA dose escalation safety and imaging feasibility study entitled "A Phase 1 ascending-dose study (NCT01229618) to assess the safety and tolerability and imaging potential of hyperpolarized pyruvate (^{13}C) injection in subjects with prostate cancer" was designed [34]. Performing the phase 1 trial required the development and optimization of a number of critical components (Fig. 6.5), including obtaining GMP quality [1-^{13}C] pyruvate and trityl radical (AH111501) and a procedure to produce a clinical grade HP probe for injection, a clinical DNP polarizer to polarize the HP [1-^{13}C]pyruvate probe, and to perform its subsequent QC, the implementation of multinuclear MR capabilities on the scanner, the construction of specialized radio frequency (RF) coils to detect both ^{1}H and ^{13}C nuclei, and the development of new MR pulse sequences to efficiently capture the dynamic hyperpolarized signal and to analyze the data (Chaps. 4 and 5).

In the phase 1 trial, a total of 31 patients received an injection of 250 mM HP [1-^{13}C] pyruvate, followed by either 1 or 2D ^{13}C dynamic or single time-point 3D ^{13}C MRSI at 3 T [3]. Safe administration was established when no dose limiting toxicities were observed for the range of doses tested (0.14–0.43 mL/kg) at an injection rate of 5 mL/s. The highest dose of HP pyruvate yielded the best pyruvate and lactate SNR in regions of cancer and the best contrast for tumor relative to normal prostate. The feasibility of imaging HP [1-^{13}C] pyruvate metabolism was established when one- and two-dimensionally localized dynamic HP data showed uptake of HP pyruvate approximately 20 s after injection, with a maximum pyruvate signal at 27 s and a maximum production of lactate at 45 s in cancer, continuing for more than 100 s (Fig. 6.5). These data demonstrated that the T_1 of HP [1-^{13}C]pyruvate (≈65 s at a magnetic field of 3 T) was more than adequate to image uptake and metabolism in the prostate of patients. In addition, the time course of uptake and metabolism corresponded well with prior data from TRAMP [38, 39, 49, 50] and human prostate tissue slice culture studies [50, 51]. The results of this trial were promising in not only confirming the safety of the agent but also showing elevated HP [1-^{13}C] lactate/[1-^{13}C] pyruvate and the apparent rate of conversion of [1-^{13}C]pyruvate to [1-^{13}C] lactate conversion in regions of biopsy-proven cancer, and in some cases, in regions of cancer that were not detected by conventional ^{1}H MRI. However, since full pathologic correlation of the prostate after surgery was not available for this study, it was not possible to fully compare the ability to detect and assess the pathologic grade of prostate cancer relative to the current standard of care imaging of localized disease,

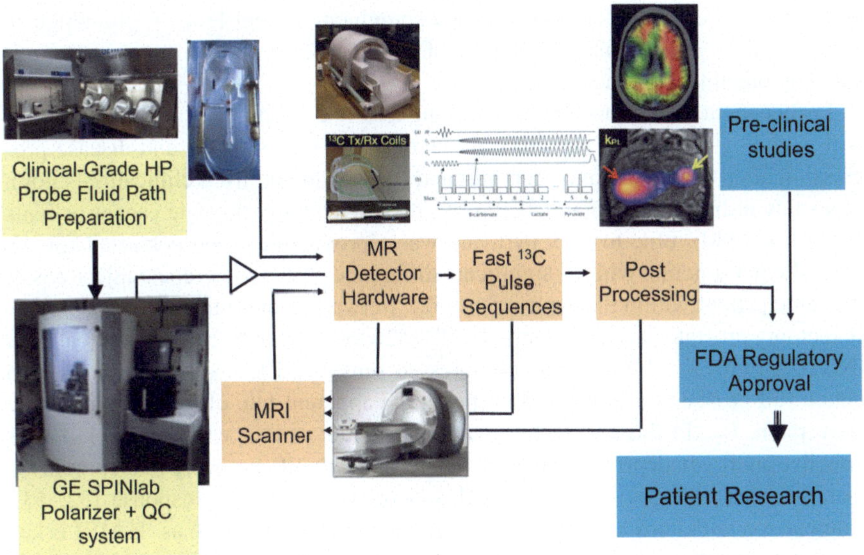

Fig. 6.5 Components required for clinical translation of hyperpolarized ^{13}C probes. Adapted from Kurhanewicz et al. [62], this figure shows the various elements that are necessary for the successful translation of a hyperpolarized ^{13}C probe for use in men

multiparametric ^{1}H MRI, thereby setting the stage for the patient validation studies discussed below.

Since the completion of the phase 1 clinical trial, significant technical improvements relevant to improving data quality and preserving safety have been implemented. These include improved technology and strategies for generating sterile-hyperpolarized material using a commercial device, the SPINlab™, new MR data acquisition sequences, and multichannel ^{1}H/^{13}C radio frequency RF coils. Pictured in Fig. 6.5 is the workflow for the current-generation SPINlab™ which is now used routinely in patient studies and provides automated polarization, dissolution, and quality assurance system. The SPINlab™ polarizer uses a 5 T magnetic field and operates at ~0.8 K. It has the capacity to polarize up to four samples simultaneously, thus opening the door for the injection of multiple HP probes within a single imaging exam. A critical part of the SPINlab™ polarization and quality assurance process is the fluid path that enables polarization in a closed system to minimize the potential for microbial contamination. Compared to the initial human study using a prototype polarizer in a clean room, the achievable polarization of [1-^{13}C]pyruvate was more than double (from ≈18% to ≈40%) and the time of delivery of HP ^{13}C pyruvate was reduced (from ≈68 s to ≈55 s). Around the world, seven sites have performed human exams using the SPINlab™ polarizer and more than 20 have been installed. For more details on the polarizer, see Chap. 2.

While initial patient studies have focused on [1-^{13}C]pyruvate, [2-^{13}C]pyruvate has recently been FDA approved and is being added to [1-^{13}C]pyruvate in ongoing clinical trials. Additionally, funded patient studies are working toward the use of ^{13}C urea, and [5-^{13}C]glutamine, [1-^{13}C]alpha-ketoglutarate, and [1,4-$^{13}C_2$]fumarate in patient studies. The rationale for the use of these probes in patient studies was motivated by preclinical studies.

Early patient validation studies involving test–retest reproducibility, correlations with pathologic findings, and clinical outcomes such as survival and disease progression after treatment have been initiated (data available from clinicaltrials.gov). In a study utilizing 2D dynamic EPSI of preprostatectomy prostate cancer patients (NCT02421380), time-to-max pyruvate was reproducible when corrected for the arterial input function with no significant difference observed in repeat injections of the same patient within the same exam [52]. In two recent clinical trials of preprostatectomy prostate cancer patients (NCT02526360) and before and after treatment (NCT02911467), a 3D dynamic compressed sensing EPSI sequence was used to obtain full prostate coverage and to model the apparent rate of pyruvate to lactate conversion, k_{PL}. In the setting of advanced prostate cancer, early (6 weeks) after effective androgen deprivation therapy, there was a significant early reduction in k_{PL} that preceded changes in mp-^{1}H MRI and predicted clinical response [52, 53]. Comparison with current state-of-the-art imaging modalities such as ^{1}H MRI is key to determining the added value of HP ^{13}C MRI, and with current HP ^{13}C MRI studies expanding to metastatic sites within the body [54], the synergy of HP MRI with currently used PET probes will also need to be investigated.

There is currently a number of single institution HP ^{13}C MRI clinical trials in the setting of breast, cervical, liver and brain cancer, and an initial multisite clinical trial for prostate cancer. Following promising preclinical studies in brain tumors, multiple sclerosis and traumatic brain injury, the feasibility of using HP ^{13}C MRI for evaluating brain metabolism in patient studies was recently demonstrated [7]. These studies demonstrated that [1-^{13}C]pyruvate is transported across the blood–brain barrier, and that conversion to [1-^{13}C]lactate can be detected in regions of both tumor and normal brain. Interestingly, the rate of conversion in normal brain was substantially higher than had previously been observed in anesthetized preclinical models. For dynamic slab-localized and 2D echo planar spectroscopic imaging data with 4–8 cm^3 spatial-resolution, it was also possible to observe the conversion of HP pyruvate to bicarbonate, which may discriminate between tumor and normal brain. Additional studies are needed to optimize the pulse sequences, RF coils, and experimental design to observe all three resonances at a finer resolution and with volumetric coverage. Single institution studies that have been funded by the NIH include protocols for obtaining data from the normal brain, in primary brain tumors prior to image-guided surgery, and for comparing that conversion to [1-^{13}C]lactate/[1-^{13}C]pyruvate in patients pre- and posttreatment with radiation and temozolomide.

There are also several clinical trials investigating noncancerous diseases, such as cardiomyopathy, hypertrophy and hypertension, and fatty liver disease (Clinicaltrials.gov). Altered cardiac energetics is known to play an important role in the progression toward heart failure, and a recent study demonstrated the feasibility of assessing pyruvate metabolism in vivo in humans [55]. The appearance of ^{13}C-bicarbonate signal after administration of hyperpolarized [1-^{13}C]pyruvate was readily detected in a cohort of healthy controls, suggesting that this technology may 1 day allow a direct measure of flux through the pyruvate dehydrogenase complex (PDC) in the in vivo human myocardium providing a means for understanding therapeutic approaches that target metabolism.

6.7 Understanding and Interpreting the Hyperpolarized Signals to Shed Light on the Underlying Biochemistry of the Pathology

Several preclinical works have shown that in the classical example of hyperpolarized pyruvate conversion to lactate, the signal is a function of not only the enzyme activity [55, 56] and the availability of NADH cofactor but also dependent on the product pool sizes. The pool size effect was clearly demonstrated by the works of Hurd et al. [57] as well as Kettunen et al. [56–59]. These findings highlight the multiple components regulating the hyperpolarized signal and the complexities of the underlying enzymatic reaction. While under in vitro conditions, it is feasible to estimate the absolute flux of the hyperpolarized probes in conjunction with thermal stable isotope-resolved metabolomics or radioisotope studies, under in vivo conditions, it becomes convolved with other confounding effects such as signal contamination from vascular component, uncertainty in the concentration of the hyperpolarized probe in the tissue of interest, intra- and extracellular compartmentalization, and the effect on the hyperpolarized signal with unknown and varying levels of enzyme and cofactor. As a result, the hyperpolarized data are represented as a ratio of the product to precursor or by an apparent first-order exchange rate constant [20, 60–62] (Chap. 6) and unable to render a metabolically pertinent flux measurement in absolute concentrations. This also implies that to thoroughly understand the enzyme system under study using hyperpolarized dDNP technique, it is imperative to do extensive preclinical studies in order to fully utilize this powerful technique to uncover kinetic mechanisms in real time.

Acknowledgements *Grants/People*: We would like to acknowledge the funding sources P41 EB013598 (NIH) and PC160630 (DoD) and the members of the Hyperpolarized MRI Technology Resource Center and the Pre-Clinical MR Imaging and Spectroscopy Core.

Problems

1. Hyperpolarized [1-^{13}C] dehydroascorbate (DHA) is a probe used to interrogate the redox status of the cell. In in vivo systems, the cyclic regeneration of the reactive oxygen species and its reductants serve to continuously reduce and oxidize ascorbic acid as shown below.
   What are all the components that will be required to estimate the apparent zero-order reaction rate of hyperpolarized DHA in a simple enzyme solution?

2. In order to test the hypothesis that increased cellular production of lactate and its efflux occur in renal cell cancers is associated with cancer aggressiveness, we designed a cell culture bioreactor study using UOK262 cells, which were established from a highly aggressive metastatic RCC. The figure below shows the bioreactor set-up containing UOK262 cells encapsulated in alginate beads with continuous flow of media and continuous infusion of hyperpolarized agent over a 90 s time period.
   The figure on the left is a schematic of the cells encapsulated in alginate microspheres being perfused in a bioreactor. The figure to the right represents hyperpolarized ^{13}C spectra under two different conditions from the bioreactor. The bottom ^{13}C spectra in the figure (below) is of alginate microspheres devoid of cells, infused with co-hyperpolarized [1-^{13}C]lactate and [1-^{13}C]pyruvate. Only one peak was observed for the [1-^{13}C]lactate signal (bottom spectra). While two peaks were observed in the alginate microspheres with UOK262 cells when infused with HP [1-^{13}C]pyruvate only (top spectra). The top inset (2.5× magnification, with black arrows) clearly shows that there are two distinct peaks for lactate (although the chemical shift difference is very small—0.031 ± 0.0005 ppm), where the chemical shift of the downfield peak coincides with that of the signal of lactate in empty alginate microspheres. A series of studies were performed to try to identify the origin of these two lactate resonances.
   (a) Increasing the cell density in the bioreactor resulted in an increase in the upfield lactate peak.

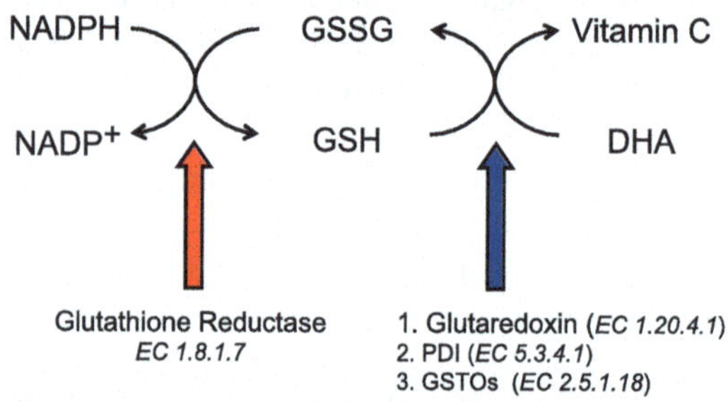

   (b) Stopping the flow of media in the bioreactor resulted in an increase in the downfield resonance.
   (c) The downfield lactate resonance decreased when the cells were pretreated with DIDS, a small molecule blocker of MCT4 transporter.
   Based on the above hyperpolarized ^{13}C studies, what could be the explanation for the two lactate peaks observed when hyperpolarized [1-^{13}C]lactate is metabolically produced by UOK262 cells after injection of hyperpolarized [1-^{13}C]pyruvate? Please make sure you explain all of the experimental findings.
3. What are some biological challenges with live animal imaging studies involving hyperpolarized MRI and interpreting the subsequent metabolites?

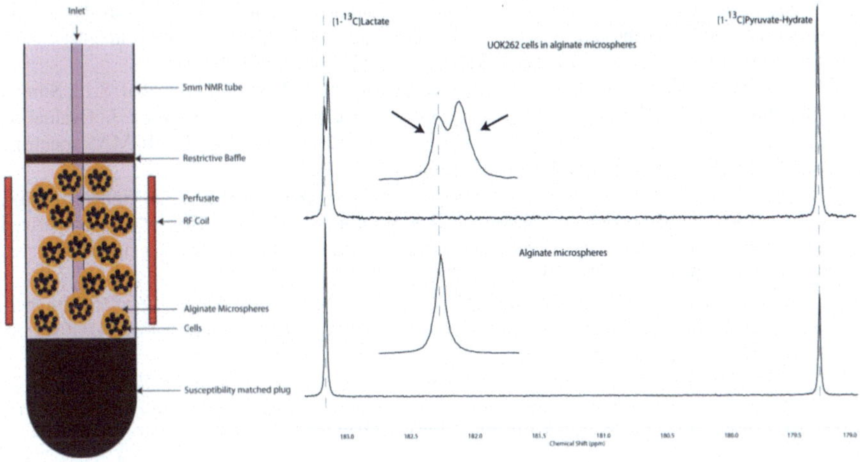

References

1. Ardenkjaer-Larsen, J.H., Fridlund, B., Gram, A., Hansson, G., Hansson, L., Lerche, M.H., Servin, R., Thaning, M., Golman, K.: Increase in signal-to-noise ratio of >10,000 times in liquid-state NMR. Proc. Natl. Acad. Sci. U. S. A. **100**, 10158–10163 (2003)
2. Golman, K.: Real-time metabolic imaging. Proc. Natl. Acad. Sci. U. S. A. **103**, 11270–11275 (2006)
3. Nelson, S.J., Kurhanewicz, J., Vigneron, D.B., Larson, P.E.Z., Harzstark, A.L., Ferrone, M., Van Criekinge, M., Chang, J.W., Bok, R., Park, I., Reed, G., Carvajal, L., Small, E.J., Munster, P., Weinberg, V.K., Ardenkjaer-Larsen, J.H., Chen, A.P., Hurd, R.E., Odegardstuen, L.-I., Robb, F.J., Tropp, J., Murray, J.A.: Metabolic imaging of patients with prostate cancer using hyperpolarized [1-13C]pyruvate. Sci. Transl. Med. **5**, 198ra108 (2013)
4. Wilson, D.M., Keshari, K.R., Larson, P.E.Z., Chen, A.P., Hu, S., Van Criekinge, M., Bok, R., Nelson, S.J., Macdonald, J.M., Vigneron, D.B., Kurhanewicz, J.: Multi-compound polarization by DNP allows simultaneous assessment of multiple enzymatic activities in vivo. J. Magn. Reson. **205**, 141–147 (2010)
5. Kurhanewicz, J., Vigneron, D.B., Brindle, K., Chekmenev, E.Y., Comment, A., Cunningham, C.H., DeBerardinis, R.J., Green, G.G., Leach, M.O., Rajan, S.S., Rizi, R.R., Ross, B.D., Warren, W.S., Malloy, C.R.: Analysis of cancer metabolism by imaging hyperpolarized nuclei: prospects for translation to clinical research. Neoplasia. **13**, 81–97 (2011)
6. McLean, M.A., Daniels, C.J., Grist, J., Schulte, R.F., Lanz, T., Chhabra, A., Earl, H.M., Basu, B., Wilkinson, I.B., Lomas, D.J., Caldas, C., Abraham, J.E., Graves, M., Gilbert, F., Brindle, K.M., Gallagher, F.A.: Feasibility of metabolic imaging of hyperpolarized 13C-pyruvate in human breast cancer European Society for Magnetic Resonance in Medicine and Biology. Springer Nature, Cham (2018)
7. Miloushev, V.Z., Granlund, K.L., Boltyanskiy, R., Lyashchenko, S.K., DeAngelis, L.M., Mellinghoff, I.K., Brennan, C.W., Tabar, V., Yang, T.J., Holodny, A.I., Sosa, R.E., Guo, Y.W., Chen, A.P., Tropp, J., Robb, F., Keshari, K.R.: Metabolic imaging of the human brain with hyperpolarized 13C pyruvate demonstrates 13C lactate production in brain tumor patients. Cancer Res. **78**(14), 3755–3760 (2018)
8. Park, I., Larson, P.E.Z., Gordon, J.W., Carvajal, L., Chen, H.-Y., Bok, R., Van Criekinge, M., Ferrone, M., Slater, J.B., Xu, D., Kurhanewicz, J., Vigneron, D.B., Chang, S., Nelson, S.J.: Development of methods and feasibility of using hyperpolarized carbon-13 imaging data for evaluating brain metabolism in patient studies. Magn. Reson. Med. **80**, 864–873 (2018)

9. Cheng, L.L., Ma, M.J., Becerra, L., Ptak, T., Tracey, I., Lackner, A., González, R.G.: Quantitative neuropathology by high resolution magic angle spinning proton magnetic resonance spectroscopy. Proc. Natl. Acad. Sci. U. S. A. **94**, 6408–6413 (1997)
10. Tessem, M.-B., Swanson, M.G., Keshari, K.R., Albers, M.J., Joun, D., Tabatabai, Z.L., Simko, J.P., Shinohara, K., Nelson, S.J., Vigneron, D.B., Gribbestad, I.S., Kurhanewicz, J.: Evaluation of lactate and alanine as metabolic biomarkers of prostate cancer using 1H HR-MAS spectroscopy of biopsy tissues. Magn. Reson. Med. **60**, 510–516 (2008)
11. Keshari, K.R., Sriram, R., Van Criekinge, M., Wilson, D.M., Wang, Z.J., Vigneron, D.B., Peehl, D.M., Kurhanewicz, J.: Metabolic reprogramming and validation of hyperpolarized 13C lactate as a prostate cancer biomarker using a human prostate tissue slice culture bioreactor. Prostate. **73**, 1171–1181 (2013)
12. Lumata, L., Jindal, A.K., Merritt, M.E., Malloy, C.R., Sherry, A.D., Kovacs, Z.: DNP by thermal mixing under optimized conditions yields >60,000-fold enhancement of 89Y NMR signal. J. Am. Chem. Soc. **133**, 8673–8680 (2011)
13. Chiavazza, E., Kubala, E., Gringeri, C.V., Düwel, S., Durst, M., Schulte, R.F., Menzel, M.I.: Earth's magnetic field enabled scalar coupling relaxation of 13C nuclei bound to fast-relaxing quadrupolar 14N in amide groups. J. Magn. Reson. **227**, 35–38 (2013)
14. Lerche, M.H., Meier, S., Jensen, P.R., Baumann, H., Petersen, B.O., Karlsson, M., Duus, J.Ø., Ardenkjaer-Larsen, J.H.: Study of molecular interactions with 13C DNP-NMR. J. Magn. Reson. **203**, 52–56 (2010)
15. Jensen, P.R., Meier, S., Ardenkjaer-Larsen, J.H., Duus, J.Ø., Karlsson, M., Lerche, M.H.: Detection of low-populated reaction intermediates with hyperpolarized NMR. Chem. Commun. (Camb.). **2009**, 5168–5170 (2009)
16. Zhang, G., Hilty, C.: Applications of dissolution dynamic nuclear polarization in chemistry and biochemistry. Magn. Reson. Chem. **56**, 566–582 (2018)
17. de Araújo, M.E.G., Lamberti, G., Huber, L.A.: Homogenization of mammalian cells. Cold Spring Harb Protoc. **2015**, 1009–1012 (2015)
18. Lumata, L., Yang, C., Ragavan, M., Carpenter, N., DeBerardinis, R.J., Merritt, M.E.: Hyperpolarized (13)C magnetic resonance and its use in metabolic assessment of cultured cells and perfused organs. Methods Enzymol. **561**, 73–106 (2015)
19. Harrison, C., Yang, C., Jindal, A., DeBerardinis, R.J., Hooshyar, M.A., Merritt, M., Dean Sherry, A., Malloy, C.R.: Comparison of kinetic models for analysis of pyruvate-to-lactate exchange by hyperpolarized 13C NMR. NMR Biomed. **25**, 1286–1294 (2012)
20. Hill, D.K., Orton, M.R., Mariotti, E., Boult, J.K.R., Panek, R., Jafar, M., Parkes, H.G., Jamin, Y., Miniotis, M.F., Al-Saffar, N.M.S., Beloueche-Babari, M., Robinson, S.P., Leach, M.O., Chung, Y.-L., Eykyn, T.R.: Model free approach to kinetic analysis of real-time hyperpolarized 13C magnetic resonance spectroscopy data. PLoS One. **8**, e71996 (2013)
21. Reineri, F., Daniele, V., Cavallari, E., Aime, S.: Assessing the transport rate of hyperpolarized pyruvate and lactate from the intra- to the extracellular space. NMR Biomed. **29**, 1022–1027 (2016)
22. Yang, C., Harrison, C., Jin, E.S., Chuang, D.T., Sherry, A.D., Malloy, C.R., Merritt, M.E., Deberardinis, R.J.: Simultaneous steady-state and dynamic 13C NMR can differentiate alternative routes of pyruvate metabolism in living cancer cells. J. Biol. Chem. **289**, 6212–6224 (2014)
23. Szwergold, B.S.: NMR spectroscopy of cells. Annu. Rev. Physiol. **54**, 775–798 (1992)
24. Sriram, R., Nguyen, J., Santos, J.D., Nguyen, L., Sun, J., Vigneron, S., Van Criekinge, M., Kurhanewicz, J., MacKenzie, J.D.: Molecular detection of inflammation in cell models using hyperpolarized 13C-pyruvate. Theranostics. **8**, 3400–3407 (2018)
25. Chaumeil, M.M., Radoul, M., Najac, C., Eriksson, P., Viswanath, P., Blough, M.D., Chesnelong, C., Luchman, H.A., Cairncross, J.G., Ronen, S.M.: Hyperpolarized 13C MR imaging detects no lactate production in mutant IDH1 gliomas: implications for diagnosis and response monitoring. NeuroImage Clin. **12**, 180–189 (2016)

26. Keshari, K.R., Wilson, D.M., Van Criekinge, M., Sriram, R., Koelsch, B.L., Wang, Z.J., VanBrocklin, H.F., Peehl, D.M., O'Brien, T., Sampath, D., Carano, R.A.D., Kurhanewicz, J.: Metabolic response of prostate cancer to nicotinamide phophoribosyltransferase inhibition in a hyperpolarized MR/PET compatible bioreactor. Prostate. **75**, 1601–1609 (2015)
27. Sriram, R., Van Criekinge, M., DeLos, S.J., Keshari, K.R., Peehl, D.M., Wang, Z.J.: Non-invasive differentiation of benign renal tumors from clear cell renal cell carcinomas using clinically translatable hyperpolarized 13C pyruvate magnetic resonance. Tomography. **2**, 35–42 (2016)
28. Sriram, R., Gordon, J., Baligand, C., Ahamed, F., DeLos, S.J., Qin, H., Bok, R.A., Vigneron, D.B., Kurhanewicz, J., Larson, P.E.Z., Wang, Z.J.: Non-invasive assessment of lactate production and compartmentalization in renal cell carcinomas using hyperpolarized 13C pyruvate MRI. Cancer. **10**, 313 (2018)
29. Merritt, M.E., Harrison, C., Storey, C., Jeffrey, F.M., Sherry, A.D., Malloy, C.R.: Hyperpolarized 13C allows a direct measure of flux through a single enzyme-catalyzed step by NMR. Proc. Natl. Acad. Sci. U. S. A. **104**, 19773–19777 (2007)
30. Josan, S., Hurd, R., Billingsley, K., Senadheera, L., Park, J.M., Yen, Y.-F., Pfefferbaum, A., Spielman, D., Mayer, D.: Effects of isoflurane anesthesia on hyperpolarized 13C metabolic measurements in rat brain. Magn. Reson. Med. **70**(4), 1117–1124 (2012)
31. Marjańska, M., Shestov, A.A., Deelchand, D.K., Kittelson, E., Henry, P.-G.: Brain metabolism under different anesthetic conditions using hyperpolarized [1- 13C]pyruvate and [2-13C]pyruvate. NMR Biomed. **31**, e4012 (2018)
32. Hu, S., Chen, A.P., Zierhut, M.L., Bok, R., Yen, Y.F., Schroeder, M.A., Hurd, R.E., Nelson, S.J., Kurhanewicz, J., Vigneron, D.B.: In vivo carbon-13 dynamic MRS and MRSI of normal and fasted rat liver with hyperpolarized 13C-pyruvate. Mol. Imaging Biol. **11**, 399–407 (2009)
33. Bastiaansen, J.A.M., Merritt, M.E., Comment, A.: Measuring changes in substrate utilization in the myocardium in response to fasting using hyperpolarized [1-(13)C]butyrate and [1-(13) C]pyruvate. Sci. Rep. **6**, 25573 (2016)
34. Schroeder, M.A., Atherton, H.J., Cochlin, L.E., Clarke, K., Radda, G.K., Tyler, D.J.: The effect of hyperpolarized tracer concentration on myocardial uptake and metabolism. Magn. Reson. Med. **61**, 1007–1014 (2009)
35. Gómez Damián, P.A., Sperl, J.I., Janich, M.A., Khegai, O., Wiesinger, F., Glaser, S.J., Haase, A., Schwaiger, M., Schulte, R.F., Menzel, M.I.: Multisite kinetic modeling of (13)C metabolic MR using [1-(13)C]pyruvate. Radiol Res Pract. **2014**, 871619 (2014)
36. Janich, M.A., Menzel, M.I., Wiesinger, F., Weidl, E., Khegai, O., Ardenkjaer-Larsen, J.H., Glaser, S.J., Haase, A., Schulte, R.F., Schwaiger, M.: Effects of pyruvate dose on in vivo metabolism and quantification of hyperpolarized ^{13}C spectra. NMR Biomed. **25**, 142–151 (2012)
37. Killion, J.J., Radinsky, R., Fidler, I.J.: Orthotopic models are necessary to predict therapy of transplantable tumors in mice | SpringerLink. Cancer Metastasis Rev. **17**, 279–284 (1998)
38. Rodrigues, T.B., Serrao, E.M., Kennedy, B.W.C., Hu, D.-E., Kettunen, M.I., Brindle, K.M.: Magnetic resonance imaging of tumor glycolysis using hyperpolarized 13C-labeled glucose. Nat. Med. **20**, 93–97 (2013)
39. Albers, M.J., Bok, R., Chen, A.P., Cunningham, C.H., Zierhut, M.L., Zhang, V.Y., Kohler, S.J., Tropp, J., Hurd, R.E., Yen, Y.-F., Nelson, S.J., Vigneron, D.B., Kurhanewicz, J.: Hyperpolarized 13C lactate, pyruvate, and alanine: noninvasive biomarkers for prostate cancer detection and grading. Cancer Res. **68**, 8607–8615 (2008)
40. Seth, P., Grant, A., Tang, J., Vinogradov, E., Wang, X., Lenkinski, R., Sukhatme, V.P.: On-target inhibition of tumor fermentative glycolysis as visualized by hyperpolarized pyruvate. Neoplasia. **13**, 60–71 (2011)
41. Thind, K., Chen, A., Friesen-Waldner, L., Ouriadov, A., Scholl, T.J., Fox, M., Wong, E., Vandyk, J., Hope, A., Santyr, G.: Detection of radiation-induced lung injury using hyperpolarized (13) C magnetic resonance spectroscopy and imaging. Magn. Reson. Med. **70**, 601–609 (2012)

42. Thind, K., Jensen, M.D., Hegarty, E., Chen, A.P., Lim, H., Martinez-Santiesteban, F., Van Dyk, J., Wong, E., Scholl, T.J., Santyr, G.E.: Mapping metabolic changes associated with early radiation induced lung injury post conformal radiotherapy using hyperpolarized (13)C-pyruvate magnetic resonance spectroscopic imaging. Radiother. Oncol. **110**, 317–322 (2014)
43. Chen, A.P., Chu, W., Gu, Y.-P., Cunnhingham, C.H.: Probing early tumor response to radiation therapy using hyperpolarized [1-(13)C]pyruvate in MDA-MB-231 Xenografts. PLoS One. **8**, e56551 (2013)
44. Brindle, K.: New approaches for imaging tumour responses to treatment. Nat. Rev. Cancer. **8**, 1–14 (2008)
45. von Morze, C., Larson, P., Hu, S., Yoshihara, H.: Investigating tumor perfusion and metabolism using multiple hyperpolarized 13 C compounds: HP001, pyruvate and urea. Magn. Reson. Imaging. **30**(3), 305–311 (2012)
46. Rider, O.J., Tyler, D.J.: Clinical implications of cardiac hyperpolarized magnetic resonance imaging. J. Cardiovasc. Magn. Reson. **15**, 93 (2013)
47. Park, J.M., Josan, S., Mayer, D., Hurd, R.E., Chung, Y., Bendahan, D., Spielman, D.M., Jue, T.: Hyperpolarized 13C NMR observation of lactate kinetics in skeletal muscle. J. Exp. Biol. **218**, 3308–3318 (2015)
48. Bastiaansen, J.A., Yoshihara, H.A., Takado, Y., Gruetter, R., Comment, A.: Hyperpolarized 13C lactate as a substrate for in vivo metabolic studies in skeletal muscle. Metabolomics. **10**, 986–994 (2014)
49. Bastiaansen, J.A.M., Cheng, T., Mishkovsky, M., Duarte, J.M.N., Comment, A., Gruetter, R.: In vivo enzymatic activity of acetylCoA synthetase in skeletal muscle revealed by (13)C turnover from hyperpolarized [1-(13)C]acetate to [1-(13)C]acetylcarnitine. Biochim. Biophys. Acta. **1830**, 4171–4178 (2013)
50. Chen, A.P., Albers, M.J., Cunningham, C.H., Kohler, S.J., Yen, Y.-F., Hurd, R.E., Tropp, J., Bok, R., Pauly, J.M., Nelson, S.J., Kurhanewicz, J., Vigneron, D.B.: Hyperpolarized C-13 spectroscopic imaging of the TRAMP mouse at 3T-initial experience. Magn. Reson. Med. **58**, 1099–1106 (2007)
51. Sriram, R., Van Criekinge, M., Santos, J.D., Vigneron, D.B., Bok, R.A., Peehl, D.M., Keshari, K.R., Kurhanewicz, J.: Hyperpolarized lactate production correlates with gleason grade in patient-derived tissues of prostate cancer. ISMRM, Toronto (2015)
52. Granlund, K.L., Vargas, H.A., Lyashchenko, S.K., DeNoble, P.J., Laudone, V.A., Eastham, J., Sosa, R.A., Kennedy, M.A., Nicholson, D., Guo, Y.W., Chen, A.P., Tropp, J., Hricak, H., Keshari, K.R.: Metabolic dynamics of hyperpolarized [1-^{13}C] pyruvate in human prostate cancer. Proceedings of the International Society of Magnetic Resonance in Medicine, Singapore (2016)
53. Aggarwal, R., Vigneron, D.B., Kurhanewicz, J.: Hyperpolarized 1-[13C]-pyruvate magnetic resonance imaging detects an early metabolic response to androgen ablation therapy in prostate cancer. Eur. Urol. **72**, 1028–1029 (2017)
54. Zhu, Z., Gordon, J.W., Chen, H.-Y., Milshteyn, E., Mammoli, D., Carvajal, L., Shin, P.J., Aggarwal, R., Bok, R., Kurhanewicz, J., Munster, P., Vigneron, D.B.: Human hyperpolarized 13C MR of liver and bone metastases using both EPSI and EPI acquisitions. Proceedings of the International Society of Magnetic Resonance in Medicine, Paris (2018)
55. von Morze, C., Tropp, J., Chen, A.P., Marco-Rius, I., Van Criekinge, M., Skloss, T.W., Mammoli, D., Kurhanewicz, J., Vigneron, D.B., Ohliger, M.A., Merritt, M.E.: Sensitivity enhancement for detection of hyperpolarized 13 C MRI probes with 1 H spin coupling introduced by enzymatic transformation in vivo. Magn. Reson. Med. **80**, 36–41 (2017)
56. Salamanca-Cardona, L., Keshari, K.R.: 13C-labeled biochemical probes for the study of cancer metabolism with dynamic nuclear polarization-enhanced magnetic resonance imaging. Cancer Metab. **3**, 9 (2015)
57. Hurd, R.E., Spielman, D., Josan, S., Yen, Y.-F., Pfefferbaum, A., Mayer, D.: Exchange-linked dissolution agents in dissolution-DNP (13) C metabolic imaging. Magn. Reson. Med. **70**, 936–942 (2012)

58. Kettunen, M.I., Hu, D.-E., Witney, T.H., Mclaughlin, R., Gallagher, F.A., Bohndiek, S.E., Day, S.E., Brindle, K.M.: Magnetization transfer measurements of exchange between hyperpolarized [1-13C]pyruvate and [1-13C]lactate in a murine lymphoma. Magn. Reson. Med. **63**, 872–880 (2010)
59. Kettunen, M.I., Kennedy, B.W.C., Hu, D.-E., Brindle, K.M.: Spin echo measurements of the extravasation and tumor cell uptake of hyperpolarized [1-(13) C]lactate and [1-(13) C]pyruvate. Magn. Reson. Med. **70**(5), 1200–1209 (2013)
60. Serrao, E.M., Kettunen, M.I., Rodrigues, T.B., Lewis, D.Y., Gallagher, F.A., Hu, D.E., Brindle, K.M.: Analysis of 13C and 14C labeling in pyruvate and lactate in tumor and blood of lymphoma-bearing mice injected with 13C- and 14C-labeled pyruvate. NMR Biomed. **50**, e3901 (2018)
61. Hill, D.K., Jamin, Y., Orton, M.R., Tardif, N., Parkes, H.G., Robinson, S.P., Leach, M.O., Chung, Y.-L., Eykyn, T.R.: ^{1}H NMR and hyperpolarized ^{13}C NMR assays of pyruvate-lactate: a comparative study. NMR Biomed. **26**, 1321–1325 (2013)
62. Qin, H., Carroll, V.N., Sriram, R., Villanueva-Meyer, J.E., von Morze, C., Wang, Z.J., Mutch, C.A., Keshari, K.R., Flavell, R.R., Kurhanewicz, J., Wilson, D.M.: Imaging glutathione depletion in the rat brain using ascorbate-derived hyperpolarized MR and PET probes. Sci. Rep. **8**, 7928 (2018)
63. Costello, L.C., Franklin, R.B.: The clinical relevance of the metabolism of prostate cancer; zinc and tumor suppression: connecting the dots. Mol. Cancer. **5**, 17 (2006)
64. Bohndiek, S.E., Kettunen, M.I., Hu, D.-E., Kennedy, B.W.C., Boren, J., Gallagher, F.A., Brindle, K.M.: Hyperpolarized [1-13C]-ascorbic and dehydroascorbic acid: vitamin C as a probe for imaging redox status in vivo. J. Am. Chem. Soc. **133**, 11795–11801 (2011)
65. Sriram, R., Van Criekinge, M., Hansen, A., Wang, Z.J., Vigneron, D.B., Wilson, D.M., Keshari, K.R., Kurhanewicz, J.: Real-time measurement of hyperpolarized lactate production and efflux as a biomarker of tumor aggressiveness in an MR compatible 3D cell culture bioreactor. NMR Biomed. **28**, 1141–1149 (2015)
66. Breukels, V., Jansen, K.C.F.J., van Heijster, F.H.A., Capozzi, A., van Bentum, P.J.M., Schalken, J.A., Comment, A., Scheenen, T.W.J.: Direct dynamic measurement of intracellular and extracellular lactate in small-volume cell suspensions with 13C hyperpolarised NMR. NMR Biomed. **28**, 1040–1048 (2015)
67. Sriram, R., Sun, J., Villanueva-Meyer, J., Mutch, C., De Los, S.J., Peters, J., Korenchan, D.E., Neumann, K., Van Criekinge, M., Kurhanewicz, J., Rosenberg, O., Wilson, D., Ohliger, M.A.: Detection of bacteria-specific metabolism using hyperpolarized [2-13C]pyruvate. ACS Infect Dis. **4**(5), 797–805 (2018)

Further Reading

Bottomley, P.A., Griffiths, J.R.: Handbook of magnetic resonance spectroscopy in vivo: MRS theory, practice and applications. Wiley, Chichester (2016)

Chapter 7
Innovating Metabolic Biomarkers for Hyperpolarized NMR

Richard L. Hesketh, Alan J. Wright, and Kevin M. Brindle

7.1 Introduction

7.1.1 What Is a Biomarker?

A biomarker "is a characteristic that is objectively measured and evaluated as an indicator of normal biological processes, pathogenic processes or responses to an exposure or therapeutic intervention" [1]. Given the requirement for an objective measurement, few imaging biomarkers in routine clinical use would qualify as almost all radiological assessment is subjective. For example, the TNM (tumor, nodes, and metastasis) classification is a biomarker widely used for staging, prognosis and response assessment of tumors but it is based on subjective measurements of tumor size, nodal and metastatic disease. Alternatively, objective measurements are frequently assessed subjectively, as is the case with measurements of the apparent diffusion coefficient (ADC) of tissue water. Therefore, a simpler, all-encompassing definition of an imaging biomarker is "any measurement derived from an image" [2].

R. L. Hesketh (✉)
Department of Radiology, University College London Hospital, London, UK

CRUK Cambridge Institute, Cambridge, UK
e-mail: richard.hesketh@nhs.net

A. J. Wright
CRUK Cambridge Institute, Cambridge, UK

K. M. Brindle
CRUK Cambridge Institute, Cambridge, UK

Department of Biochemistry, University of Cambridge, Cambridge, UK

T. Jue, D. Mayer (eds.), *Dynamic Hyperpolarized Nuclear Magnetic Resonance*, Handbook of Modern Biophysics, https://doi.org/10.1007/978-3-030-55043-1_7

Biomarkers of disease are used at every step in the clinical decision-making process from screening, diagnosis, prognosis, stratifying patients for treatment and monitoring response to therapy [3]. Imaging is an essential part of all these steps for most cancers and for many inflammatory, infectious, and degenerative diseases.

7.1.2 What Makes a Good Biomarker?

Biomarkers in their most simplistic form have long existed in the field of medicine. Hippocrates, the father of modern medicine, recognized that the frequency of fever was a clinical biomarker of prognosis in patients with malaria, albeit the discovery of the parasite species behind the differences in fever periodicity would wait until the work of Alphonse Laveran, Camillo Golgi and others over 2000 years later [4, 5]. Our increased understanding of disease has coincided with huge advances in technology that has led to the discovery of over 100,000 biomarkers. However, only around 100 are in routine clinical use and, for the most part, the simplest of measurements remains the most accurate and widely used [6]. For example, as a predictor of morbidity and mortality following traumatic brain injury (TBI), bedside clinical assessment of conscious level using the Glasgow Coma Scale (GCS) performs as well as computed tomography (CT)-based classification [7]. The most successful and common approach to improving biomarker performance is to use multivariate models of varying complexity to incorporate data from multiple biomarkers. Most diseases have some sort of calculable, multiple variable scores for one or more of prognostication, patient stratification, and treatment response monitoring. From the previous TBI example, CT or GCS alone perform similarly. However, combining GCS, CT, and other clinical and biochemical metrics improves prognostication [7, 8].

There are notable examples where comprehension of the underlying biology has led to the development and clinical use of novel biomarkers. In the 1960s, tamoxifen, an estrogen antagonist, was synthesized by scientists trying to develop the emergency contraceptive pill. Unfortunately, it was not effective as a contraceptive. Instead, it increased fertility. However, breast cancer had been recognized to be estrogen sensitive following the cases of remission induced by bilateral oophorectomy first reported in 1895 by British surgeon George Beatson. That understanding led to an alternative usage. The initial results in treating breast cancer were slightly disappointing. Tamoxifen only induced a positive response in 22% of breast cancer patients [9]. Four years later in 1975, however, Elwood Jensen discovered the low response rate was partly explained by tumor estrogen receptor (ER) expression. A response to tamoxifen was seen in 63% of patients with ER+ tumors versus only 2% of ER− tumors [10]. Jensen's work was the foundation for the first biomarker for targeted therapy in cancer. Along with progesterone receptor (PR) and human epidermal growth factor receptor 2 (HER2) status, ER status has become an essential determinant of the clinical management of breast cancer today.

The advances in medical imaging over the last 50 years represent simply astounding feats of technology and have revolutionized and replaced many aspects of traditional clinical medicine. The first CT acquired in 1971 took around 4 min per slice to acquire [11]. Today, a 2-ton CT gantry rotating at approximately three revolutions per second can acquire 0.5 mm isotropic voxels of an entire patient in just a few seconds, a truly remarkable and perhaps under-appreciated accomplishment of medical physics. The development of magnetic resonance imaging (MRI) over a similar timescale is equally noteworthy, from the blurry proton images of two tubes of water generated by Lauterbur in 1972 [12] to the wide array of high-resolution anatomical and multinuclear functional imaging techniques that are in use today. However, the translation of functional imaging biomarkers for clinical use has not been efficient. Many techniques, probes, and metrics have been proposed from preclinical studies but only a handful has been translated into clinical application. Clinical translation is hindered by the logistically complex and expensive translational pathway, a process that can take decades [2].

For any new functional imaging biomarker to be useful clinically, it must provide information that (a) is not available from standard-of-care anatomical imaging and (b) have the potential to change clinical management. Using cancer as an example, the treatment options for solid tumors have traditionally been surgery or radiotherapy for localized tumors and cytotoxic chemotherapy for tumors that have spread beyond the organ they originated in (metastatic tumors). This information is provided by anatomical imaging to a high level of accuracy, with only limited additional information being provided by imaging of tissue function for most solid tumors. Over the last 20 years, tumor genotype and molecular phenotype have been used increasingly to determine what type of treatment to use and the potential of functional imaging to determine these phenotypes noninvasively is a highly attractive proposition. Furthermore, for many cancers there are now multiple targeted therapies available, several of which have paradoxically made anatomical imaging harder to interpret (discussed in "Tumor Staging" below), where detection of response to therapy in a timely manner is becoming increasingly important.

To achieve this a biomarker must demonstrate sufficient repeatability and reproducibility for a given application. Depending on the application the stringency of test-retest requirements may vary considerably. As part of validation a new biomarker should be able to demonstrate a test sensitivity and specificity that is not achievable using current standard-of-care imaging. However, unlike biochemical biomarkers, imaging biomarkers do not necessarily measure a specific biological process [2]. Even the most targeted imaging biomarkers invariably report on multiple aspects of in vivo biology. For example, the widely used positron emission tomography (PET) tracer, the glucose analog 2-[^{18}F]fluoro-2-deoxy-D-glucose ([^{18}F]FDG), is often erroneously reported to be a surrogate marker of glycolytic activity. In reality, it is a complex biomarker that depends on tissue perfusion, endogenous glucose concentration and the expression and activity of glucose transporters and several enzymes, most notably hexokinase and glucose-6-phosphatase, and is not necessarily reflective of downstream glycolytic activity [13, 14]. Similar limitations will be true for all injected hyperpolarized ^{13}C-labelled biomarkers of metabolic processes, with perfusion limitations being unavoidable and active membrane transporters being required for observation of intracellular metabolism. Interpreting the fate of even the simplest metabolic probes depends on numerous factors.

7.2 Desirable Characteristics of a Hyperpolarized Substrate

7.2.1 Polarizable Agents

Magnetic resonance spectroscopic imaging with hyperpolarized isotope-labeled cell substrates has revolutionized our ability to image metabolism in vivo, including in the clinic (reviewed in [15, 16]). The pathway to clinical translation of hyperpolarized substrates bears many similarities to that of PET tracers as does the ultimate clinical goal of providing novel data that can influence patient management. However, many specific considerations must be addressed during hyperpolarized substrate development. Most substrates proposed for translation are ^{13}C labelled. Several methods can be used to hyperpolarize ^{13}C labelled substrates including brute-force hyperpolarization, parahydrogen induced polarization (PHIP), signal amplification by reversible exchange (SABRE), and dissolution dynamic nuclear polarization (DNP). As DNP currently offers the highest signal amplification and has progressed to clinical use, this chapter will refer to DNP as the polarization method unless stated otherwise.

7.2.2 Toxicity

Toxicity poses a significant concern because hyperpolarized agents need to be administered in much larger quantities than radiotracers to reach the threshold of detection. Although the injected amount is similar to gadolinium-based contrast agents, most hyperpolarized substrates are not biologically inert and significant transient disruption to in vivo metabolism is possible. The injection of hyperpolarized [1-^{13}C]pyruvate at a concentration of approximately 0.1 mmol/kg has not resulted in any adverse effects in patients. Mice can tolerate quantities in excess of 1 mmol/kg, although an enhanced oxygen consumption rate and subsequent transient hypoxia have been observed [17–19]. However, in anesthetized mice the administration of ~0.4 mmol/kg [1-^{13}C]dehydroascorbic acid (the oxidized form of vitamin C) resulted in transient cardiorespiratory arrest [20].

7.2.3 Sample Preparation and DNP

DNP experiments require a stable free radical, which has unpaired electrons, to be mixed with the molecule of interest and rapidly frozen to form a glass. At low temperatures (~1 K) and high magnetic field (typically >3 T), the spins of unpaired electrons, which have a gyromagnetic ratio > 2600 times that of ^{13}C nuclei, are almost 100% polarized. Microwave irradiation close to the resonance frequency of the electron induces electron-nuclear spin transitions and a gradual build-up of nuclear spin polarization in susceptible nuclei that are in close proximity to the unpaired electrons, a process that typically takes between 15 min and 2 h depending on the polarization conditions and substrate [21, 22]. Following polarization, the sample is rapidly dissolved in a super-heated solvent and flushed out of the system using pressurized gas in a process known as dissolution [23].

Polarization can be further enhanced by up to 300% by the addition of a Gd^{3+} chelate, an effect that diminishes at a higher field strength of 4.6 T [22, 24]. Gd^{3+} chelates are not used in clinical samples to avoid administration of Gd^{3+}, which can accumulate in tissue, albeit without credible evidence of a clinical effect thus far [25]. A few substrates are self-glassing, notably pyruvic acid, but others require the addition of glassing agents, for example, glycerol or dimethyl sulfoxide, to prevent crystallization and inhomogeneous distribution of the radical. Furthermore, the sample to be polarized must be highly concentrated prior to freezing as the dissolution step results in significant dilution. The dilution is of particular concern in small animal experiments that can use only limited injection volumes. For example, neat [1-^{13}C]pyruvic acid, a liquid at room temperature, has a concentration of 14 M but in most animal experiments dissolution dilutes the concentration nearly 200-fold to 75 mM. Without high solubility in aqueous solution, the DNP substrate will not reach the concentration required for clinical use. Furthermore, any glassing agents or solvents added to the sample must also be nontoxic, limiting the clinically applicable molecules.

In the clinical polarizer sterility and purity of the sample is of utmost importance. The small molecule imaging agent is mixed with the free radical in the sample cup of a fluid path, a single-use sterilizable unit that is prepared and sealed in a pharmacy. This is installed in the polarizer, located close to the scanner, where the sample is polarized and an automated dissolution process is performed. Quality control (QC) checks are used to confirm that the radical has been largely removed by filtration and that the pH is suitable for intravenous injection. The imaging agent can then be transferred to the MRI scanner room and injected *via* an automated injection pump. Currently, the QC steps prolong the time between the start of dissolution to injection into the patient by approximately a minute in clinical studies compared to <15 s in preclinical studies.

7.2.4 T₁

The rate of hyperpolarization decay is determined by the T_1 of the nucleus that is less than 1 min for most ^{13}C nuclei. Rapid dissolution, transfer of the substrate to the subject and rapid imaging are essential to maximize available signal. Several other strategies exist to prolong the T_1. Although the addition of Gd^{3+} to the sample can increase the polarization, the presence of paramagnetic ions in the injected sample will usually decrease the T_1. The T_1 can also decrease substantially through a variety of mechanisms at the low magnetic fields that can occur during transfer between the hyperpolarizer and scanner magnet, which has led to attempts to maintain a sufficiently high magnetic field during the transfer [26].

Most hyperpolarized molecules are small. For example, [1-^{13}C]pyruvic acid is just 89 Da. Small molecules are generally more soluble and are more likely to be substrates in rapid metabolic reactions. Small molecules also tend to have longer T_1 values. As molecular size increases, correlation time (τ_c) increases. T_1 relaxation is optimal when $\tau_c = 1/\omega$ (where ω = Larmor frequency). For small molecules in solution at room temperature, τ_c is typically several orders of magnitude below $1/\omega$, which results in long T_1.

In addition to τ_c, molecular interactions affect the T_1 of a nucleus. The most important interaction is dipole–dipole interactions that occur primarily with protons due to their abundance in biological molecules and their large magnetic moment. The strength of the dipolar coupling is distance dependent; therefore, nuclei with attached protons generally have a T_1 too short to be useful in hyperpolarization experiments. For this reason, ^{13}C hyperpolarized substrates containing carbonyl groups have been the most utilized. The effect of dipolar coupling can be reduced by the substitution of protons with deuterons. For example, the universal deuteration of glucose carbons was reported to increase the combined in vivo T_1 from <1 s to nearly 9 s [27]. For clinical applications, a long T_1 is necessary. From the point of dissolution to injection, there are inevitable signal losses that are dictated by this first order process. Additionally, for in vivo experiments the time required for injection and perfusion to the tissue of interest is an unavoidable source of T_1-dependent signal decay. Miloushev et al. found that [1-^{13}Cpyruvate levels in the human brain peaked at 12 s after intravenous injection [28].

7.2.5 *Chemical Shift*

The enhanced sensitivity provided by hyperpolarization allows imaging of a labelled substrate and its conversion into a metabolic product. With the exception of a few substrates developed to measure tissue extracellular pH or perfusion, most substrates target fast, mostly catabolic reactions that occur on the timescale of the hyperpolarization lifetime, typically <2 min. The substrate and product signals must be spectrally resolved. For example, glutamine labelled on carbons 1 and 5 has been

Table 7.1 Chemical shifts of ^{13}C-labelled glutamine and relevant metabolic products

Compound	Chemical shift (ppm)
[1-^{13}C]Glutamine	174.8
[1-^{13}C]Glutamate	175.5
[1-^{13}C]2-Hydroxyglutarate	180.8
[5-^{13}C]Glutamine	178.5
[5-^{13}C]Glutamate	181.5
[5-^{13}C]2-Hydroxyglutarate	180

successfully hyperpolarized. The chemical shift difference between [5-^{13}C]glutamine and the reaction product [5-^{13}C]glutamate is 3 ppm [29]. In contrast, only 0.7 ppm separates the [1-^{13}C]glutamine and [1-^{13}C]glutamate signals, which is difficult to resolve in vivo meaning that [1-^{13}C]glutamine is poorly suited to probe glutaminase activity. Conversely, [1-^{13}C]2-hydroxyglutarate (2-HG), a downstream product of glutamine metabolism in tumors that have isocitrate dehydrogenase (IDH) mutations, has a chemical shift approximately 5 ppm greater than [1-^{13}C]glutamine, compared to only 1.5 ppm between [5-^{13}C]glutamine and [5-^{13}C]2-HG. Therefore, to determine IDH mutational status in vivo [1-^{13}C]glutamine is a more appropriate substrate than [5-^{13}C]glutamine [30] (Table 7.1).

7.3 Functional Imaging and Clinical Applications

7.3.1 Glycolysis

Glycolytic flux can be assessed directly using hyperpolarized [U-^{13}C,U-^{2}H]glucose; however, the short hyperpolarization lifetime of this molecule means that, even though glycolysis is a rapid process, very little of the ^{13}C hyperpolarization survives the ten steps in the glycolytic pathway to result in lactate labelling [27]. Moreover, given the indirect relationship between signal intensity and concentration in a hyperpolarized experiment, it would be difficult, if not impossible, to quantify the rate of glycolytic flux from these data. Recent studies of oral administration of [6,6′-$^{2}H_2$]glucose in glioma patients have shown that tumor glycolysis can be assessed using ^{2}H MR imaging [31]. Preclinical studies have shown that the imaging technique can be used to produce quantitative maps of tumor glycolysis in vivo [32].

[1-^{13}C]pyruvic acid has a relatively long T_1 (about 60 s at 3 T) and is highly soluble in water. Injected [1-^{13}C]pyruvate exchanges its hyperpolarized ^{13}C label with the endogenous lactate pool, which is often present at millimolar concentrations and is elevated in many pathologies, such as cancer. The endogenous pyruvate pool exists at submillimolar concentration. The exchange is catalyzed by lactate dehydrogenase (LDH) in a reaction that is close to chemical equilibrium in vivo, resulting in rapid exchange of the hyperpolarized ^{13}C label with relatively little net conversion of pyruvate to lactate [33]. The amount of lactate labelling is largely dependent on lactate pool size that in turn is partially determined by the rate of glycolysis. Therefore, the amount of lactate labelling provides an indirect measure of glycolytic activity. However, complicating factors temper the interpretation of lactate labelling as a measure of glycolysis. The rate of [1-^{13}C]lactate formation is dependent on circulation, tissue perfusion, expression of the mono-carboxylate transporters, the concentration of LDH, NADH (the substrate and coenzyme of LDH), and the rate of flux from pyruvate into other competing metabolic pathways, for example, dehydrogenation and decarboxylation at entry points into the tricarboxylic acid cycle [18, 33–35].

Due to its position at a metabolic crossroads, favorable hyperpolarization levels, and T_1, [1-^{13}C]pyruvate has been the most widely investigated hyperpolarized substrate in preclinical studies and the only hyperpolarized ^{13}C probe to be used in clinical trials so far. Cancer has also been the dominant focus for research with hyperpolarized substrates. As a probe of upregulated glycolysis, it was proposed that hyperpolarized [1-^{13}C]pyruvate could be used for localizing and diagnosing solid tumors in vivo as elevated glycolysis is a feature of most solid tumors [36]. Subsequently, [1-^{13}C]pyruvate has also been investigated as an imaging biomarker of tumor grade and response to treatment, perhaps the two most pertinent and complex clinical questions.

In conjunction with prostate specific antigen (PSA), multiparametric ^{1}H imaging (mpMRI) consisting primarily of T_2-weighted, diffusion-weighted and dynamic contrast enhanced T_1-weighted sequences, is now the first line test for diagnosis and targeted biopsy planning for suspected prostate cancer. However, mpMRI can be misleading particularly in younger patients, with a false negative rate of up to 27% [37]. Imaging with hyperpolarized substrates, improved diffusion-weighted imaging techniques, including the VERDICT (Vascular, Extracellular, and Restricted Diffusion for Cytometry in Tumors) model and [^{68}Ga]-PSMA PET have all been proposed as improvements on the current mpMRI techniques [38, 39]. After injection of hyperpolarized [1-^{13}C]pyruvate, increased lactate labelling was associated with tumors of higher grade in a mouse model of prostate adenocarcinoma (TRAMP) [40]. Initial results in prostate cancer patients have similarly reported that increased lactate labelling following [1-^{13}C]pyruvate injection was associated with a higher Gleason score [41].

7.3.2 Tumor Staging

An imaging biomarker of a disease state is not just capable of localizing pathology but can potentially follow, with multiple measurements over time, tumor growth, metastasis and response to therapy. Currently, tumor staging is usually performed using anatomical cross-sectional imaging, with radiologists often employing rudimentary measurements of size to stage and determine tumor responses to treatment. The increasing use of immunotherapy in cancer treatment has brought further challenges to image interpretation. Pseudoprogression, a phenomenon where tumors transiently increase before reducing in size in response to treatment, is rare in tumors treated with cytotoxic chemotherapy but in patients treated with immunotherapy pseudoprogression can lead to the misdiagnosis of progressive disease in as many as 14% of patients using traditional, size-based response criteria (RECIST) [42, 43]. Furthermore, hyperprogression, a doubling of tumor growth rate following initiation of treatment, has been reported in between 4–29% of patients treated with immunotherapy [44]. Differentiation of response, progression, and pseudoprogression is therefore more complex than ever before and improved imaging biomarkers are required.

Changes in metabolic flux may be used as a surrogate for cell viability and may precede changes in tumor volume. By imaging before and after therapy a decrease in lactate labelling following injection of [1-^{13}C]pyruvate has been shown in multiple animal models and in a single prostate cancer patient to be a marker of treatment response [14, 18, 45, 46]. This means that ineffective treatments can be stopped and changed much sooner, which is of great potential benefit when used in conjunction with personalized medicine and targeted cancer therapies. Using an imaging biomarker should improve treatment outcome.

As pyruvate is the product of the glycolytic pathway, hyperpolarized [1-^{13}C] pyruvate has frequently been compared to the widely used PET tracer [^{18}F]FDG, which provides a surrogate marker for glucose uptake (Fig. 7.1). Several studies have shown that the two probes measure different aspects of the glycolytic pathway and that measurements of [^{18}F]FDG uptake do not provide an accurate assessment of glycolytic activity [14, 35, 47].

7.3.3 Tumor Phenotyping

Over the last 20 years tumor biology has been studied to a depth that was previously inconceivable, primarily driven by publication of the Human Genome Project in 2001 and the accelerated development of sequencing technology that accompanied it [48]. It is now widely recognized that the genomic, epigenomic, and proteomic signatures of individual cancers can vary greatly, even for tumors arising from the same tissue. This recognition has led to significant improvements in prognostication and treatment. It is now common to perform targeted sequencing, rtPCR (real-time polymerase chain reaction) or at the very least immunohistochemistry to identify mutations or expressed proteins that present potential targets for therapy. For example, breast cancers are routinely categorized based on their HER2, ER, and PR status. Genomic and transcriptomic profiling has revealed ten distinct genomic subtypes [49, 50]. However, cancer is a disease driven by genomic instability and there is often significant intratumoral, intertumoral and temporal heterogeneity, partially explaining why a disparate response to treatment is so frequently seen in different metastatic deposits in the same patient [51]. This heterogeneity poses a significant problem for biopsy driven management, almost invariably performed on

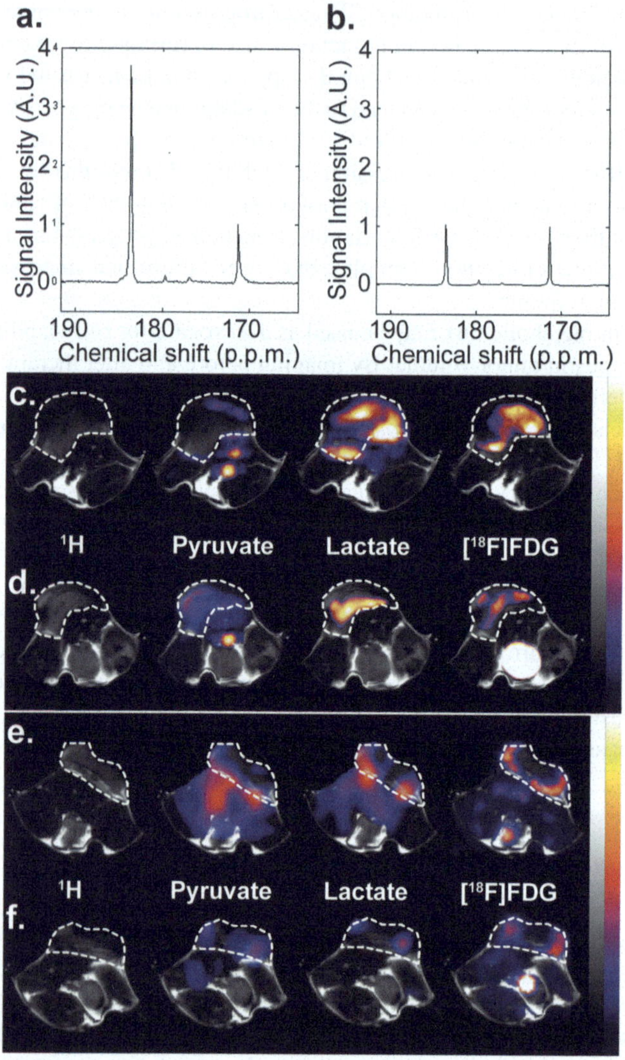

Fig. 7.1 Imaging of EL4 murine T-cell lymphomas before and 24 h after treatment with etoposide showing a reduction in the tumor [1-^{13}C]lactate signal and [^{18}F]FDG signal after treatment. Magnetic resonance spectroscopy (**a**) before and (**b**) after treatment. The resonance at 171 ppm is from [1-^{13}C]pyruvate and at 183 p.p.m. is that of [1-^{13}C]lactate. ^{13}C MR images and [^{18}F]FDG-PET images overlaid on ^{1}H MRI reference images before (**c**) and (**d**) after treatment. The false color scale is the same in the pre- and post-treatment images. The ^{13}C MR images were acquired using a 2D EPI sequence [111]

the most accessible tumor. Even the novel concept of the "liquid biopsy," based on circulating DNA or circulating tumor cells isolated from blood, has demonstrated significant discrepancies with samples taken from the tumor or tumors [52]. Deriving functional signatures from imaging, where whole tumors and metastases can be sampled noninvasively, is therefore an attractive alternative to invasive biopsy. One of the most promising targets for determining tumor phenotype using metabolic imaging is isocitrate dehydrogenase (IDH). It is mutated in the majority of low-grade gliomas, and mutation is associated with a better prognosis. In contrast to wild-type IDH, which catalyzes the oxidative decarboxylation of isocitrate to alpha-ketoglutarate (α-KG), the most common isoform of mutant IDH reduces α-KG to D-2-hydroxyglutate (2-HG), which is present at low levels in normal cells. This neomorphic function of mutant IDH leads to an accumulation of 2-HG, termed an oncometabolite because of its tumorigenic effects [53]. Several methods for imaging 2-HG accumulation using hyperpolarized ^{13}C-labelled substrates have been proposed. Elevated levels of 2HG result in decreased PDH activity, which leads to decreased hyperpolarized ^{13}C label flux between [2-^{13}C]pyruvate and [5-^{13}C]glutamate [54]. [1-^{13}C]2-HG has also been imaged directly following the injection of [1-^{13}C]glutamine or [1-^{13}C]α-KG [30, 55]. The reported T_1 of both substrates (52 s at 3 T and 34 s at 1 T for [1-^{13}C]α-KG and [1-^{13}C]glutamine, respectively) are sufficiently long to make these substrates usable in the clinic. However, the higher polarization levels and faster transport into cells make [1-^{13}C]glutamine the more likely candidate for clinical translation.

7.3.4 Inflammation

Increased hyperpolarized ^{13}C label exchange between pyruvate and lactate is not only observed in tumor metabolism but can also be used to detect the presence of inflammation. In a rat model of arthritis there was elevated conversion of [1-^{13}C]pyruvate to lactate [56], which was in part due to the glycolytic nature of leukocytes and also due to upregulated glycolysis in other cells in the inflamed tissue. This increased local lactate concentration, however, does not necessarily lead to a lowering of tissue pH (Fig. 7.2) [57]. The localization of inflammatory cells can be used to identify other diseases. A high density of pro-inflammatory mononuclear phagocytes was imaged with [1-^{13}C]pyruvate in a model of multiple sclerosis [58]. Imaging leukocyte infiltration has been proposed as an early biomarker of organ transplant rejection. Following injection of hyperpolarized [1-^{13}C]pyruvate an increased lactate/pyruvate signal ratio was observed in a lung allograft undergoing acute tissue rejection, which was ascribed to an influx of CD8+ T cells [59].

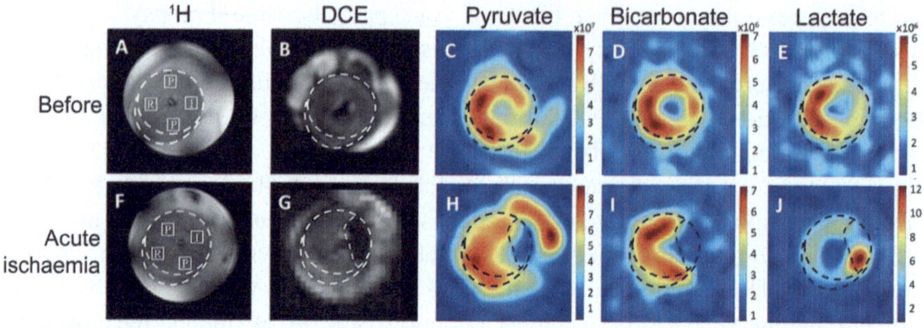

Fig. 7.2 Imaging metabolic activity and pH in inflamed tissue using [1-^{13}C]pyruvate and ^{13}C bicarbonate. (**a**) Chemical Shift Imaging (CSI) of hyperpolarized [1-^{13}C]pyruvate and [1-^{13}C]lactate in an inflamed mouse foot 7 days after injection of Freund's adjuvant. An anatomical image is overlaid with a lactate/pyruvate signal intensity ratio map. CSI data were collected 20 s after intravenous injection of [1-^{13}C] pyruvate with an example spectrum (purple box) showing elevated resonances of [1-^{13}C]lactate in regions of inflammation. (**b**) A coil localized NMR spectrum of an inflamed foot 12 s after administration of [^{13}C]bicarbonate. The ratio of the [^{13}C]bicarbonate and [^{13}C]carbon dioxide signals indicates a pH of 7.3. Inflamed paws were found not to have a significantly lower pH than normal [57]

7.3.5 Immune Cells

Imaging with hyperpolarized [1-^{13}C]pyruvate can be used for a wide range of potential clinical applications, given the variety of cell types that show elevated glycolysis under different pathological conditions. Increased glycolytic flux may be associated with the recruitment of immune cells to tumors. Imaging biomarkers of tumor associated macrophages or T-cell recruitment would be useful for assessing tumor responses to immune therapies (such as anti-PDL-1 antibodies). The relative contribution to tumor glycolysis by tumor-associated leukocytes has long been an open question when using [^{18}F]FDG-PET, since macrophages are highly glycolytic. However, if they constitute a minor cell population in the tumor, then their contribution to overall tumor glycolysis may be very small [14]. Measuring changes in tumor immune cell infiltration in response to therapy could be a very challenging application for hyperpolarized [1-^{13}C]pyruvate as the effect of tumor cell loss may confound any increase in tumor glycolysis resulting from an influx of leukocytes. However, immune cells might be differentiated by imaging a metabolic activity that is unique to these cells. For example, the arginase activity of myeloid derived suppressor cells (MDSCs) can be monitored using the labelled urea produced from hyperpolarized [6-^{13}C]arginine [60]. It is this arginase activity that gives MDSCs their T-cell suppressing capabilities and tumor promoting activity, making it an attractive target for cancer immunotherapy.

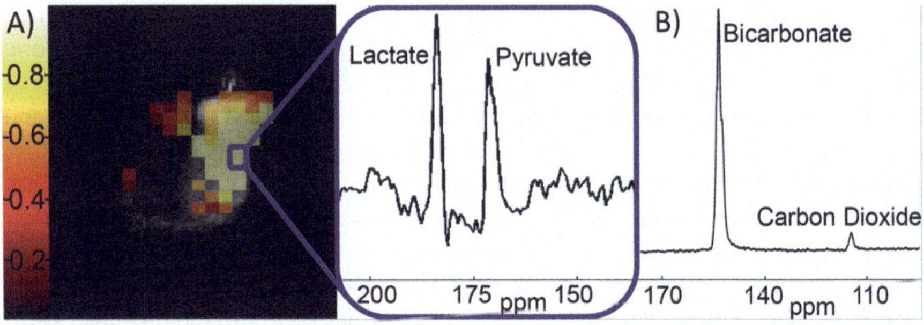

Fig. 7.3 MRI of perfused rat hearts before and within 10 min of the onset of acute ischemia by occlusion of the LAD artery. (**a**, **f**) high resolution axial ^{1}H MR images—the approximate infarct region is labelled **i**; (**b**, **g**) dynamic contrast enhanced (DCE) images showing the hypoperfused segment after LAD artery occlusion in (**g**); (**c**–**e** and **h**–**i**) chemical shift imaging of the distribution of hyperpolarized metabolites, (**c**, **h**) [1-^{13}C]pyruvate; (**d**, **i**) [^{13}C]bicarbonate; (**e**, **j**) [1-^{13}C]lactate. Adapted from Ball et al. 2013 and used under the Creative Commons Attribution License (CC BY) [62]

7.3.6 Ischemia

Metabolic changes that occur secondary to ischemia can also potentially be imaged using hyperpolarized [1-^{13}C]pyruvate. Acute cardiac ischemia is characterized by a switch to anaerobic glycolysis and the accumulation of lactate, therefore [1-^{13}C]pyruvate has been proposed as an early biomarker of acute ischemia. In the normal heart [1-^{13}C]lactate and ^{13}C bicarbonate signals are both readily detectable following injection of [1-^{13}C]pyruvate, the bicarbonate signal being generated from the CO_2 released when pyruvate is converted to acetyl-CoA by the pyruvate dehydrogenase (PDH) complex [61, 62]. Within 10 min of occlusion of the left anterior descending artery (LAD) in perfused rat hearts, an increase in [1-^{13}C]lactate signal and a corresponding decrease in the [^{13}C]bicarbonate signal was observed following [1-^{13}C]pyruvate injection (Fig. 7.3) [62]. In a separate study a similar switch to anaerobic glycolysis was observed in a model of right ventricular heart failure [63]. However, Ball et al. also demonstrated in a separate cohort of rats that 4 weeks after LAD occlusion, the [1-^{13}C]lactate and [^{13}C]bicarbonate signals were both decreased in the infarcted territory [62]. Together, these results demonstrate that hyperpolarized [1-^{13}C]pyruvate may be able to influence clinical management by differentiating between acute infarction, chronic infarction and heart failure in vivo.

7.3.7 Necrosis

Hyperpolarized [1-^{13}C]pyruvate is a negative contrast agent for detecting tumor treatment response since this results in a decrease in lactate labelling. If pretreatment label flux to lactate is low then the dynamic range available for detecting treatment response may be too small for accurate assessment. An alternative is to have a hyperpolarized molecule that acts as a functional imaging biomarker of a metabolic reaction that is increased in the presence of cell death. Hyperpolarized [1,4-$^{13}C_2$] fumarate, which is converted to malate by the enzyme fumarase in the cytosol and mitochondria of cells and in the cytoplasm of red blood cells, has been proposed as just such a marker. Malate production from hyperpolarized [1,4-$^{13}C_2$]fumarate is not normally observed in vivo, as the transport of fumarate across the cell and mitochondrial membranes is too slow when compared to the lifetime of the hyperpolarization. However, if there is cellular necrosis, which results in increased plasma membrane permeability, then fumarate can gain rapid access to the enzyme. Since malate production results from hydration of fumarate, the enzyme has no coenzyme requirement and will be active in the extracellular space or in a necrotic cell. Hyperpolarized [1,4-$^{13}C_2$]fumarate was first demonstrated as an early detector of cell death in tumors following chemotherapy [64]. Hyperpolarized fumarate has been shown to distinguish between different kidney pathologies in animal models of disease [65]. Mouse kidneys with acute tubular necrosis showed an increase in [1,4-$^{13}C_2$]malate production, while those with glomerulonephritis did not. Recent studies have demonstrated that [1-^{13}C]pyruvate and [1,4-$^{13}C_2$]fumarate can be co-polarized and used in conjunction with other imaging agents [66, 67]. The potential to image simultaneously different metabolic processes could present a significant advantage of using hyperpolarized MRI instead of PET.

7.3.8 pH

A good target for a new functional imaging method is a fundamental biological process that is dysregulated in pathology. One such target, suitable for hyperpolarized small molecules, is pH. A lower extracellular pH is a feature of both solid tumors and ischemia, where lactate production from the Warburg effect, anaerobiosis or poor perfusion as well as changes to extracellular buffering will lead to a proton build-up. In acute inflammation, such as in infection, there may also be a transient lowering of extracellular pH due to the presence of highly glycolytic leukocytes [68]. Hyperpolarized [^{13}C]bicarbonate can be used as a probe of extracellular pH, since it converts rapidly to $^{13}CO_2$ in the near-equilibrium reaction catalyzed by carbonic anhydrase [69, 70]. The ratio of the $H^{13}CO_3^-$ and $^{13}CO_2$ resonances can be used to calculate the local pH using the Henderson-Hasselbalch equation and assuming a value for the bicarbonate pKa in vivo [71]).

The drawbacks of using [^{13}C]bicarbonate as a pH probe are its relatively short T_1 in vivo of ~10 s, the requirement that the ^{13}C label equilibrates between the $H^{13}CO_3^-$ and $^{13}CO_2$ pools within the lifetime of the hyperpolarization and that the pH estimation is based on measurements of signal intensities. A ratio of peak areas can be very sensitive to errors caused by noisy data. Alternatively, several hyperpolarized probes have been developed that exhibit pH-dependent chemical shifts. Measuring pH from a chemical shift only requires sufficient signal to observe the resonance and the accuracy is then mostly dependent on the resonance linewidth rather than the signal-to-noise ratio. $^{15}N_2$-imidazole and several ^{15}N-pyridine derivatives show chemical shifts of up to 60 ppm per pH unit and T_1s in excess of 1 min. However, these large chemical shifts will require high bandwidth RF excitation pulses that could create chemical shift artifacts and may make in vivo imaging challenging [72, 73]. An alternative probe for measuring extracellular pH is [1,5-$^{13}C_2$] zymonic acid that forms from dimerization of [1-^{13}C]pyruvic acid and was discovered accidentally during experiments with this substrate [74]. As such it is a good substrate for dissolution DNP, polarizing to over 20% at 3.35 T with a $T_1 > 40$ s in aqueous solution. Deuteration, to yield [3,6,6,6-$^2H_4$, 1,5-$^{13}C_2$] zymonic acid led to further increases in T_1 and an increase in signal-to-noise ratio (SNR) of up to 46% in vivo [75]. The exchangeable proton of the enolic hydroxyl group at the second carbon position has a pK_a of 6.9. Dissociation changes the chemical shift of the first carbon, which exhibits an increased chemical shift of approximately 4 ppm as the pH increases from 6 to 8 [74].

7.3.9 *Redox State*

Hyperpolarized agents designed to assess tissue redox status are limited by their lack of sensitivity and the fact that they perturb the systems they are measuring by reacting with NADPH or NADH. They are, rather, markers of available reducing power and cannot probe the pre-existing redox state. For example, dehydroascorbic acid (DHA) is rapidly transported into cells where it is reduced to ascorbic acid with concomitant oxidation of NADPH [76, 77]. The NADPH balance is then restored by the abundant cellular antioxidant, glutathione. Timm et al. showed that the rate of [1-^{13}C]DHA reduction in a tumor was dependent on both the concentration of reduced glutathione and the rate of NADPH production in the cell. NADPH production in this tumor was determined largely by pentose phosphate pathway flux [20].

[1,3-$^{13}C_2$]acetoacetate is reduced to β-hydroxybutyrate (BHB) in the mitochondria by the NADH-dependent enzyme β-hydroxybutyrate dehydrogenase. Measuring the conversion of acetoacetate to BHB provides a sensitive probe of mitochondrial function and its reducing power, which may be useful in assessing liver and kidney pathologies. In rat kidneys, hyperpolarized [1,3-$^{13}C_2$]acetoacetate was rapidly converted to [1,3-$^{13}C_2$]β-hydroxybutyrate, and this flux was increased when the mitochondrial NADH/NAD ratio was increased by metformin inhibition of complex I [78].

7.3.10 *Perfusion*

Gd^{3+} chelates that enter the extravascular space are used routinely to improve tissue contrast in MRI studies. However, it has been extensively publicized recently that Gd^{3+} chelates can accumulate in small quantities in tissues, particularly in the basal ganglia of the brain [79]. Although there is as yet no compelling evidence that this has a clinical effect, patients are sometimes reluctant to have Gd^{3+} chelates injected [80]. Several hyperpolarized substrates have been proposed as markers of perfusion. HP001 (*bis*-1,1-(hydroxymethyl)-[1-^{13}C]cyclopropane-d_8) has a long T_1 and polarizes well [81]. However, it is an exogenous substance and would have to undergo extensive safety testing prior to use. An alternative hyperpolarized perfusion agent is ^{13}C urea, an endogenous waste product of the urea cycle that is biologically inert in humans and also has a reported T_1 of 43 s in aqueous solution [67]. Urea is present in high concentration in blood (2.5–7 mM) and is well tolerated with sustained levels >50 mmol/L required for symptomatic uremia, far exceeding what would be injected in a hyperpolarized study [82]. Urea rapidly equilibrates with the extravascular space with slower movement into cells and across the blood–brain barrier, mostly by facilitated diffusion via urea transporters. Hyperpolarized [^{13}C]urea has been used as a perfusion marker in studies on tumors and the heart with the additional benefit that it can be co-polarized and used in combination with other substrates for the simultaneous assessment of perfusion and metabolism [67, 81, 83]. An increase in signal-to-noise has been reported by using [^{13}C, $^{15}N_2$]urea, which eliminates ^{13}C–^{14}N scalar coupling and significantly prolongs the ^{13}C T_2 [84]. Alternatively, transferring ^{15}N polarization to spin-coupled protons increased the sensitivity of detection of hyperpolarized [$^{15}N_2$]urea compared to direct ^{15}N detection and has a theoretical SNR improvement over direct ^{13}C detection, although the feasibility of performing this experiment in vivo has yet to be demonstrated [85].

7.3.11 Pulmonary Imaging with Hyperpolarized Gases

Current clinical imaging in pulmonary medicine is heavily reliant upon CT to produce high-resolution anatomical images of the lungs and airways and, to a lesser extent, scintigraphy or single photon emission computed tomography (SPECT). However, many of the airway and interstitial lung diseases are chronic conditions, requiring regular, repeated imaging for which the use of imaging techniques that rely on ionizing radiation is undesirable. Furthermore, CT provides minimal functional information, and scintigraphy/SPECT has limited spatial resolution. The lungs have poor suitability for MRI due to their low proton-density, large number of air–tissue interfaces and constant movement, although various advances in technology and imaging techniques have started to address these shortcomings [86]. MRI with hyperpolarized gases has the potential to address some of the shortcomings of current imaging techniques by avoiding ionizing radiation, improving the spatial resolution, and reducing the acquisition time of SPECT. So far, ^{3}He and ^{129}Xe have been the most investigated hyperpolarized gases. It is possible to hyperpolarize gases using DNP but spin-exchange optical pumping is the more common technique [87, 88]. ^{3}He has a three times greater gyromagnetic ratio than ^{129}Xe and a higher level of polarization. ^{3}He poses no clinically significant side effects. However, ^{3}He has a low natural abundance and is extremely expensive to produce. Consequently,

^{129}Xe is more attractive. Unlike ^{3}He, ^{129}Xe can diffuse across the alveolar wall and permits the measurement of gaseous exchange. It does pose a small anesthesia risk in large inhaled doses [89, 90]. Both gases have been used in clinical trials to investigate diseases such as chronic obstructive pulmonary disease (COPD), asthma, cystic fibrosis, and pulmonary fibrosis. The largest clinical trial of hyperpolarized MRI to date demonstrated that hyperpolarized ^{3}He outperformed CT in correlating with pulmonary function tests (PFTs) [91]. Similarly, hyperpolarized ^{129}Xe experiments have been correlated with pulmonary function in COPD [92]. Interstitial uptake and red blood cell transfer of hyperpolarized ^{129}Xe has been correlated with diffusion capacity for carbon monoxide in idiopathic pulmonary fibrosis [93].

7.4 Clinical Translation

7.4.1 Clinical Relevance

Developing new hyperpolarized biomarkers of metabolism for clinical translation will require applications that can change clinical practice while at the same time justifying the cost of such a complicated molecular imaging method. Developing functional biomarkers of fundamental biological processes may have the greatest potential. [1-^{13}C]pyruvate has been shown to be safe and has been used in prostate cancer patients to localize cancerous foci and provide staging and prognostic information for targeted therapies [17, 41]. Hyperpolarized pyruvate labelled in both the C1 [94] and C2 [95] positions has been studied in the normal human brain. These studies have demonstrated that both lactate and bicarbonate are detectable when the blood–brain barrier is intact (Fig. 7.4). This may limit the specificity of [1-^{13}C]lactate for detecting for pathological processes in the brain, such as tumors.

A large array of clinical trials with [1-^{13}C]pyruvate has started and other substrates are likely to follow (reviewed in [96]). Developments in DNP technology as well as improved MR hardware and acquisition sequences could significantly increase signal-to-noise ratio as well as reduce the financial cost of clinical hyperpolarized studies. The first clinical studies have demonstrated a remarkably high repeatability and this will need to continue to be demonstrated in multicenter studies [17, 41]. If these technical aspects can be successfully addressed, then it is possible that one of the many promising preclinical hyperpolarized probes can translate to become a biomarker in routine clinical use.

7.4.2 Challenges to Clinical Translation and Potential Solutions

It is important to remember that dissolution DNP was only invented in 2003 [23]. Within 20 years the first hyperpolarized biomarker to be translated to the clinic, [1-^{13}C]pyruvate, has been shown to work in principle and has entered multiple clini-

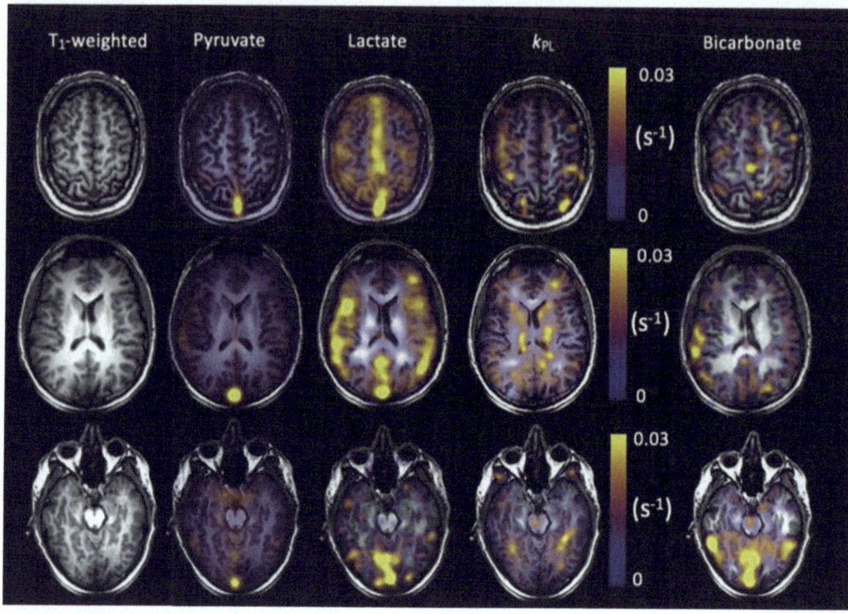

Fig. 7.4 Summed images from the brain of a healthy volunteer demonstrating [1-^{13}C]pyruvate, [1-^{13}C]lactate and [^{13}C]bicarbonate signal from three axial slices. Adapted from Grist et al. 2019 and used under the Creative Commons Attribution License (CC BY) [94]

cal trials at several sites in North America and Europe. Over 60 hyperpolarized imaging agents have been developed, many of which have potential for future clinical translation [96, 97]. This impressive progress has also brought numerous challenges. Future developments to facilitate clinical translation of hyperpolarized MRI can broadly be grouped into three categories; (a) maximize the available signal, (b) reduce the financial cost of the technique, and (c) find a compelling clinical application for the technology.

The development of new polarizers at higher field strengths [98] and lower temperatures can improve polarization levels [99]. Beyond improvements in polarizer technology, there are other improvements that could increase signal-to-noise in clinical experiments. Current quality control (QC) measures delay patient injection by around a minute after dissolution. Polarization decays to thermal equilibrium with the longitudinal relaxation constant T_1. This process is rapid, for example, the T_1 of [1-^{13}C]pyruvate is approximately 45 s in a solution at room temperature and can be significantly less in vivo [100]. Although for most of this time, the sample is within the stray fields of the polarizer and MRI magnets, which reduces the rate of relaxation, this still means that for [1-^{13}C]pyruvate at least one T_1 has elapsed prior to injection, that is, 63% of signal has been lost. If substrates with a shorter T_1 are to be translated, reducing the time between dissolution and injection will be of paramount importance.

Removal of the radical used to polarize the substrate would speed up the dissolution process by eliminating the filtration step and reducing the QC requirements. A number of methods have been proposed to achieve this. Photogenerated radicals can be removed by thermal annihilation at temperatures below the freezing point of the substrate [101, 102]. Hydrophobic radicals can be more easily separated during dissolution [103]. Once the radical has been removed, polarization is long-lasting provided the sample is kept at low temperature (~4 K) and in a magnetic field [103]. This could facilitate centralized production of hyperpolarized probes followed by transport to peripheral clinical sites for dissolution and injection, in a similar fashion to how most radiotracers for PET are produced and distributed.

Further gains in signal should be possible through optimization of the MRI hardware and pulse sequences. The transient nature of hyperpolarization places high demands on hardware and pulse sequences in order to acquire 3D data while minimizing excitation pulses that further deplete the signal. Dual-tuned, parallel-array coils would permit ^{1}H imaging to be carried out without changing the coil and could significantly improve sensitivity and coverage while also reducing the number of required RF pulses. Improved gradients would increase the image resolution and speed of acquisition; the low gyromagnetic ratio of ^{13}C means gradient strengths four times greater than those needed for ^{1}H are required to achieve the same effect. Numerous pulse sequences have been designed based upon magnetic resonance spectroscopic imaging (MRSI), balanced-steady state free precession (bSSFP) [104], echo-planar spectroscopic imaging (EPSI) [105], spectral-spatial EPI [106], spiral [107] and simultaneous auto-calibrating and k-space estimation (SAKE) acquisitions [106, 108]. A range of compressed sensing techniques have also been employed to further maximize spatial and temporal resolution while minimizing the number of RF pulses [109]. An alternative approach is to transfer magnetization from hyperpolarized ^{13}C to adjacent spin-coupled protons and detect hyperpolarization in the proton signal, as was described above for ^{1}H detection of hyperpolarized ^{15}N-labeled urea. This may be particularly advantageous at low field strengths where the longer ^{1}H and ^{13}C T_2 relaxation times will improve the efficiency of polarization transfer [85, 110]. While pulse sequence development is important, collaboration and standardization between different clinical sites will be essential for the field to produce comparable data and move towards multisite clinical trials.

7.5 Summary

MRI of hyperpolarized substrates has provided novel insights into in vivo biology, which is impossible using any other imaging modality. Over 60 hyperpolarized substrates have been used in preclinical studies of cancer, inflammation, ischemia, and cell death. Numerous aspects of dysregulated metabolism have been targeted including glycolysis, the TCA cycle, and redox balance. Other applications include imaging tissue pH and pulmonary function. Uses for hyperpolarized imaging have been proposed for almost every stage of patient care, from screening and diagnosis, staging, and prognostication to monitoring response to treatment. One substrate, [1-^{13}C]pyruvate, has entered clinical trials and initial results have been promising. [1-^{13}C]pyruvate has many of the attributes required for clinical translation—a relatively long T_1, high levels of polarization and a nontoxic, naturally occurring metabolite that is rapidly converted to downstream metabolic products, most notably [1-^{13}C]lactate, which can be detected by increased signal at a different chemical shift.

The main technical barrier to clinical translation is improving the signal-to-noise ratio, where a multifaceted potential solution includes improved hardware, imaging protocols, and pulse sequences. Optimization of the technology will pave the way for the critical question to be answered—can biomarkers derived from imaging with hyperpolarized substrates influence clinical management by providing data that are not available using other modalities? Studies in patient populations over the next decade should go a long way to answering whether the broad range of potential applications of the technique that have been identified in preclinical studies, many of which were discussed here, can be replicated in patients and have a role in routine clinical imaging to guide patient care.

Acknowledgments This work was supported by grants from Cancer Research UK A28289, A17242, and A16465.

Problems

1. [1-^{13}C]pyruvate is the perfect metabolite for dissolution dynamic nuclear polarization. Discuss how true this statement is with regard to the hyperpolarization and clinical applications of [1-^{13}C] pyruvate as an imaging agent.
2. How might DNP imaging agents be used in the management of cancer patients in the future?

References

1. Biomarkers Definitions Working Group: Biomarkers and surrogate endpoints: preferred definitions and conceptual framework. Clin. Pharmacol. Ther. **69**(3), 89–95 (2001). https://doi.org/10.1067/mcp.2001.113989
2. O'Connor, J.P., Aboagye, E.O., Adams, J.E., Aerts, H.J., Barrington, S.F., Beer, A.J., Boellaard, R., Bohndiek, S.E., Brady, M., Brown, G., Buckley, D.L., Chenevert, T.L., Clarke, L.P., Collette, S., Cook, G.J., De Souza, N.M., Dickson, J.C., Dive, C., Evelhoch, J.L., Faivre-Finn, C., Gallagher, F.A., Gilbert, F.J., Gillies, R.J., Goh, V., Griffiths, J.R., Groves, A.M., Halligan, S., Harris, A.L., Hawkes, D.J., Hoekstra, O.S., Huang, E.P., Hutton, B.F., Jackson, E.F., Jayson, G.C., Jones, A., Koh, D.M., Lacombe, D., Lambin, P., Lassau, N., Leach, M.O., Lee, T.Y., Leen, E.L., Lewis, J.S., Liu, Y., Lythgoe, M.F., Manoharan, P., Maxwell, R.J., Miles, K.A., Morgan, B., Morris, S., Ng, T., Padhani, A.R., Parker, G.J., Partridge, M., Pathak, A.P., Peet, A.C., Punwani, S., Reynolds, A.R., Robinson, S.P., Shankar, L.K., Sharma, R.A., Soloviev, D., Stroobants, S., Sullivan, D.C., Taylor, S.A., Tofts, P.S., Tozer, G.M.,

van Herk, M., Walker-Samuel, S., Wason, J., Williams, K.J., Workman, P., Yankeelov, T.E., Brindle, K.M., McShane, L.M., Jackson, A., Waterton, J.C.: Imaging biomarker roadmap for cancer studies. Nat. Rev. Clin. Oncol. **14**(3), 169–186 (2017). https://doi.org/10.1038/nrclinonc.2016.162

3. Mayeux, R.: Biomarkers: potential uses and limitations. NeuroRx. **1**(2), 182–188 (2004). https://doi.org/10.1602/neurorx.1.2.182
4. Hippocrates: English translation: Adams F. The Genuine Works of Hippocrates. On Airs, Waters, and Places., vol. 7, 2 ed. Sydenham Society, London (400 BCE)
5. Cox, F.E.: History of the discovery of the malaria parasites and their vectors. Parasit. Vectors. **3**(1), 5 (2010). https://doi.org/10.1186/1756-3305-3-5
6. Poste, G.: Bring on the biomarkers. Nature. **469**(7329), 156–157 (2011). https://doi.org/10.1038/469156a
7. Majdan, M., Brazinova, A., Rusnak, M., Leitgeb, J.: Outcome prediction after traumatic brain injury: comparison of the performance of routinely used severity scores and multivariable prognostic models. J. Neurosci. Rural. Pract. **8**(1), 20–29 (2017). https://doi.org/10.4103/0976-3147.193543
8. Steyerberg, E.W., Mushkudiani, N., Perel, P., Butcher, I., Lu, J., McHugh, G.S., Murray, G.D., Marmarou, A., Roberts, I., Habbema, J.D., Maas, A.I.: Predicting outcome after traumatic brain injury: development and international validation of prognostic scores based on admission characteristics. PLoS Med. **5**(8), e165 (2008).; discussion e165). https://doi.org/10.1371/journal.pmed.0050165
9. Cole, M.P., Jones, C.T., Todd, I.D.: A new anti-oestrogenic agent in late breast cancer. An early clinical appraisal of ICI46474. Br. J. Cancer. **25**(2), 270–275 (1971). https://doi.org/10.1038/bjc.1971.33
10. Jensen, E.V.: Estrogen receptors in hormone-dependent breast cancers. Cancer Res. **35**(11 Pt. 2), 3362–3364 (1975)
11. Hounsfield, G.N.: Computerized transverse axial scanning (tomography). 1. Description of system. Br. J. Radiol. **46**(552), 1016–1022 (1973). https://doi.org/10.1259/0007-1285-46-552-1016
12. Lauterbur, P.C.: Image formation by induced local interactions: examples employing nuclear magnetic resonance. Nature. **242**(5394), 190–191 (1973). https://doi.org/10.1038/242190a0
13. Avril, N.: GLUT1 expression in tissue and (18)F-FDG uptake. J. Nucl. Med. **45**(6), 930–932 (2004)
14. Hesketh, R.L., Wang, J., Wright, A.J., Lewis, D.Y., Denton, A.E., Grenfell, R., Miller, J.L., Bielik, R., Gehrung, M., Fala, M., Ros, S., Xie, B., Hu, D.E., Brindle, K.M.: Magnetic resonance imaging is more sensitive than PET for detecting treatment-induced cell death-dependent changes in glycolysis. Cancer Res. **79**(14), 3557–3569 (2019). https://doi.org/10.1158/0008-5472.CAN-19-0182
15. Comment, A., Merritt, M.E.: Hyperpolarized magnetic resonance as a sensitive detector of metabolic function. Biochemistry. **53**(47), 7333–7357 (2014). https://doi.org/10.1021/bi501225t
16. Serrao, E.M., Brindle, K.M.: Potential clinical roles for metabolic imaging with hyperpolarized [1-(13)C]pyruvate. Front. Oncol. **6**, 59 (2016). https://doi.org/10.3389/fonc.2016.00059
17. Nelson, S.J., Kurhanewicz, J., Vigneron, D.B., Larson, P.E., Harzstark, A.L., Ferrone, M., van Criekinge, M., Chang, J.W., Bok, R., Park, I., Reed, G., Carvajal, L., Small, E.J., Munster, P., Weinberg, V.K., Ardenkjaer-Larsen, J.H., Chen, A.P., Hurd, R.E., Odegardstuen, L.I., Robb, F.J., Tropp, J., Murray, J.A.: Metabolic imaging of patients with prostate cancer using hyperpolarized [1-(1)(3)C]pyruvate. Sci. Transl. Med. **5**(198), 198ra108 (2013). https://doi.org/10.1126/scitranslmed.3006070
18. Day, S.E., Kettunen, M.I., Gallagher, F.A., Hu, D.E., Lerche, M., Wolber, J., Golman, K., Ardenkjaer-Larsen, J.H., Brindle, K.M.: Detecting tumor response to treatment using hyperpolarized 13C magnetic resonance imaging and spectroscopy. Nat. Med. **13**(11), 1382–1387 (2007). https://doi.org/10.1038/nm1650

19. Takakusagi, Y., Matsumoto, S., Saito, K., Matsuo, M., Kishimoto, S., Wojtkowiak, J.W., DeGraff, W., Kesarwala, A.H., Choudhuri, R., Devasahayam, N., Subramanian, S., Munasinghe, J.P., Gillies, R.J., Mitchell, J.B., Hart, C.P., Krishna, M.C.: Pyruvate induces transient tumor hypoxia by enhancing mitochondrial oxygen consumption and potentiates the anti-tumor effect of a hypoxia-activated prodrug TH-302. PLoS One. **9**(9), e107995 (2014). https://doi.org/10.1371/journal.pone.0107995
20. Timm, K.N., Hu, D.E., Williams, M., Wright, A.J., Kettunen, M.I., Kennedy, B.W., Larkin, T.J., Dzien, P., Marco-Rius, I., Bohndiek, S.E., Brindle, K.M.: Assessing oxidative stress in tumors by measuring the rate of hyperpolarized [1-13C]dehydroascorbic acid reduction using 13C magnetic resonance spectroscopy. J. Biol. Chem. **292**(5), 1737–1748 (2017). https://doi.org/10.1074/jbc.M116.761536
21. Ardenkjaer-Larsen, J.H., Macholl, S., Jóhannesson, H.: Dynamic nuclear polarization with trityls at 1.2 K. Appl. Magn. Reson. **34**(3–4), 509–522 (2008). https://doi.org/10.1007/s00723-008-0134-4
22. Lumata, L., Merritt, M.E., Malloy, C.R., Sherry, A.D., Kovacs, Z.: Impact of Gd3+ on DNP of [1-13C]pyruvate doped with trityl OX063, BDPA, or 4-oxo-TEMPO. J. Phys. Chem. A. **116**(21), 5129–5138 (2012). https://doi.org/10.1021/jp302399f
23. Ardenkjaer-Larsen, J.H., Fridlund, B., Gram, A., Hansson, G., Hansson, L., Lerche, M.H., Servin, R., Thaning, M., Golman, K.: Increase in signal-to-noise ratio of > 10,000 times in liquid-state NMR. Proc. Natl. Acad. Sci. U. S. A. **100**(18), 10158–10163 (2003). https://doi.org/10.1073/pnas.1733835100
24. Johannesson, H., Macholl, S., Ardenkjaer-Larsen, J.H.: Dynamic nuclear polarization of [1-13C]pyruvic acid at 4.6 tesla. J. Magn. Reson. **197**(2), 167–175 (2009). https://doi.org/10.1016/j.jmr.2008.12.016
25. Choi, J.W., Moon, W.J.: Gadolinium deposition in the brain: current updates. Korean J. Radiol. **20**(1), 134–147 (2019). https://doi.org/10.3348/kjr.2018.0356
26. Milani, J., Vuichoud, B., Bornet, A., Mieville, P., Mottier, R., Jannin, S., Bodenhausen, G.: A magnetic tunnel to shelter hyperpolarized fluids. Rev. Sci. Instrum. **86**(2), 024101 (2015). https://doi.org/10.1063/1.4908196
27. Rodrigues, T.B., Serrao, E.M., Kennedy, B.W., Hu, D.E., Kettunen, M.I., Brindle, K.M.: Magnetic resonance imaging of tumor glycolysis using hyperpolarized 13C-labeled glucose. Nat. Med. **20**(1), 93–97 (2014). https://doi.org/10.1038/nm.3416
28. Miloushev, V.Z., Granlund, K.L., Boltyanskiy, R., Lyashchenko, S.K., DeAngelis, L.M., Mellinghoff, I.K., Brennan, C.W., Tabar, V., Yang, T.J., Holodny, A.I., Sosa, R.E., Guo, Y.W., Chen, A.P., Tropp, J., Robb, F., Keshari, K.R.: Metabolic imaging of the human brain with hyperpolarized (13)C pyruvate demonstrates (13)C lactate production in brain tumor patients. Cancer Res. **78**(14), 3755–3760 (2018). https://doi.org/10.1158/0008-5472.CAN-18-0221
29. Gallagher, F.A., Kettunen, M.I., Day, S.E., Lerche, M., Brindle, K.M.: 13C MR spectroscopy measurements of glutaminase activity in human hepatocellular carcinoma cells using hyperpolarized 13C-labeled glutamine. Magn. Reson. Med. **60**(2), 253–257 (2008). https://doi.org/10.1002/mrm.21650
30. Salamanca-Cardona, L., Shah, H., Poot, A.J., Correa, F.M., Di Gialleonardo, V., Lui, H., Miloushev, V.Z., Granlund, K.L., Tee, S.S., Cross, J.R., Thompson, C.B., Keshari, K.R.: In vivo imaging of glutamine metabolism to the oncometabolite 2-hydroxyglutarate in IDH1/2 mutant tumors. Cell Metab. **26**(6), 830–841. e833 (2017). https://doi.org/10.1016/j.cmet.2017.10.001
31. De Feyter, H.M., Behar, K.L., Corbin, Z.A., Fulbright, R.K., Brown, P.B., McIntyre, S., Nixon, T.W., Rothman, D.L., de Graaf, R.A.: Deuterium metabolic imaging (DMI) for MRI-based 3D mapping of metabolism in vivo. Sci. Adv. **4**(8), eaat7314 (2018). https://doi.org/10.1126/sciadv.aat7314
32. Kreis, F., Wright, A.J., Hesse, F., Fala, M., Hu, D.E., Brindle, K.M.: Measuring tumor glycolytic flux in vivo by using fast deuterium MRI. Radiology. **294**, 191242 (2019). https://doi.org/10.1148/radiol.2019191242

33. Witney, T.H., Kettunen, M.I., Brindle, K.M.: Kinetic modeling of hyperpolarized 13C label exchange between pyruvate and lactate in tumor cells. J. Biol. Chem. **286**(28), 24572–24580 (2011). https://doi.org/10.1074/jbc.M111.237727
34. Keshari, K.R., Sriram, R., Koelsch, B.L., Van Criekinge, M., Wilson, D.M., Kurhanewicz, J., Wang, Z.J.: Hyperpolarized 13C-pyruvate magnetic resonance reveals rapid lactate export in metastatic renal cell carcinomas. Cancer Res. **73**(2), 529–538 (2013). https://doi.org/10.1158/0008-5472.CAN-12-3461
35. Witney, T.H., Kettunen, M.I., Day, S.E., Hu, D.E., Neves, A.A., Gallagher, F.A., Fulton, S.M., Brindle, K.M.: A comparison between radiolabeled fluorodeoxyglucose uptake and hyperpolarized (13)C-labeled pyruvate utilization as methods for detecting tumor response to treatment. Neoplasia. **11**(6), 574–582, 571 p following 582 (2009). https://doi.org/10.1593/neo.09254
36. Golman, K., Zandt, R.I., Lerche, M., Pehrson, R., Ardenkjaer-Larsen, J.H.: Metabolic imaging by hyperpolarized 13C magnetic resonance imaging for in vivo tumor diagnosis. Cancer Res. **66**(22), 10855–10860 (2006). https://doi.org/10.1158/0008-5472.CAN-06-2564
37. Ahmed, H.U., El-Shater Bosaily, A., Brown, L.C., Gabe, R., Kaplan, R., Parmar, M.K., Collaco-Moraes, Y., Ward, K., Hindley, R.G., Freeman, A., Kirkham, A.P., Oldroyd, R., Parker, C., Emberton, M., Group, P.S.: Diagnostic accuracy of multi-parametric MRI and TRUS biopsy in prostate cancer (PROMIS): a paired validating confirmatory study. Lancet. **389**(10071), 815–822 (2017). https://doi.org/10.1016/S0140-6736(16)32401-1
38. Demirci, E., Kabasakal, L., Sahin, O.E., Akgun, E., Gultekin, M.H., Doganca, T., Tuna, M.B., Obek, C., Kilic, M., Esen, T., Kural, A.R.: Can SUVmax values of Ga-68-PSMA PET/CT scan predict the clinically significant prostate cancer? Nucl. Med. Commun. **40**(1), 86–91 (2019). https://doi.org/10.1097/MNM.0000000000000942
39. Johnston, E.W., Bonet-Carne, E., Ferizi, U., Yvernault, B., Pye, H., Patel, D., Clemente, J., Piga, W., Heavey, S., Sidhu, H.S., Giganti, F., O'Callaghan, J., Brizmohun Appayya, M., Grey, A., Saborowska, A., Ourselin, S., Hawkes, D., Moore, C.M., Emberton, M., Ahmed, H.U., Whitaker, H., Rodriguez-Justo, M., Freeman, A., Atkinson, D., Alexander, D., Panagiotaki, E., Punwani, S.: VERDICT MRI for prostate cancer: intracellular volume fraction versus apparent diffusion coefficient. Radiology. **291**(2), 391–397 (2019). https://doi.org/10.1148/radiol.2019181749
40. Chen, H.Y., Larson, P.E.Z., Bok, R.A., von Morze, C., Sriram, R., Delos Santos, R., Delos Santos, J., Gordon, J.W., Bahrami, N., Ferrone, M., Kurhanewicz, J., Vigneron, D.B.: Assessing prostate Cancer aggressiveness with hyperpolarized dual-agent 3D dynamic imaging of metabolism and perfusion. Cancer Res. **77**(12), 3207–3216 (2017). https://doi.org/10.1158/0008-5472.CAN-16-2083
41. Granlund, K.L., Tee, S.S., Vargas, H.A., Lyashchenko, S.K., Reznik, E., Fine, S., Laudone, V., Eastham, J.A., Touijer, K.A., Reuter, V.E., Gonen, M., Sosa, R.E., Nicholson, D., Guo, Y.W., Chen, A.P., Tropp, J., Robb, F., Hricak, H., Keshari, K.R.: Hyperpolarized MRI of human prostate Cancer reveals increased lactate with tumor grade driven by Monocarboxylate transporter 1. Cell Metab. **31**, 105 (2019). https://doi.org/10.1016/j.cmet.2019.08.024
42. Eisenhauer, E.A., Therasse, P., Bogaerts, J., Schwartz, L.H., Sargent, D., Ford, R., Dancey, J., Arbuck, S., Gwyther, S., Mooney, M., Rubinstein, L., Shankar, L., Dodd, L., Kaplan, R., Lacombe, D., Verweij, J.: New response evaluation criteria in solid tumours: revised RECIST guideline (version 1.1). Eur. J. Cancer. **45**(2), 228–247 (2009). https://doi.org/10.1016/j.ejca.2008.10.026
43. Hodi, F.S., Hwu, W.J., Kefford, R., Weber, J.S., Daud, A., Hamid, O., Patnaik, A., Ribas, A., Robert, C., Gangadhar, T.C., Joshua, A.M., Hersey, P., Dronca, R., Joseph, R., Hille, D., Xue, D., Li, X.N., Kang, S.P., Ebbinghaus, S., Perrone, A., Wolchok, J.D.: Evaluation of immune-related response criteria and RECIST v1.1 in patients with advanced melanoma treated with pembrolizumab. J. Clin. Oncol. **34**(13), 1510–1517 (2016). https://doi.org/10.1200/JCO.2015.64.0391

44. Frelaut, M., Le Tourneau, C., Borcoman, E.: Hyperprogression under immunotherapy. Int. J. Mol. Sci. **20**(11), 2674 (2019). https://doi.org/10.3390/ijms20112674
45. Witney, T.H., Kettunen, M.I., Hu, D.E., Gallagher, F.A., Bohndiek, S.E., Napolitano, R., Brindle, K.M.: Detecting treatment response in a model of human breast adenocarcinoma using hyperpolarised [1-13C]pyruvate and [1,4-13C2]fumarate. Br. J. Cancer. **103**(9), 1400–1406 (2010). https://doi.org/10.1038/sj.bjc.6605945
46. Aggarwal, R., Vigneron, D.B., Kurhanewicz, J.: Hyperpolarized 1-[(13)C]-pyruvate magnetic resonance imaging detects an early metabolic response to androgen ablation therapy in prostate Cancer. Eur. Urol. **72**(6), 1028–1029 (2017). https://doi.org/10.1016/j.eururo.2017.07.022
47. Gutte, H., Hansen, A.E., Henriksen, S.T., Johannesen, H.H., Ardenkjaer-Larsen, J., Vignaud, A., Hansen, A.E., Borresen, B., Klausen, T.L., Wittekind, A.M., Gillings, N., Kristensen, A.T., Clemmensen, A., Hojgaard, L., Kjaer, A.: Simultaneous hyperpolarized (13)C-pyruvate MRI and (18)F-FDG-PET in cancer (hyperPET): feasibility of a new imaging concept using a clinical PET/MRI scanner. Am. J. Nucl. Med. Mol. Imaging. **5**(1), 38–45 (2015)
48. Lander, E.S., Linton, L.M., Birren, B., Nusbaum, C., Zody, M.C., Baldwin, J., Devon, K., Dewar, K., Doyle, M., FitzHugh, W., Funke, R., Gage, D., Harris, K., Heaford, A., Howland, J., Kann, L., Lehoczky, J., LeVine, R., McEwan, P., McKernan, K., Meldrim, J., Mesirov, J.P., Miranda, C., Morris, W., Naylor, J., Raymond, C., Rosetti, M., Santos, R., Sheridan, A., Sougnez, C., Stange-Thomann, Y., Stojanovic, N., Subramanian, A., Wyman, D., Rogers, J., Sulston, J., Ainscough, R., Beck, S., Bentley, D., Burton, J., Clee, C., Carter, N., Coulson, A., Deadman, R., Deloukas, P., Dunham, A., Dunham, I., Durbin, R., French, L., Grafham, D., Gregory, S., Hubbard, T., Humphray, S., Hunt, A., Jones, M., Lloyd, C., McMurray, A., Matthews, L., Mercer, S., Milne, S., Mullikin, J.C., Mungall, A., Plumb, R., Ross, M., Shownkeen, R., Sims, S., Waterston, R.H., Wilson, R.K., Hillier, L.W., McPherson, J.D., Marra, M.A., Mardis, E.R., Fulton, L.A., Chinwalla, A.T., Pepin, K.H., Gish, W.R., Chissoe, S.L., Wendl, M.C., Delehaunty, K.D., Miner, T.L., Delehaunty, A., Kramer, J.B., Cook, L.L., Fulton, R.S., Johnson, D.L., Minx, P.J., Clifton, S.W., Hawkins, T., Branscomb, E., Predki, P., Richardson, P., Wenning, S., Slezak, T., Doggett, N., Cheng, J.F., Olsen, A., Lucas, S., Elkin, C., Uberbacher, E., Frazier, M., Gibbs, R.A., Muzny, D.M., Scherer, S.E., Bouck, J.B., Sodergren, E.J., Worley, K.C., Rives, C.M., Gorrell, J.H., Metzker, M.L., Naylor, S.L., Kucherlapati, R.S., Nelson, D.L., Weinstock, G.M., Sakaki, Y., Fujiyama, A., Hattori, M., Yada, T., Toyoda, A., Itoh, T., Kawagoe, C., Watanabe, H., Totoki, Y., Taylor, T., Weissenbach, J., Heilig, R., Saurin, W., Artiguenave, F., Brottier, P., Bruls, T., Pelletier, E., Robert, C., Wincker, P., Smith, D.R., Doucette-Stamm, L., Rubenfield, M., Weinstock, K., Lee, H.M., Dubois, J., Rosenthal, A., Platzer, M., Nyakatura, G., Taudien, S., Rump, A., Yang, H., Yu, J., Wang, J., Huang, G., Gu, J., Hood, L., Rowen, L., Madan, A., Qin, S., Davis, R.W., Federspiel, N.A., Abola, A.P., Proctor, M.J., Myers, R.M., Schmutz, J., Dickson, M., Grimwood, J., Cox, D.R., Olson, M.V., Kaul, R., Raymond, C., Shimizu, N., Kawasaki, K., Minoshima, S., Evans, G.A., Athanasiou, M., Schultz, R., Roe, B.A., Chen, F., Pan, H., Ramser, J., Lehrach, H., Reinhardt, R., McCombie, W.R., de la Bastide, M., Dedhia, N., Blocker, H., Hornischer, K., Nordsiek, G., Agarwala, R., Aravind, L., Bailey, J.A., Bateman, A., Batzoglou, S., Birney, E., Bork, P., Brown, D.G., Burge, C.B., Cerutti, L., Chen, H.C., Church, D., Clamp, M., Copley, R.R., Doerks, T., Eddy, S.R., Eichler, E.E., Furey, T.S., Galagan, J., Gilbert, J.G., Harmon, C., Hayashizaki, Y., Haussler, D., Hermjakob, H., Hokamp, K., Jang, W., Johnson, L.S., Jones, T.A., Kasif, S., Kaspryzk, A., Kennedy, S., Kent, W.J., Kitts, P., Koonin, E.V., Korf, I., Kulp, D., Lancet, D., Lowe, T.M., McLysaght, A., Mikkelsen, T., Moran, J.V., Mulder, N., Pollara, V.J., Ponting, C.P., Schuler, G., Schultz, J., Slater, G., Smit, A.F., Stupka, E., Szustakowki, J., Thierry-Mieg, D., Thierry-Mieg, J., Wagner, L., Wallis, J., Wheeler, R., Williams, A., Wolf, Y.I., Wolfe, K.H., Yang, S.P., Yeh, R.F., Collins, F., Guyer, M.S., Peterson, J., Felsenfeld, A., Wetterstrand, K.A., Patrinos, A., Morgan, M.J., de Jong, P., Catanese, J.J., Osoegawa, K., Shizuya, H., Choi, S., Chen, Y.J., Szustakowki, J.: International human genome sequencing, C.: initial sequencing and analysis of the human genome. Nature. **409**(6822), 860–921 (2001). https://doi.org/10.1038/35057062

49. Onitilo, A.A., Engel, J.M., Greenlee, R.T., Mukesh, B.N.: Breast cancer subtypes based on ER/PR and Her2 expression: comparison of clinicopathologic features and survival. Clin. Med. Res. **7**(1–2), 4–13 (2009). https://doi.org/10.3121/cmr.2009.825
50. Mukherjee, A., Russell, R., Chin, S.F., Liu, B., Rueda, O.M., Ali, H.R., Turashvili, G., Mahler-Araujo, B., Ellis, I.O., Aparicio, S., Caldas, C., Provenzano, E.: Associations between genomic stratification of breast cancer and centrally reviewed tumour pathology in the METABRIC cohort. NPJ Breast Cancer. **4**, 5 (2018). https://doi.org/10.1038/s41523-018-0056-8
51. Angus, L., Smid, M., Wilting, S.M., van Riet, J., Van Hoeck, A., Nguyen, L., Nik-Zainal, S., Steenbruggen, T.G., Tjan-Heijnen, V.C.G., Labots, M., van Riel, J., Bloemendal, H.J., Steeghs, N., Lolkema, M.P., Voest, E.E., van de Werken, H.J.G., Jager, A., Cuppen, E., Sleijfer, S., Martens, J.W.M.: The genomic landscape of metastatic breast cancer highlights changes in mutation and signature frequencies. Nat. Genet. **51**(10), 1450–1458 (2019). https://doi.org/10.1038/s41588-019-0507-7
52. Aktas, B., Muller, V., Tewes, M., Zeitz, J., Kasimir-Bauer, S., Loehberg, C.R., Rack, B., Schneeweiss, A., Fehm, T.: Comparison of estrogen and progesterone receptor status of circulating tumor cells and the primary tumor in metastatic breast cancer patients. Gynecol. Oncol. **122**(2), 356–360 (2011). https://doi.org/10.1016/j.ygyno.2011.04.039
53. Cohen, A.L., Holmen, S.L., Colman, H.: IDH1 and IDH2 mutations in gliomas. Curr. Neurol. Neurosci. Rep. **13**(5), 345 (2013). https://doi.org/10.1007/s11910-013-0345-4
54. Izquierdo-Garcia, J.L., Viswanath, P., Eriksson, P., Cai, L., Radoul, M., Chaumeil, M.M., Blough, M., Luchman, H.A., Weiss, S., Cairncross, J.G., Phillips, J.J., Pieper, R.O., Ronen, S.M.: IDH1 mutation induces reprogramming of pyruvate metabolism. Cancer Res. **75**(15), 2999–3009 (2015). https://doi.org/10.1158/0008-5472.CAN-15-0840
55. Chaumeil, M.M., Larson, P.E., Yoshihara, H.A., Danforth, O.M., Vigneron, D.B., Nelson, S.J., Pieper, R.O., Phillips, J.J., Ronen, S.M.: Non-invasive in vivo assessment of IDH1 mutational status in glioma. Nat. Commun. **4**, 2429 (2013). https://doi.org/10.1038/ncomms3429
56. MacKenzie, J.D., Yen, Y.F., Mayer, D., Tropp, J.S., Hurd, R.E., Spielman, D.M.: Detection of inflammatory arthritis by using hyperpolarized 13C-pyruvate with MR imaging and spectroscopy. Radiology. **259**(2), 414–420 (2011). https://doi.org/10.1148/radiol.10101921
57. Wright, A.J., Husson, Z.M.A., Hu, D.E., Callejo, G., Brindle, K.M., Smith, E.S.J.: Increased hyperpolarized [1-(13) C] lactate production in a model of joint inflammation is not accompanied by tissue acidosis as assessed using hyperpolarized (13) C-labelled bicarbonate. NMR Biomed. **31**(4), e3892 (2018). https://doi.org/10.1002/nbm.3892
58. Guglielmetti, C., Najac, C., Didonna, A., Van der Linden, A., Ronen, S.M., Chaumeil, M.M.: Hyperpolarized (13)C MR metabolic imaging can detect neuroinflammation in vivo in a multiple sclerosis murine model. Proc. Natl. Acad. Sci. U. S. A. **114**(33), E6982–E6991 (2017). https://doi.org/10.1073/pnas.1613345114
59. Siddiqui, S., Habertheuer, A., Xin, Y., Pourfathi, M., Tao, J.Q., Hamedani, H., Kadlecek, S., Duncan, I., Vallabhajosyula, P., Naji, A., Chatterjee, S., Rizi, R.: Detection of lung transplant rejection in a rat model using hyperpolarized [1-(13) C] pyruvate-based metabolic imaging. NMR Biomed. **32**(8), e4107 (2019). https://doi.org/10.1002/nbm.4107
60. Najac, C., Chaumeil, M.M., Kohanbash, G., Guglielmetti, C., Gordon, J.W., Okada, H., Ronen, S.M.: Detection of inflammatory cell function using (13)C magnetic resonance spectroscopy of hyperpolarized [6-(13)C]-arginine. Sci. Rep. **6**, 31397 (2016). https://doi.org/10.1038/srep31397
61. Cunningham, C.H., Lau, J.Y., Chen, A.P., Geraghty, B.J., Perks, W.J., Roifman, I., Wright, G.A., Connelly, K.A.: Hyperpolarized 13C metabolic MRI of the human heart: initial experience. Circ. Res. **119**(11), 1177–1182 (2016). https://doi.org/10.1161/CIRCRESAHA.116.309769
62. Ball, D.R., Cruickshank, R., Carr, C.A., Stuckey, D.J., Lee, P., Clarke, K., Tyler, D.J.: Metabolic imaging of acute and chronic infarction in the perfused rat heart using hyperpolarised [1-13C]pyruvate. NMR Biomed. **26**(11), 1441–1450 (2013). https://doi.org/10.1002/nbm.2972

63. Agger, P., Hyldebrandt, J.A., Hansen, E.S.S., Omann, C., Bogh, N., Waziri, F., Nielsen, P.M., Laustsen, C.: Magnetic resonance hyperpolarization imaging detects early myocardial dysfunction in a porcine model of right ventricular heart failure. Eur. Heart J. Cardiovasc. Imaging. **21**, 93 (2019). https://doi.org/10.1093/ehjci/jez074
64. Gallagher, F.A., Kettunen, M.I., Hu, D.E., Jensen, P.R., Zandt, R.I., Karlsson, M., Gisselsson, A., Nelson, S.K., Witney, T.H., Bohndiek, S.E., Hansson, G., Peitersen, T., Lerche, M.H., Brindle, K.M.: Production of hyperpolarized [1,4-13C2]malate from [1,4-13C2]fumarate is a marker of cell necrosis and treatment response in tumors. Proc. Natl. Acad. Sci. U. S. A. **106**(47), 19801–19806 (2009). https://doi.org/10.1073/pnas.0911447106
65. Clatworthy, M.R., Kettunen, M.I., Hu, D.E., Mathews, R.J., Witney, T.H., Kennedy, B.W., Bohndiek, S.E., Gallagher, F.A., Jarvis, L.B., Smith, K.G., Brindle, K.M.: Magnetic resonance imaging with hyperpolarized [1,4-(13)C2]fumarate allows detection of early renal acute tubular necrosis. Proc. Natl. Acad. Sci. U. S. A. **109**(33), 13374–13379 (2012). https://doi.org/10.1073/pnas.1205539109
66. Eldirdiri, A., Clemmensen, A., Bowen, S., Kjaer, A., Ardenkjaer-Larsen, J.H.: Simultaneous imaging of hyperpolarized [1,4-(13) C2]fumarate, [1-(13) C]pyruvate and (18) F-FDG in a rat model of necrosis in a clinical PET/MR scanner. NMR Biomed. **30**(12), e3803 (2017). https://doi.org/10.1002/nbm.3803
67. Wilson, D.M., Keshari, K.R., Larson, P.E., Chen, A.P., Hu, S., Van Criekinge, M., Bok, R., Nelson, S.J., Macdonald, J.M., Vigneron, D.B., Kurhanewicz, J.: Multi-compound polarization by DNP allows simultaneous assessment of multiple enzymatic activities in vivo. J. Magn. Reson. **205**(1), 141–147 (2010). https://doi.org/10.1016/j.jmr.2010.04.012
68. Scholz, D.J., Janich, M.A., Kollisch, U., Schulte, R.F., Ardenkjaer-Larsen, J.H., Frank, A., Haase, A., Schwaiger, M., Menzel, M.I.: Quantified pH imaging with hyperpolarized (13) C-bicarbonate. Magn. Reson. Med. **73**(6), 2274–2282 (2015). https://doi.org/10.1002/mrm.25357
69. Gallagher, F.A., Kettunen, M.I., Day, S.E., Hu, D.E., Ardenkjaer-Larsen, J.H., Zandt, R., Jensen, P.R., Karlsson, M., Golman, K., Lerche, M.H., Brindle, K.M.: Magnetic resonance imaging of pH in vivo using hyperpolarized 13C-labelled bicarbonate. Nature. **453**(7197), 940–943 (2008). https://doi.org/10.1038/nature07017
70. Gallagher, F.A., Sladen, H., Kettunen, M.I., Serrao, E.M., Rodrigues, T.B., Wright, A., Gill, A.B., McGuire, S., Booth, T.C., Boren, J., McIntyre, A., Miller, J.L., Lee, S.H., Honess, D., Day, S.E., Hu, D.E., Howat, W.J., Harris, A.L., Brindle, K.M.: Carbonic anhydrase activity monitored in vivo by hyperpolarized 13C-magnetic resonance spectroscopy demonstrates its importance for pH regulation in tumors. Cancer Res. **75**(19), 4109–4118 (2015). https://doi.org/10.1158/0008-5472.CAN-15-0857
71. Gallagher, F.A., Kettunen, M.I., Brindle, K.M.: Imaging pH with hyperpolarized 13C. NMR Biomed. **24**(8), 1006–1015 (2011). https://doi.org/10.1002/nbm.1742
72. Jiang, W., Lumata, L., Chen, W., Zhang, S., Kovacs, Z., Sherry, A.D., Khemtong, C.: Hyperpolarized 15N-pyridine derivatives as pH-sensitive MRI agents. Sci. Rep. **5**, 9104 (2015). https://doi.org/10.1038/srep09104
73. Shchepin, R.V., Barskiy, D.A., Coffey, A.M., Theis, T., Shi, F., Warren, W.S., Goodson, B.M., Chekmenev, E.Y.: (15)N hyperpolarization of imidazole-(15)N2 for magnetic resonance pH sensing via SABRE-SHEATH. ACS Sens. **1**(6), 640–644 (2016). https://doi.org/10.1021/acssensors.6b00231
74. Duwel, S., Hundshammer, C., Gersch, M., Feuerecker, B., Steiger, K., Buck, A., Walch, A., Haase, A., Glaser, S.J., Schwaiger, M., Schilling, F.: Imaging of pH in vivo using hyperpolarized (13)C-labelled zymonic acid. Nat. Commun. **8**, 15126 (2017). https://doi.org/10.1038/ncomms15126
75. Hundshammer, C., Duwel, S., Kocher, S.S., Gersch, M., Feuerecker, B., Scheurer, C., Haase, A., Glaser, S.J., Schwaiger, M., Schilling, F.: Deuteration of hyperpolarized (13) C-labeled Zymonic acid enables sensitivity-enhanced dynamic MRI of pH. ChemPhysChem. **18**(18), 2422–2425 (2017). https://doi.org/10.1002/cphc.201700779

76. Keshari, K.R., Kurhanewicz, J., Bok, R., Larson, P.E., Vigneron, D.B., Wilson, D.M.: Hyperpolarized 13C dehydroascorbate as an endogenous redox sensor for in vivo metabolic imaging. Proc. Natl. Acad. Sci. U. S. A. **108**(46), 18606–18611 (2011). https://doi.org/10.1073/pnas.1106920108
77. Bohndiek, S.E., Kettunen, M.I., Hu, D.E., Kennedy, B.W., Boren, J., Gallagher, F.A., Brindle, K.M.: Hyperpolarized [1-13C]-ascorbic and dehydroascorbic acid: vitamin C as a probe for imaging redox status in vivo. J. Am. Chem. Soc. **133**(30), 11795–11801 (2011). https://doi.org/10.1021/ja2045925
78. von Morze, C., Ohliger, M.A., Marco-Rius, I., Wilson, D.M., Flavell, R.R., Pearce, D., Vigneron, D.B., Kurhanewicz, J., Wang, Z.J.: Direct assessment of renal mitochondrial redox state using hyperpolarized (13) C-acetoacetate. Magn. Reson. Med. **79**(4), 1862–1869 (2018). https://doi.org/10.1002/mrm.27054
79. Kanda, T., Fukusato, T., Matsuda, M., Toyoda, K., Oba, H., Kotoku, J., Haruyama, T., Kitajima, K., Furui, S.: Gadolinium-based contrast agent accumulates in the brain even in subjects without severe renal dysfunction: evaluation of autopsy brain specimens with inductively coupled plasma mass spectroscopy. Radiology. **276**(1), 228–232 (2015). https://doi.org/10.1148/radiol.2015142690
80. Guo, B.J., Yang, Z.L., Zhang, L.J.: Gadolinium deposition in brain: current scientific evidence and future perspectives. Front. Mol. Neurosci. **11**, 335 (2018). https://doi.org/10.3389/fnmol.2018.00335
81. von Morze, C., Larson, P.E., Hu, S., Yoshihara, H.A., Bok, R.A., Goga, A., Ardenkjaer-Larsen, J.H., Vigneron, D.B.: Investigating tumor perfusion and metabolism using multiple hyperpolarized (13)C compounds: HP001, pyruvate and urea. Magn. Reson. Imaging. **30**(3), 305–311 (2012). https://doi.org/10.1016/j.mri.2011.09.026
82. Mackenzie, J., Chacko, B.: An isolated elevation in blood urea level is not 'uraemia' and not an indication for renal replacement therapy in the ICU. Crit. Care. **21**(1), 275 (2017). https://doi.org/10.1186/s13054-017-1868-x
83. Lau, A.Z., Miller, J.J., Robson, M.D., Tyler, D.J.: Simultaneous assessment of cardiac metabolism and perfusion using copolarized [1-(13) C]pyruvate and (13) C-urea. Magn. Reson. Med. **77**(1), 151–158 (2017). https://doi.org/10.1002/mrm.26106
84. Reed, G.D., von Morze, C., Bok, R., Koelsch, B.L., Van Criekinge, M., Smith, K.J., Hong, S., Larson, P.E., Kurhanewicz, J., Vigneron, D.B.: High resolution (13)C MRI with hyperpolarized urea: in vivo T(2) mapping and (15)N labeling effects. IEEE Trans. Med. Imaging. **33**(2), 362–371 (2014). https://doi.org/10.1109/TMI.2013.2285120
85. Kreis, F., Wright, A.J., Somai, V., Katz-Brull, R., Brindle, K.M.: Increasing the sensitivity of hyperpolarized [(15) N2]urea detection by serial transfer of polarization to spin-coupled protons. Magn. Reson. Med. **84**, 1844 (2020). https://doi.org/10.1002/mrm.28241
86. Biederer, J., Mirsadraee, S., Beer, M., Molinari, F., Hintze, C., Bauman, G., Both, M., Van Beek, E.J., Wild, J., Puderbach, M.: MRI of the lung (3/3)-current applications and future perspectives. Insights Imaging. **3**(4), 373–386 (2012). https://doi.org/10.1007/s13244-011-0142-z
87. Pourfathi, M., Clapp, J., Kadlecek, S.J., Keenan, C.D., Ghosh, R.K., Kuzma, N.N., Rizi, R.R.: Low-temperature dynamic nuclear polarization of gases in frozen mixtures. Magn. Reson. Med. **76**(3), 1007–1014 (2016). https://doi.org/10.1002/mrm.26002
88. Albert, M.S., Cates, G.D., Driehuys, B., Happer, W., Saam, B., Springer Jr., C.S., Wishnia, A.: Biological magnetic resonance imaging using laser-polarized 129Xe. Nature. **370**(6486), 199–201 (1994). https://doi.org/10.1038/370199a0
89. Roos, J.E., McAdams, H.P., Kaushik, S.S., Driehuys, B.: Hyperpolarized gas MR imaging: technique and applications. Magn. Reson. Imaging Clin. N. Am. **23**(2), 217–229 (2015). https://doi.org/10.1016/j.mric.2015.01.003
90. Kern, A.L., Vogel-Claussen, J.: Hyperpolarized gas MRI in pulmonology. Br. J. Radiol. **91**(1084), 20170647 (2018). https://doi.org/10.1259/bjr.20170647

91. van Beek, E.J., Dahmen, A.M., Stavngaard, T., Gast, K.K., Heussel, C.P., Krummenauer, F., Schmiedeskamp, J., Wild, J.M., Sogaard, L.V., Morbach, A.E., Schreiber, L.M., Kauczor, H.U.: Hyperpolarised 3He MRI versus HRCT in COPD and normal volunteers: PHIL trial. Eur. Respir. J. **34**(6), 1311–1321 (2009). https://doi.org/10.1183/09031936.00138508
92. Doganay, O., Matin, T., Chen, M., Kim, M., McIntyre, A., McGowan, D.R., Bradley, K.M., Povey, T., Gleeson, F.V.: Time-series hyperpolarized xenon-129 MRI of lobar lung ventilation of COPD in comparison to V/Q-SPECT/CT and CT. Eur. Radiol. **29**(8), 4058–4067 (2019). https://doi.org/10.1007/s00330-018-5888-y
93. Wang, J.M., Robertson, S.H., Wang, Z., He, M., Virgincar, R.S., Schrank, G.M., Smigla, R.M., O'Riordan, T.G., Sundy, J., Ebner, L., Rackley, C.R., McAdams, P., Driehuys, B.: Using hyperpolarized (129)Xe MRI to quantify regional gas transfer in idiopathic pulmonary fibrosis. Thorax. **73**(1), 21–28 (2018). https://doi.org/10.1136/thoraxjnl-2017-210070
94. Grist, J.T., McLean, M.A., Riemer, F., Schulte, R.F., Deen, S.S., Zaccagna, F., Woitek, R., Daniels, C.J., Kaggie, J.D., Matys, T., Patterson, I., Slough, R., Gill, A.B., Chhabra, A., Eichenberger, R., Laurent, M.C., Comment, A., Gillard, J.H., Coles, A.J., Tyler, D.J., Wilkinson, I., Basu, B., Lomas, D.J., Graves, M.J., Brindle, K.M., Gallagher, F.A.: Quantifying normal human brain metabolism using hyperpolarized [1-(13)C]pyruvate and magnetic resonance imaging. NeuroImage. **189**, 171–179 (2019). https://doi.org/10.1016/j.neuroimage.2019.01.027
95. Chung, B.T., Chen, H.Y., Gordon, J., Mammoli, D., Sriram, R., Autry, A.W., Le Page, L.M., Chaumeil, M.M., Shin, P., Slater, J., Tan, C.T., Suszczynski, C., Chang, S., Li, Y., Bok, R.A., Ronen, S.M., Larson, P.E.Z., Kurhanewicz, J., Vigneron, D.B.: First hyperpolarized [2-(13)C]pyruvate MR studies of human brain metabolism. J. Magn. Reson. **309**, 106617 (2019). https://doi.org/10.1016/j.jmr.2019.106617
96. Kurhanewicz, J., Vigneron, D.B., Ardenkjaer-Larsen, J.H., Bankson, J.A., Brindle, K., Cunningham, C.H., Gallagher, F.A., Keshari, K.R., Kjaer, A., Laustsen, C., Mankoff, D.A., Merritt, M.E., Nelson, S.J., Pauly, J.M., Lee, P., Ronen, S., Tyler, D.J., Rajan, S.S., Spielman, D.M., Wald, L., Zhang, X., Malloy, C.R., Rizi, R.: Hyperpolarized (13)C MRI: path to clinical translation in oncology. Neoplasia. **21**(1), 1–16 (2019). https://doi.org/10.1016/j.neo.2018.09.006
97. Keshari, K.R., Wilson, D.M.: Chemistry and biochemistry of 13C hyperpolarized magnetic resonance using dynamic nuclear polarization. Chem. Soc. Rev. **43**(5), 1627–1659 (2014). https://doi.org/10.1039/c3cs60124b
98. Capozzi, A., Patel, S., Wenckebach, W.T., Karlsson, M., Lerche, M.H., Ardenkjaer-Larsen, J.H.: Gadolinium effect at high-magnetic-field DNP: 70% (13)C polarization of [U-(13)C] glucose using Trityl. J. Phys. Chem. Lett. **10**(12), 3420–3425 (2019). https://doi.org/10.1021/acs.jpclett.9b01306
99. Cheng, T., Gaunt, A.P., Marco-Rius, I., Gehrung, M., Chen, A.P., van der Klink, J.J., Comment, A.: A multisample 7 T dynamic nuclear polarization polarizer for preclinical hyperpolarized MR. NMR Biomed. **33**, e4264 (2020). https://doi.org/10.1002/nbm.4264
100. Chattergoon, N., Martinez-Santiesteban, F., Handler, W.B., Ardenkjaer-Larsen, J.H., Scholl, T.J.: Field dependence of T1 for hyperpolarized [1-13C]pyruvate. Contrast Media Mol. Imaging. **8**(1), 57–62 (2013). https://doi.org/10.1002/cmmi.1494
101. Eichhorn, T.R., Takado, Y., Salameh, N., Capozzi, A., Cheng, T., Hyacinthe, J.N., Mishkovsky, M., Roussel, C., Comment, A.: Hyperpolarization without persistent radicals for in vivo real-time metabolic imaging. Proc. Natl. Acad. Sci. U. S. A. **110**(45), 18064–18069 (2013). https://doi.org/10.1073/pnas.1314928110
102. Marco-Rius, I., Cheng, T., Gaunt, A.P., Patel, S., Kreis, F., Capozzi, A., Wright, A.J., Brindle, K.M., Ouari, O., Comment, A.: Photogenerated radical in Phenylglyoxylic acid for in vivo hyperpolarized (13)C MR with photosensitive metabolic substrates. J. Am. Chem. Soc. **140**(43), 14455–14463 (2018). https://doi.org/10.1021/jacs.8b09326

103. Ji, X., Bornet, A., Vuichoud, B., Milani, J., Gajan, D., Rossini, A.J., Emsley, L., Bodenhausen, G., Jannin, S.: Transportable hyperpolarized metabolites. Nat. Commun. **8**, 13975 (2017). https://doi.org/10.1038/ncomms13975
104. von Morze, C., Sukumar, S., Reed, G.D., Larson, P.E., Bok, R.A., Kurhanewicz, J., Vigneron, D.B.: Frequency-specific SSFP for hyperpolarized (1)(3)C metabolic imaging at 14.1 T. Magn. Reson. Imaging. **31**(2), 163–170 (2013). https://doi.org/10.1016/j.mri.2012.06.037
105. Chen, H.Y., Larson, P.E.Z., Gordon, J.W., Bok, R.A., Ferrone, M., van Criekinge, M., Carvajal, L., Cao, P., Pauly, J.M., Kerr, A.B., Park, I., Slater, J.B., Nelson, S.J., Munster, P.N., Aggarwal, R., Kurhanewicz, J., Vigneron, D.B.: Technique development of 3D dynamic CS-EPSI for hyperpolarized (13) C pyruvate MR molecular imaging of human prostate cancer. Magn. Reson. Med. **80**(5), 2062–2072 (2018). https://doi.org/10.1002/mrm.27179
106. Gordon, J.W., Hansen, R.B., Shin, P.J., Feng, Y., Vigneron, D.B., Larson, P.E.Z.: 3D hyperpolarized C-13 EPI with calibrationless parallel imaging. J. Magn. Reson. **289**, 92–99 (2018). https://doi.org/10.1016/j.jmr.2018.02.011
107. Wang, J., Hesketh, R.L., Wright, A.J., Brindle, K.M.: Hyperpolarized (13) C spectroscopic imaging using single-shot 3D sequences with unpaired adiabatic refocusing pulses. NMR Biomed. **31**(11), e4004 (2018). https://doi.org/10.1002/nbm.4004
108. Shin, P.J., Larson, P.E., Ohliger, M.A., Elad, M., Pauly, J.M., Vigneron, D.B., Lustig, M.: Calibrationless parallel imaging reconstruction based on structured low-rank matrix completion. Magn. Reson. Med. **72**(4), 959–970 (2014). https://doi.org/10.1002/mrm.24997
109. Hu, S., Lustig, M., Chen, A.P., Crane, J., Kerr, A., Kelley, D.A., Hurd, R., Kurhanewicz, J., Nelson, S.J., Pauly, J.M., Vigneron, D.B.: Compressed sensing for resolution enhancement of hyperpolarized 13C flyback 3D-MRSI. J. Magn. Reson. **192**(2), 258–264 (2008). https://doi.org/10.1016/j.jmr.2008.03.003
110. Wang, J., Kreis, F., Wright, A.J., Hesketh, R.L., Levitt, M.H., Brindle, K.M.: Dynamic (1) H imaging of hyperpolarized [1-(13) C]lactate in vivo using a reverse INEPT experiment. Magn. Reson. Med. **79**(2), 741–747 (2018). https://doi.org/10.1002/mrm.26725
111. Wang, J., Wright, A.J., Hesketh, R.L., Hu, D.E., Brindle, K.M.: A referenceless Nyquist ghost correction workflow for echo planar imaging of hyperpolarized [1-(13) C]pyruvate and [1-(13) C]lactate. NMR Biomed. **31**(2), e3866 (2018). https://doi.org/10.1002/nbm.3866

Further Reading

Imaging biomarker roadmap for cancer studies.

O'Connor, J.P., Aboagye, E.O., Adams, J.E., et al.: Imaging biomarker roadmap for cancer studies. Nat. Rev. Clin. Oncol. **14**(3), 169–186 (2017). https://doi.org/10.1038/nrclinonc.2016.162

Hyperpolarized magnetic resonance as a sensitive detector of metabolic function.

Comment, A., Merritt, M.E.: Hyperpolarized magnetic resonance as a sensitive detector of metabolic function. Biochemistry. **53**(47), 7333–7357 (2014). https://doi.org/10.1021/bi501225t

Chemistry and biochemistry of 13C hyperpolarized magnetic resonance using dynamic nuclear polarization.

Keshari, K.R., Wilson, D.M.: Chemistry and biochemistry of 13C hyperpolarized magnetic resonance using dynamic nuclear polarization. Chem. Soc. Rev. **43**(5), 1627–1659 (2014). https://doi.org/10.1039/c3cs60124b

Hyperpolarized ^{13}C MRI: Path to clinical translation in oncology.

Kurhanewicz, J., Vigneron, D.B., Ardenkjaer-Larsen, J.H., et al.: Hyperpolarized (13)C MRI: path to clinical translation in oncology. Neoplasia. **21**(1), 1–16 (2019). https://doi.org/10.1016/j.neo.2018.09.006

Chapter 8
New Insights into Metabolic Regulation from Hyperpolarized ^{13}C MRS/MRI Studies

A. Dean Sherry and Craig R. Malloy

Abbreviations

GNG	Gluconeogenesis
HP	Hyperpolarized
NMR	Nuclear magnetic resonance
PC	Pyruvate carboxylase
PDH	Pyruvate dehydrogenase
PEP	Phosphoenolpyruvate
PEPCK	Phosphoenolpyruvate carboxykinase
TCA cycle	Tricarboxylic acid cycle or Krebs cycle

A. D. Sherry (✉)
Advanced Imaging Research Center and the Department of Radiology, University of Texas Southwestern Medical Center, Dallas, TX, USA

Department of Chemistry and Biochemistry, University of Texas at Dallas, Richardson, TX, USA
e-mail: Dean.Sherry@utsouthwestern.edu

C. R. Malloy
Advanced Imaging Research Center and the Department of Radiology, University of Texas Southwestern Medical Center, Dallas, TX, USA

Department of Internal Medicine, University of Texas Southwestern Medical Center, Dallas, TX, USA

Veterans Affairs North Texas Health Care System, Dallas, TX, USA

T. Jue, D. Mayer (eds.), *Dynamic Hyperpolarized Nuclear Magnetic Resonance*, Handbook of Modern Biophysics, https://doi.org/10.1007/978-3-030-55043-1_8

8.1 Introduction

Disruption of intermediary metabolism plays a central role in many high-impact diseases such as cancer, diabetes, heart disease, and genetic disorders. The growing appreciation of complex interactions among disease, genetics, intermediary metabolism, and cell signaling has reawakened intense interest in the classical metabolic pathways. Although the details of transporter kinetics and enzyme regulation continue to be refined, the focus of metabolic research is no longer the discovery of enzymes. In fact, the details of pathways are well described in maps and databases with extensive information about genes, enzymes, substrates, products, and cell compartments. Instead, the focus of research has shifted to a better understanding of the integrated function of these systems. The detailed mechanism of the Warburg phenomenon, for example, is only beginning to emerge. The role of intermediates in common metabolic pathways also continues to be refined. The last two decades have seen the surprising development of new concepts such as "oncometabolites," small molecules such as fumarate, succinate, and 2-hydroxyglutarate that are thought to drive the development of some malignancies [1, 2].

Recent studies of intermediary metabolism have emphasized measurement of gene transcripts, enzyme expression, and large-scale metabolomics to assess relative changes in the concentration of small metabolites. Although these methods provide considerable insight, information about flux in specific pathways and reactions is essential to link metabolism and the physiology of functioning tissue. Historically, fluxes were measured in individual pathways using appropriate mathematical models to interpret the distribution of tracers such as ^{14}C and ^{3}H. Radioactive tracers were preferred because of high detection sensitivity, but these methods are limited by practical constraints, particularly for studies in human subjects. Perhaps less appreciated is the fact that the information yield from radiotracer experiments is inherently quite poor compared to the complexity of metabolism in vivo. For these reasons, the field has moved to the use of stable isotopes, particularly ^{13}C, with detection by mass spectrometry and NMR spectroscopy. Both technologies provide information about the labeling patterns in product molecules, which are inherently difficult or impossible to determine by radiotracer methods. If sample mass is adequate and the tissue can be conveniently biopsied, then analysis by either ^{13}C NMR or mass spectrometry can be considered. Both techniques offer valuable, quantitative insights into intermediary metabolism, an approach, which has recently proven valuable in cancer studies [3–7]. Mass spectrometry is more sensitive than NMR and can be done on smaller samples, while ^{13}C NMR is less sensitive and requires more sample but offers higher information yield.

The requirement of obtaining a tissue biopsy is a critical limitation, so there is great interest in obtaining truly noninvasively metabolic data from a tracer experiment. The advent of hyperpolarization (HP) technologies, described in detail elsewhere, enables real-time monitoring of metabolism of ^{13}C-enriched substrates in vivo, at least through a few enzyme-catalyzed steps. It is less comprehensive than steady-state isotopomer methods but as technical improvements come, it is likely that noninvasive imaging with hyperpolarized ^{13}C coupled with metabolomics and mathematical analysis of metabolic pathways will enable fundamentally new concepts in disease pathophysiology. In this chapter, we summarize how HP ^{13}C has provided new insights into four important metabolic pathways and suggest possible further advances and limitations of the technique.

8.2 Pyruvate Dehydrogenase and the Krebs Cycle

Pyruvate dehydrogenase (PDH) catalyzes the conversion of pyruvate to acetyl-CoA and CO_2. PDH is a multienzyme complex located on the inner surface of the inner mitochondrial membrane. Flux through PDH is sensitive to mass-action and is regulated by both phosphorylation–dephosphorylation and the allosteric regulators, nicotinamide adenine dinucleotide hydrogen (NADH), and acetyl-CoA. The regulatory and phosphorylation sites are well established but the activity of the enzyme in vivo is not always known or easily predicted. [1-^{13}C]pyruvate is widely used in hyperpolarized ^{13}C studies, and it is generally assumed that the appearance of HP-bicarbonate reflects PDH activity and may provide an index of mitochondrial integrity. In highly oxidative tissues such as the heart, this may be true but two important factors should be considered when interpreting HP-bicarbonate production from [1-^{13}C]pyruvate.

The first issue is that PDH flux is very sensitive to competing substrates. Even in an isolated perfused heart where available substrates can be precisely controlled, the heart is still able to oxidize stored glycogen and triglycerides. In vivo, the heart is exposed to complex mixtures of short-, medium-, and long-chain fatty acids, glucose, lactate, and other oxidizable compounds that may vary widely in concentration and directly or indirectly influence PDH activity. The biochemical environment of PDH is difficult to monitor in vivo. Moreno et al. [8] demonstrated that the generation of HP-bicarbonate after exposure to HP-[1-^{13}C]pyruvate is very sensitive to competing substrates and that pyruvate polarization is altered by weak binding to albumin. In the presence of physiological levels of pyruvate (0.12 mM), lactate (1.2 mM), glucose (5 mM), ketones (0.17 mM), and fatty acids (0.4 mM), fatty acids and ketones overwhelmingly supply the majority of acetyl-CoA to the TCA cycle. However, with the addition of a bolus of pyruvate (3 mM to mimic HP conditions), pyruvate then provides ~80% of acetyl-CoA, while the contributions of ketones and fatty acids are reduced to near zero (Fig. 8.1). Under metabolic conditions mimicking a fast, meaning higher concentrations of ketones and fatty acids, HP-[^{13}C]bicarbonate production was much reduced. This indicates that the amount of HP-bicarbonate formed in an HP experiment will depend upon (1) dietary state and (2) the amount of HP-pyruvate reaching the tissue. Thus, under certain metabolic conditions, PDH could be fully active (dephosphorylated) even though production of HP-bicarbonate is relatively low because other substrates contribute more to the pool of acetyl-CoA entering the TCA cycle than HP-pyruvate. This means that it is possible to have a condition where the signal of HP [^{13}C]bicarbonate

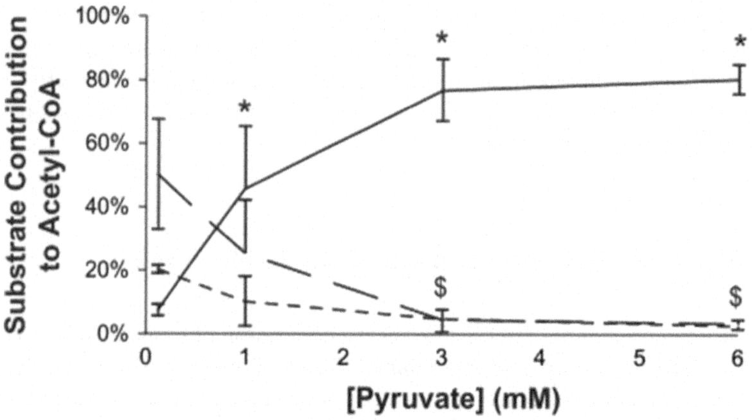

Fig. 8.1 Influence of [pyruvate] on sources of acetyl-CoA and flux through pyruvate dehydrogenase under fed conditions. Isolated hearts were supplied with a physiological mixture of fatty acids, ketones, and other substrates designed to replicate concentrations under fed, resting conditions. The relative rates of oxidation of ketones (dotted line), long-chain fatty acids (broken line), or pyruvate (solid line) are shown as a function of the [pyruvate]. At 3 mM pyruvate, oxidation of competing substrates is largely suppressed. Reproduced with permission from [8]. [APS statement: "Authors may republish parts of their final-published work (e.g., figures, tables), without charge and without requesting permission, provided that full citation of the source is given in the new work"]

approaches zero even when the activity of PDH is fully functional. It has also been shown that low concentrations of added propionate restore PDH flux even when the concentration of competing substrates is significant [9]. Thus, the interplay among normal regulation of PDH, physiological state of the animal or patient, and delivery must play an important role in a proper interpretation of an HP experiment.

A second important factor is the alternative pathway for HP [1-^{13}C]pyruvate to enter the TCA cycle via pyruvate carboxylase (PC) to form HP-[1-^{13}C]oxaloacetate which would also produce HP-bicarbonate after three or more forward steps in the TCA cycle (OAA → citrate → isocitrate → α-KG). The entry of pyruvate into the TCA cycle via PC, the dominant anaplerotic reaction in heart, is known to be small compared to forward flux through the TCA cycle (y_{pc} ~ 3–5% of TCA cycle flux) [10–13] so the production of HP-bicarbonate via this pathway should be relatively small. However, under metabolic conditions where fatty acids and ketones dominate production of acetyl-CoA, any production of HP-bicarbonate via entry of HP-pyruvate into the cycle via PC could be misinterpreted as reflecting flux through PDH.

Notwithstanding these potential limitations, the production of HP-bicarbonate from HP-[1-^{13}C]pyruvate has been reported to reflect PDH activity in cardiac tissue both in perfused hearts [14] and in vivo [15]. In isolated hearts, the addition of octanoate to compete with pyruvate for production of acetyl-CoA increased production of HP-lactate and reduced HP-bicarbonate production to near zero [14]. Nonpolarized ^{13}C isotopomer studies were used to verify that PDH flux was indeed lower in the presence of octanoate while TCA cycle flux was unchanged. Thus, in a simple isolated perfused heart preparation where substrate availability is easily controlled, the production of HP-bicarbonate does appear to accurately reflect PDH activity. One might anticipate this may not be true in the in vivo myocardium because substrate availability (fed versus fasted) and pyruvate delivery could be added variables. Nonetheless, in rat hearts in vivo, Schroeder et al. [15] observed less production of HP-bicarbonate from HP-[1-^{13}C]pyruvate in both diabetic and 24 h starved rats compared to controls and ascribed these differences to reduced flux through PDH due to competition from higher plasma levels of fatty acids and ketone bodies. These physiological conditions also promote the phosphorylation of PDH via PDH kinase, which reduces the activity of PDH, consistent with their observations. Nevertheless, this study illustrates a common problem when faced with the interpretation of metabolic imaging studies based only on HP data. In this case, the question is, if less HP-bicarbonate is produced, does this reflect lower PDH activity or simply more competition from fatty acids, and ketone bodies for the production of acetyl-CoA? This question cannot be answered without additional supporting information.

Given that PDH flux is likely a good biomarker of intact mitochondria, there is a great deal of clinical interest in developing HP-[1-^{13}C]pyruvate to assess common heart diseases such as ischemia [16] or heart failure or to detect responses to drug interventions. Merritt et al. [17] evaluated the utility of HP [1-^{13}C]pyruvate to detect changes in PDH activity early vs. late after reperfusion of ischemic heart tissue, and compared the HP ^{13}C results with conventional ^{31}P NMR spectroscopic measures of PCr (phosphocreatine) and adenosine triphosphate (ATP) levels. Interestingly after brief ischemia followed by reperfusion, cardiac energetics and oxygen consumption were not sensitive to the earlier episode of ischemia, that is, these measures recovered quickly to preischemic levels. However, the appearance of HP [^{13}C]bicarbonate was impaired early after ischemia but recovered later, perhaps reflecting a highly reduced metabolic state immediately after the ischemic injury.

In a study designed to test whether a sudden increase in cardiac activity would stimulate flux of HP-[1-^{13}C]pyruvate through PDH to yield more HP-bicarbonate, isoproterenol was presented to perfused rat hearts ~30 s after each heart was exposed to HP-[1-^{13}C]pyruvate [18]. β-adrenergic stimulation resulted in a significant increase in heart rate from 290 to 420 beats/min but, surprisingly, no increase in the production of HP-bicarbonate from HP-[1-^{13}C]pyruvate. Those results are illustrated in Fig. 8.2. Instead, the signal of HP-lactate, which had already reached an apex in intensity ~20 s after the addition of HP-pyruvate, began to increase in intensity for a second time. This effect was later traced to the rapid breakdown of glycogen to form additional nonpolarized lactate, which was then available to exchange with the remaining HP-[1-^{13}C]pyruvate to generate a second lactate apex. This observation is interesting for two reasons. First, adrenergic stimulation of cardiac work did not increase flux through PDH as reported by the production of HP-bicarbonate. Instead, glycogen was rapidly converted to pyruvate and lactate to supply additional energy in the form of carbohydrates. Presumably, any HP-pyruvate remaining in the tissue after the addition of isoproterenol was quickly diluted by the

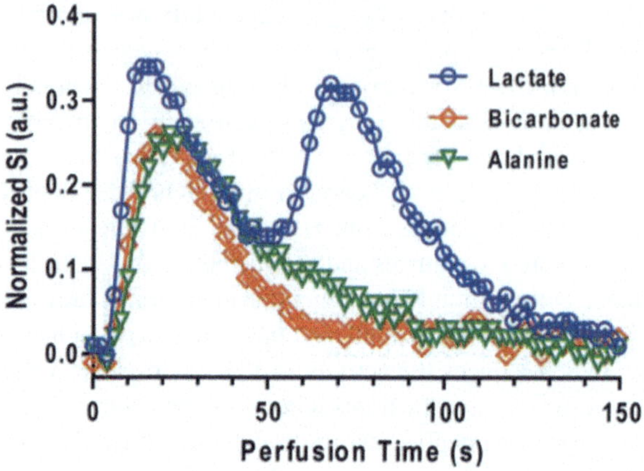

Fig. 8.2 Normalized ^{13}C resonance intensities versus time after the addition of HP-[1-^{13}C]pyruvate in isolated rat hearts. These data show the time-dependent signals of HP-lactate, alanine, and bicarbonate with additional isoproterenol at 30 s after the addition of HP [1-^{13}C]pyruvate. An elevated cardiac workload (as measured by heart rate and developed pressure) was achieved approximately 15 s after the addition of isoproterenol. Almost immediately thereafter, the signal intensity of HP-lactate which had begun to decay after reaching the first apex sharply increased in intensity and reached a second apex. Stimulation of cardiac function did not significantly alter the total HP bicarbonate signal. Reproduced with permission from [18]. [MRM statement: Authors—If you wish to reuse your own article (or an amended version of it) in a new publication of which you are the author, editor, or coeditor, prior permission is not required (with the usual acknowledgments)]

nonpolarized pyruvate generated from glycogenolysis and this resulted in no change in the appearance of HP-bicarbonate. This was further verified by isotopomer analysis of ^{13}C NMR spectra of tissue extracts. These results are consistent with the "glycogen shunt" hypothesis put forth by Shulman and Rothman [19] who suggested that the rapid breakdown of glycogen to lactate is absolutely required to buffer the short-term energy requirements of exercising muscle. This study further illustrates that the production of HP-bicarbonate from HP-[1-^{13}C]pyruvate is not necessarily an accurate biomarker of PDH flux.

Schroeder et al. [20] paced porcine hearts at a high rate over a period of several weeks to induce cardiomyopathy and periodically collected ^{31}P NMR spectra to monitor high-energy phosphates and ^{13}C NMR spectra after injection of either HP-[1-^{13}C]pyruvate or HP-[2-^{13}C]pyruvate to monitor downstream metabolic products (see Fig. 8.3). Unlike the use of [1-^{13}C]pyruvate where TCA cycle flux is assumed proportional to bicarbonate appearance (assuming appropriate control of competing substrates) [2-^{13}C]pyruvate is metabolized to [1-^{13}C]acetyl-CoA and eventually [5-^{13}C]glutamate, thereby directly interrogating the first span of the TCA cycle [21]. In this cardiomyopathy model, the myocardial PCr/ATP ratio progressively dropped throughout the pacing protocol from 2.3 ± 0.2 at baseline to 1.7 ± 0.1 at early pacing, and 1.3 ± 0.2 at moderate subclinical dysfunction. The amount of HP-[5-^{13}C]glutamate produced from HP-[2-^{13}C]pyruvate, considered an index of combined PDH activity and TCA cycle flux, also decreased in parallel with loss of

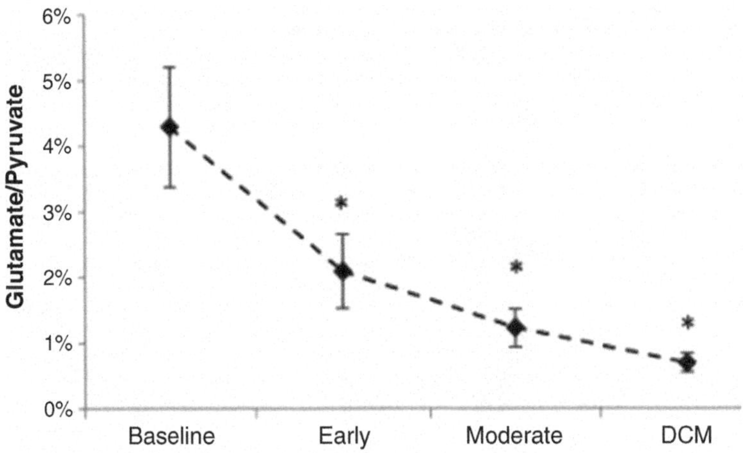

Fig. 8.3 Hyperpolarized ^{13}C magnetic resonance spectroscopy (MRS) showing altered [5-^{13}C]glutamate production during the development of dilated cardiomyopathy. Congestive heart failure was induced by long-term rapid pacing. Metabolism of HP-[2-^{13}C]pyruvate was repeatedly examined. Under these conditions, HP-[2-^{13}C]pyruvate is metabolized to HP-[5-^{13}C] α-ketoglutarate in the first span of the TCA cycle and subsequently to HP-[5-^{13}C]glutamate. The [5-^{13}C]glutamate/[2-^{13}C]pyruvate ratio is shown, measured for all pigs during development of heart failure. Reproduced with permission from [20]. [Permission granted from Eur. J. Heart Failure through RightsLink Copyright Clearance Center (https://s100.copyright.com/AppDispatchServlet)]

heart function while production of HP-bicarbonate remained equal to baseline levels over a much longer period, finally dropping at the end point of dilated cardiomyopathy. Although there may other interpretations of these data, this exciting study demonstrated for the first time that metabolic changes can be detected by HP ^{13}C metabolic imaging long before heart function actually deteriorates to a point where it is easily detected using standard clinical measures of left ventricular function. If such studies could be collected in the human myocardium well before irreversible heart failure sets in, then the administration of drugs or some other metabolic intervention could potentially save human lives.

The first hyperpolarized ^{13}C metabolic imaging study of the human heart using HP-[1-^{13}C]pyruvate was reported by Cunningham, et al. [22]. The metabolic images collected in that study, Fig. 8.4, show that HP-bicarbonate was mainly confined to the LV myocardium, whereas HP-[1-^{13}C]pyruvate appeared largely in the blood pool with little in the muscle itself while HP-[1-^{13}C]lactate produced via exchange with pyruvate appeared both within the chambers and in the myocardium. This nicely illustrates that information about myocardial metabolism and perhaps PDH flux can be obtained regionally within the tissue. One interesting feature of this study was that the amount of HP-bicarbonate produced from HP-[1-^{13}C]pyruvate appears to be considerably higher in human hearts compared to levels observed in rodent studies. This again illustrates that PDH flux is highly dependent upon dietary state, substrate delivery, and likely other yet to be determined factors. Most recently, hyperpolarized ^{13}C spectroscopy studies of the human heart were acquired in

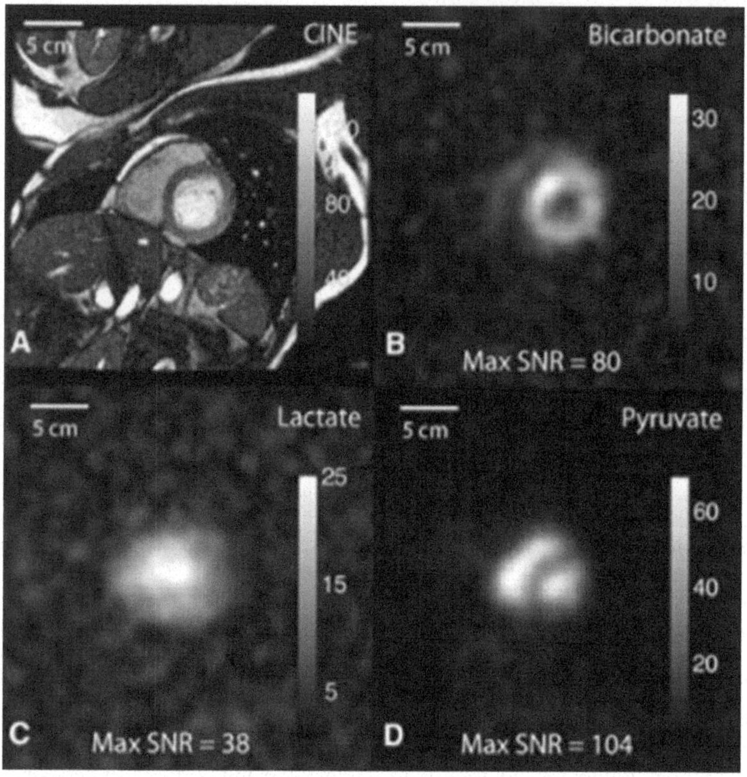

Fig. 8.4 Short axis anatomic and ^{13}C images from the heart of a healthy human subject. Panel (**a**) shows the anatomical image. Panel (**b**) shows the HP-bicarbonate image with an intense signal in the left ventricular myocardium and much less, although detectable, signal from the right ventricular myocardium. Panel (**c**) shows the HP [1-^{13}C]lactate image which appears to contain signal from both the blood pool and myocardium. Panel (**d**) is the HP[1-^{13}C]pyruvate image with, as expected, signal confined largely to the left and right ventricular cavities. Reproduced with permission from [22]. [Permission requested from AHA journals on 2/5/2019]

patients with type 2 diabetes [23], finding that flux through PDH was reduced among patients with type 2 diabetes. The effects of a cardiotoxic cancer therapy, doxorubicin, was studied in women with breast cancer [24]. A small decrease in flux through PDH but not LDH was observed in association with chemotherapy. In both studies, reduced PDH flux may be an indicator of subtle mitochondrial injury but again competition from other substrates should be considered.

The potential clinical significance of understanding PDH flux in the heart and other highly oxidative tissues is evident. Cancers are commonly thought to rely only on glycolysis for energy production (the Warburg phenomenon) but it has become clear that the TCA cycle may contribute significantly as well [5, 25, 26]. In isolated cancer cells [27] metabolism of HP-[1-^{13}C]pyruvate to HP bicarbonate was readily observed, but HP bicarbonate production in cancers in vivo is not commonly observed.

8.3 Pyruvate Carboxylase and Gluconeogenesis

Pyruvate carboxylase (PC) catalyzes the conversion of pyruvate to the 4-carbon TCA cycle intermediate, oxaloacetate. In some tissues, the dominant pathway for HP-[1-^{13}C]pyruvate could be the production of HP-[1-^{13}C]oxaloacetate rather than flux through PDH. PC is ubiquitous in virtually all tissues and is highly regulated. It requires biotin and one equivalent of ATP for activation of CO_2, and its activity is regulated by mitochondrial levels of acetyl-CoA:

$$\text{Pyruvate} + HCO_3^- + \text{ATP} \rightarrow \text{Oxaloacetate} + \text{ADP} + \text{Pi}.$$

PC is a critical anaplerotic reaction in the liver because it provides oxaloacetate in the first step of gluconeogenesis, in the brain where it is essential for the biosynthesis of neurotransmitters in astroglia, and in other tissues for net production of TCA cycle intermediates. In liver tissues where anaplerotic flux can be four- to sevenfold higher than TCA cycle flux [28, 29], drawing metabolic conclusions solely on the basis of HP-bicarbonate produced from HP-[1-^{13}C]pyruvate is tenuous. In this case (see Fig. 8.5), production of HP-bicarbonate could arise from flux through PDH (single step), carboxylation of HP-[1-^{13}C]pyruvate via PC followed by (a) forward flux in the TCA cycle (three steps) or (b) backward scrambling of 4-carbon intermediates followed by decarboxylation via PEPCK or the malic enzyme (at least three steps). Thus, it is not surprising to find little consensus in the literature as to which pathway dominates in producing $^{14}CO_2$ from [1-^{14}C]pyruvate. Some models of hepatic metabolism neglect flux through PDH entirely [30] or conclude that flux through PDH is small and even zero under some experimental conditions [31]. Others [32] assign the appearance of labeled bicarbonate only to flux through PDH.

In an early HP study of metabolism in isolated, perfused mouse livers, signals from HP-[1-^{13}C]malate and HP-[1-^{13}C]aspartate were evident within seconds after HP-[1-^{13}C]pyruvate was introduced into the perfusate [29, 33]. This provided direct evidence for the flux of HP-[1-^{13}C]pyruvate through PC as illustrated in Fig. 8.6. In addition to those signals, HP-bicarbonate was also observed. So this invoked the obvious question, does this bicarbonate reflect flux through PDH, the TCA cycle, or PEPCK? In an attempt to answer this question, mice lacking liver PEPCK were also examined, and in those livers, little HP-bicarbonate was detected. This suggested that the HP-bicarbonate seen in control mice might largely reflect flux of TCA cycle intermediates through PEPCK, an exciting result that, if true, suggests the bicarbonate signal could possibly be used as an index of gluconeogenesis. However, additional nonpolarized ^{13}C experiments using [1,2,3-$^{13}C_3$]pyruvate showed that pyruvate contributed ~7% of all acetyl-CoA in this preparation so clearly some HP-bicarbonate reflects PDH activity.

In the first in vivo liver study [34], the flux of HP-[1-^{13}C]pyruvate into TCA cycle intermediates via PC was also evident by the appearance of HP-[1-^{13}C]malate and HP-[1-^{13}C]aspartate resonances in the spectra. The kinetic rates of appearance of

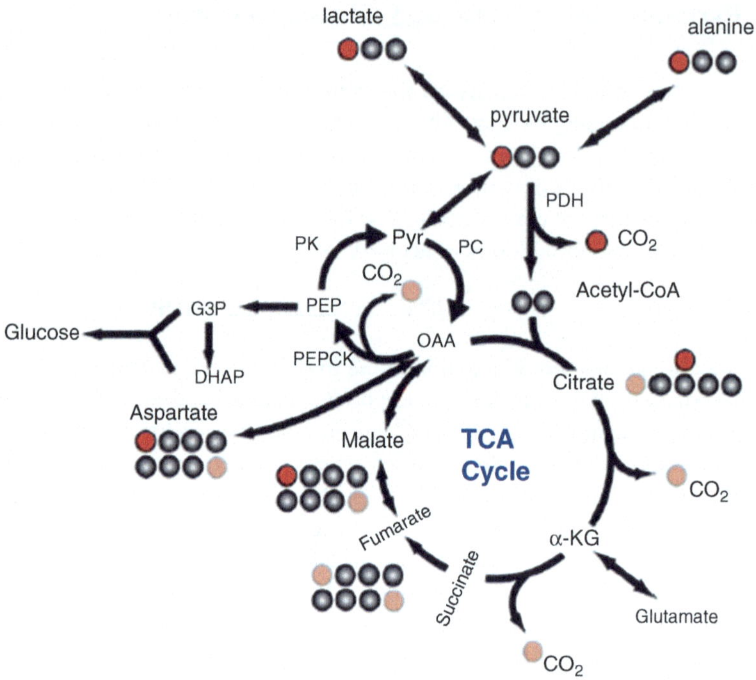

Fig. 8.5 Schematic of carbon flow in the liver originating from [1-^{13}C]pyruvate. The introduction of [1-^{13}C]pyruvate leads to labeling at the C1 of lactate and alanine (red circles). If PDH flux is present, CO_2 can also be labeled. If pyruvate carboxylase flux is significant, the label can appear at the C1 of malate. If the label is symmetrically distributed by the exchange between malate and fumarate, then it can also appear at the four positions of various metabolites (red, semitransparent circles). CO_2 can be eliminated by the action of either PEPCK on the C4 position of oxaloacetate or by enzymatic reactions in the forward span of the TCA cycle. Labeling of the CO_2 in these cases depends upon the level of symmetrization achieved by the malate to fumarate exchange. Reproduced with permission from [29]. [PNAS copyright: Anyone may, without requesting permission, use original figures or tables published in PNAS for noncommercial and educational use (i.e., in a review article, in a book that is not for sale), provided that the full journal reference is cited]

those signals ($k_{pyr \rightarrow mal}$ and $k_{pyr \rightarrow asp}$), presumably imaging biomarkers of PC flux, were higher in the liver of mice fed a high-fat diet. This correlated well with increased PC activity directly measured in tissue homogenate. This exciting study demonstrated for the first time that HP metabolic imaging could directly measure enzyme fluxes specific to liver in a type 2 model of diabetes. Nonetheless, the fact that little HP-bicarbonate was observed in these livers begs the question, why does one not detect upstream gluconeogenic products from these labeled TCA cycle intermediates? In principle, a relatively large amount of HP-bicarbonate should have been generated from these HP-intermediates, yet only a small HP-bicarbonate signal was observed. In this case, the appearance of HP-bicarbonate was ascribed entirely to flux through PDH without further direct experimental verification.

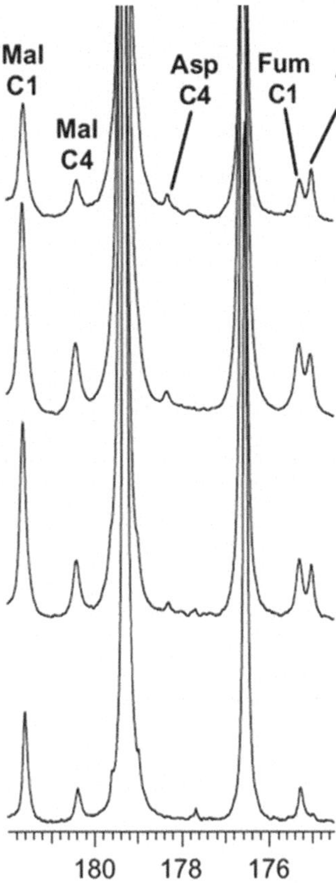

Fig. 8.6 ^{13}C NMR spectra during metabolism of hyperpolarized [2-^{13}C]- and [1-^{13}C]pyruvate. Signals of [1-^{13}C]lactate, [1-^{13}C]alanine and [1-^{13}C]pyruvate hydrate, [5-^{13}C]glutamate, [1-^{13}C]malate, [4-^{13}C]malate, [1-^{13}C]aspartate, [4-^{13}C]aspartate, and [1 or 4-^{13}C]fumarate were observed. Each panel represents livers perfused with different nonpolarized substrates prior to exposure to HP-pyruvate: (**a**) (1.5 mM) lactate and (0.15 mM) pyruvate; (**b**) (0.2 mM) octanoate; (**c**) (1.5 mM) lactate, (0.15 mM) pyruvate and (0.2 mM) octanoate; (**d**) (4 mM) pyruvate and (0.2 mM) octanoate. Reproduced with permission from Fig. 8.5 in ref. [33]. [Permission granted from Metabolomics through RightsLink Copyright Clearance Center (https://s100.copyright.com/AppDispatchServlet)]

In a second in vivo liver study, HP-bicarbonate was produced from HP-[1-^{13}C]pyruvate in the liver of fed rats but not in 21 h fasted rats [35]. In separate bench infusion studies in fed versus fasted animals, nonpolarized [2,3-$^{13}C_2$]pyruvate was used under identical conditions (bolus injection of 2.5 mL of 80 mM pyruvate) to validate and interpret the HP results. High-resolution ^{13}C NMR spectra of those liver extracts are shown in Fig. 8.7. The large doublet (labeled D45) in the glutamate C4 resonance of the liver extract from the fed animal shows that PDH is active (Fig. 8.7b). The very small D45 resonance in the spectrum of the fasted animal shows that PDH is relatively inactive under these feeding conditions. This is entirely

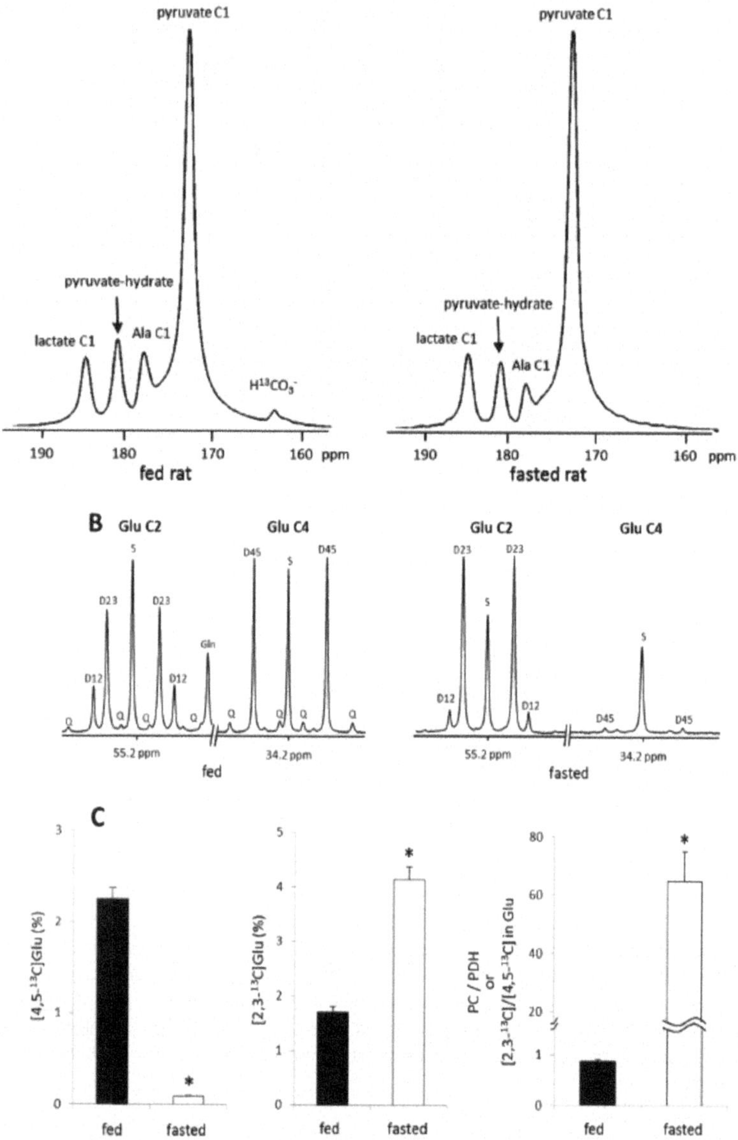

Fig. 8.7 Effect of nutritional state on integrated metabolic analysis of pyruvate carboxylation by conventional and hyperpolarized ^{13}C NMR. The upper panel shows ^{13}C NMR spectra of rat liver in vivo after intravenous injection of HP [1-^{13}C]pyruvate. Livers of both fed (left) and fasted (right) rats show HP signals from [1-^{13}C]lactate, [1-^{13}C]alanine, [1-^{13}C]pyruvate, and the hydrate of [1-^{13}C]pyruvate, but only livers of fed animals show the signal from HP [^{13}C]bicarbonate. The middle and bottom panels show results from conventional ^{13}C NMR isotopomer analysis after injection of [2,3-$^{13}C_2$]pyruvate and extraction at approximately the same time as the HP data. The entry of [2,3-$^{13}C_2$]pyruvate to the TCA cycle through PDH yields [4,5-^{13}C]glutamate while the entry through PC produces [2,3-$^{13}C_2$]glutamate after the condensation of [2,3$^{13}C_2$]oxaloacetate and acetyl-CoA followed by the forward turn through the TCA cycle. In ^{13}C NMR spectra of C2

consistent with the appearance of HP-bicarbonate in the liver of the fed animal and the absence of HP-bicarbonate in the liver from the fasted animal. Similarly, if [2,3-$^{13}C_2$]pyruvate enters the TCA cycle via PC to yield [2,3-$^{13}C_2$]oxaloacetate, then these labeled carbons would appear as D23 in the C2 resonance of glutamate. In the spectrum from the fed animal, the D23 is present but smaller than the C2 singlet while in the spectrum of the fasted animal, the D23 is larger than the C2 singlet. This directly shows that anaplerosis via PC is higher in the liver of fasted animals than in the liver from fed animals. Finally, the left bar graph in Fig. 8.7c shows that the relative PC/PDH flux ratio in the fasted animals is much higher than in fed animals, consistent with higher glucose production (GNG = gluconeogenesis) in the fasted state. Thus, the two in vivo liver HP studies report essentially the same thing: in mice fed a high-fat diet, the flux of HP-[1-^{13}C]pyruvate through PC is stimulated versus controls. In fasted rats where liver GNG is known to be high, PC flux as reported by ^{13}C isotopomer analysis is greatly stimulated overfed animals. Yet neither model provided evidence for the production of HP-bicarbonate from HP-oxaloacetate via PEPCK, the first committed step in GNG. This is rather unexpected given that HP-malate, HP-aspartate, and HP-glutamate have all been detected in various tissues. Other studies in isolated livers where a detailed analysis of pathway fluxes based on direct measurements of oxygen consumption and ketone production also concluded that bicarbonate production did not correlate well with pyruvate carboxylation [33] in spite of high flux through PEPCK. Might there be a biological reason for the loss of spin polarization at the level of PEPCK?

One can only speculate at this point but a potential source for depolarization can be gleaned from the crystal structure of PEPCK [36, 37]. PEPCK is a GTP-dependent enzyme that catalyzes decarboxylation and concomitant phosphorylation of oxaloacetate to form PEP:

$$\text{Oxaloacetate} + \text{GTP} \rightarrow \text{CO}_2 + \text{GDP} + \text{PEP}\left(\text{Phosphoenolpyruvate}\right).$$

The rat liver cytosolic form of the enzyme contains a paramagnetic Mn^{2+} ion in the active site that binds to an oxygen atom on the C1 carboxyl group plus the carbonyl oxygen atom on C2 [37]. Binding of Mn^{2+} to the C2 carbonyl oxygen stabilizes the enol form of OAA, which weakens the C3–C4 bond and promotes release of the C4 carboxyl group as CO_2, as illustrated in Fig. 8.8. Simultaneous with the release of CO_2, the nucleophilic enol oxygen on C2 attacks the γ-phosphate of GTP to form PEP and GDP. A divalent Mg^{2+} bound to the α- and β-phosphates of GTP

Fig. 8.7 (continued) (~55.2 ppm) and C4 (~34.2 ppm) regions of glutamate isolated from liver, the signal from [4,5-$^{13}C_2$]glutamate is strong in fed animals, but negligible in fasted rats. The signal from [2,3-$^{13}C_2$]glutamate is easily detected in fed animals and very strong in fasted animals. The ratio of [2,3-$^{13}C_2$]/[4,5-$^{13}C_2$] in glutamate, reflecting the ratio of PC/PDH, was dramatically higher in fasted rats compared with fed rats. Reproduced with permission from [35]. [NMR in Biomedicine statement: Authors—If you wish to reuse your own article (or an amended version of it) in a new publication of which you are the author, editor, or coeditor, prior permission is not required (with the usual acknowledgments)]

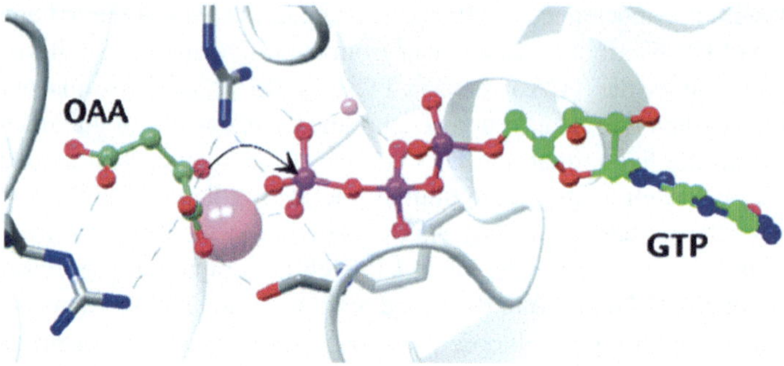

Fig. 8.8 Structure of the Mn^{2+}-OAA-GTP complex in the active site of PEPCK. The large pink sphere represents Mn^{2+} coordinated to both the C1 and C2 oxygen atoms of OAA and the γ-phosphate of GTP. The other active site ligands coordinated to the Mn^{2+} ions were omitted for clarity. Reproduced with permission from [37]. [Permission granted from Biochemistry through RightsLink Copyright Clearance Center (https://s100.copyright.com/AppDispatchServlet)]

sits nearby the Mn^{2+} and aids in the release of the γ-phosphate. Given that Mn^{2+} has five unpaired electrons and is quite efficient at relaxing nearby NMR nuclei (r^{-6} dependence), it is reasonable to assume that all highly polarized OAA ^{13}C atoms could be rapidly relaxed by the paramagnetic Mn^{2+} ion when OAA binds in the active site of PEPCK. This may be the reason why we have not been able to conclusively assign production of HP-bicarbonate to flux through PEPCK even under conditions when highly polarized 4-carbon TCA cycle intermediates can be detected and gluconeogenesis is active.

8.4 Glycolysis

Given the importance of glucose as an energy source and a source of carbons for many biosynthetic processes, the possibility of imaging intermediates along those many pathways using HP ^{13}C probes is very appealing. Unfortunately, carbohydrates are not great candidates for HP ^{13}C imaging because most carbons have at least one directly attached proton and this reduces the T_1 of a directly bonded ^{13}C atom via dipolar interactions. Those directly bonded protons can be substituted with deuterium to lengthen the carbon T_1 values [38] but even with this modification, the T_1s are rather short (~10–12 s). Nevertheless, [U-$^2H_7$, U-$^{13}C_6$]glucose has been polarized and investigated in suspensions of *Escherichia coli* [39], yeast [40–42], and breast cancer cells [43] and in an in vivo lymphoma mouse model [44]. In *E. coli*, resonances characteristic of gluconate-6-phosphate (6-PGA), fructose-1,6-bisphosphate (FBP), dihydroxyacetone phosphate (DHAP), pyruvate, acetyl-CoA, lactate, alanine, acetate, formate, CO_2, HCO_3^-, ribulose-5P, and ethanol where all observed within a few seconds after delivery of HP-[U-$^2H_7$, U-$^{13}C_6$]glucose to a cell suspension. This demonstrates that multiple pathway intermediates can be detected in living cells, and the kinetics of the appearance of HP-carbons in those intermediates can provide few new insights into metabolic control. Nonetheless, a comparison of labeled intermediates in exponentially growing cells versus stationary phase cells showed that more glucose is converted to δ-6-phosphogluconolactone (five-fold) in the former, consistent with greater flux through the oxidative branch of the pentose phosphate pathway during the cell growth phase [39]. Similarly, in fermenting yeast cells, FBP, 6-phosphogluconate, and pyruvate along with the catabolic end products, ethanol, and CO_2, were rapidly produced from HP-[U-$^2H_7$, U-$^{13}C_6$]glucose in a real-time DNP-NMR assay. The appearance of HP-pyruvate and HP-ethanol versus lack of appearance of HP-acetaldehyde was ascribed to the well-known rate-limiting pyruvate decarboxylase reaction [45] and rapid conversion of acetaldehyde to ethanol.

Although HP studies in isolated cells have not revealed anything substantially new about metabolism or metabolic flux compared to more traditional tracer experiments, they do offer the huge advantage of rapid detection so that physiological responses to any specific metabolic intervention can be assayed quickly, an important advantage in biotechnical optimization of cellular networks [39]. HP-[U-$^2H_7$, U-$^{13}C_6$]glucose has also been exposed to T47D breast cancer cells [43] and, in this case, fewer intermediates (DHAP, 3PG, lactate) were directly visible by ^{13}C NMR in comparison to those detected in yeast and *E. coli*. Nevertheless, given that the chemical shift of most glycolytic intermediates is well established, a unique radio-frequency pulse saturation method was used to indirectly monitor relative flux from several "invisible" metabolites to lactate [43]. It was shown that the appearance of HP-lactate from HP-glucose followed Michaelis-Menten-like kinetics with a $K_m = 3.5 \pm 1.5$ mM and $V_{max} = 34 \pm 4$ fmol/cell/min [43], values completely consistent with earlier reported constants. One question that arises whenever perdeuterated glucose is used for polarization studies is the possible impact of a deuterium kinetic isotope effect on measured kinetic rate constants. Funk et al. [46] recently demonstrated using isotopomer methods that the rate of conversion of [U-$^{13}C_6$]glucose to lactate versus [U-$^{13}C_6$,U-$^2H_7$]glucose to lactate does not differ in isolated perfused rat hearts so this suggests that kinetic constants measured by the appearance of HP-lactate from HP-deuterated glucose should accurately reflect glycolysis.

HP-[U-$^2H_7$, U-$^{13}C_6$]glucose has also been used to image glycolysis in an in vivo lymphoma EL4 tumor model [44]. This was the first report of the detection of HP-lactate, the end-product of glycolysis, from HP-glucose in vivo. Unlike in cell suspensions where a sample of highly polarized [U-$^2H_7$, U-$^{13}C_6$]glucose can be mixed quickly with cells already suspended in an NMR tube in a magnet, in vivo detection requires blood circulation and cell uptake of HP-[U-$^2H_7$, U-$^{13}C_6$]glucose before entering glycolysis so this first report stimulated great excitement among the metabolic imaging community. Indeed, low levels of DHAP, 6-phosphogluconate (6PG), and bicarbonate (HCO_3^-) were also detected in those tumors, but the signal from HP-lactate demonstrated for the first time that the entire glycolytic pathway can potentially be monitored in real time. Given the importance of the Warburg effect in tumors and the possibility of imaging the impact of cancer drugs on glycolysis, this was an important scientific observation. Unfortunately, other tumors are much less glycolytic than the EL4 tumor so, unlike HP-pyruvate, HP-[U-$^2H_7$, U-$^{13}C_6$]glucose has yet to become a widely used probe in metabolic imaging.

In an attempt to further lengthen the T_1 of glucose, [U-$^2H_7$, 3,4-$^{13}C_2$]glucose was prepared and its relaxation properties compared with those of HP-[U-$^2H_7$, U-$^{13}C_6$]glucose [47]. The idea in creating [U-$^2H_7$, 3,4-$^{13}C_2$]glucose was to reduce the number of carbon–carbon spin–spin interactions within glucose such that when [U-$^2H_7$, 3,4-$^{13}C_2$]glucose undergoes glycolysis, it produces singly C1-labeled, largely nonprotonated glycolytic intermediates from [1-^{13}C]1,3-*bis*-P-glycerate to [1-^{13}C]lactate in the glycolytic pathway. Any HP-[1-^{13}C]lactate formed from HP-[U-$^2H_7$, 3,4-$^{13}C_2$]glucose would also appear as a singlet instead of a doublet and this would essentially improve *S*/*N* by an added factor of 2. Although the differences in carbon T_1s between the two labeled glucose molecules were small, the added advantage of having a singlet in [1-^{13}C]lactate did prove beneficial in that HP-lactate was clearly detected in the brain of a live mouse within about 10 s after tail vein delivery of HP-[U-$^2H_7$, 3,4-$^{13}C_2$]glucose [47]. Although this was once again a nice demonstration of the potential of HP ^{13}C for imaging the glycolytic pathway, the T_1s of glucose will likely limit applications to mouse models of cancer and perhaps traumatic brain injury. The prospects of translating this technology to humans seem remote at this point in technology development.

Given that some glycolytic intermediates contain nonprotonated carbons, other longer T_1 hyperpolarized probes of glycolysis have been investigated. Park et al. [48] polarized [1-^{13}C]glycerate and observed its biochemical conversion to [1-^{13}C]pyruvate and [1-^{13}C]lactate in vivo in rat liver. The T_1 of this probe (~60 s at 3 T) was sufficiently long for the molecule to be transported into hepatocytes via monocarboxylate transporters, phosphorylated to 3-phospho-[1-^{13}C]glycerate (3PG), and converted to the glycolytic end products, HP-[1-^{13}C]pyruvate and HP-[1-^{13}C]lactate with excellent signal-to-noise. Although small differences were observed in the amount of HP-[1-^{13}C]lactate formed in fed versus fasted animals, the most interesting observation of the study was that both HP-pyruvate and HP-lactate were directly detected, thereby offering possible use of this probe as an indicator of cytosolic redox state.

A second small molecule containing a nonprotonated carbon that has been considered as an HP probe of glycolysis is [2-^{13}C]dihydroxyacetone [49]. This neutral molecule also quickly enters hepatocytes, becomes phosphorylated, and enters the glycolytic pathway at the level of triosephosphate isomerase (TPI), slightly upstream of 3PG. Although the T_1 of this probe is about 50% shorter (~32 s at 9.4 T) than 3PG, the phosphorylated product, HP-[2-^{13}C]DHAP, showed many more detectable glycolytic intermediates in the liver, both downstream (glyceraldehyde-3-P, 3PG, PEP, pyruvate, lactate, and alanine) and upstream (glycerol-3-P, glucose-6-P, and glucose) of TPI. This indicates that HP-[2-^{13}C]dihydroxyacetone, unlike HP-[1-^{13}C]glycerate, provides a broader metabolic snapshot of both glycolysis and gluconeogenesis from triose to glucose. These probe differences suggest that there is an important point of regulation between these two entry points, perhaps at the level of glyceraldehyde-3-P dehydrogenase. A second feature of the [2-^{13}C]dihydroxyacetone study that was not evident in the [1-^{13}C]glycerate study was the detection of both

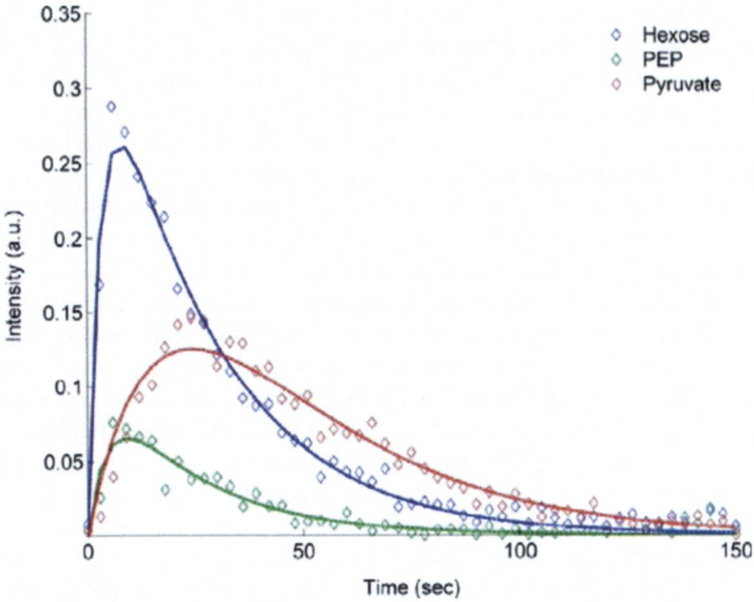

Fig. 8.9 Time course for hyperpolarized metabolites of dihydroxyacetone in the gluconeogenic state. Hexose denotes the sum of all the hexose resonances, while PEP and pyruvate refer only to the individual metabolites. Pyruvate kinase is the only enzyme of the glycolytic–gluconeogenic pathway that produces a delay in the kinetic curves. Units are arbitrary. Reproduced with permission from [49]. [JBC: Authors of manuscripts, submitted at any time, need not contact the journal to request permission to reuse their own material]

HP-PEP and HP-pyruvate when using [2-^{13}C]dihydroxyacetone. The kinetic curves (see Fig. 8.9) show a rather substantial delay in the appearance of HP-pyruvate compared to HP-PEP, which suggests that the activity of pyruvate kinase can be readily monitored using this most interesting new HP probe.

8.5 Pentose Phosphate Pathway

Reduced nicotinamide adenine dinucleotide phosphate (NADPH) plays a key role in metabolism in at least two respects. It is an essential cofactor in reductive biosynthetic pathways such as de novo lipogenesis from acetyl-CoA and is required for the regeneration of glutathione, a reaction catalyzed by glutathione reductase (GSSG → 2GSH) in response to oxidative stress. Two equivalents of NADPH are produced in the first two enzymes in the pentose phosphate pathway, glucose-6-phosphate dehydrogenase (G6PD) and 6-phosphogluconolactonase (6PGDH). The generation of NADPH might be particularly important for studies related to nutritional excess—very important in western societies—because the PPP is assumed to be essential for both de novo lipogenesis to store excess calories and glutathione reduction to protect against oxidative stress with excess nutrition [50]. The PPP is upregulated in tissues with oxidative stress related to brain trauma and in cancer to provide ribose for nucleotide biosynthesis. However, very little is known about the activity of the PPP in the evolution of these diseases so noninvasive methods to monitor this pathway flux could have a substantial impact in future studies of human patients.

In the irreversible oxidative phase of the pentose phosphate pathway, glucose-6-phosphate is oxidized to glucono-δ-lactone-6-phosphate, which is hydrolyzed and decarboxylated to ribulose-5-phosphate plus CO_2. We recently reported that the production of hyperpolarized bicarbonate can be detected in isolated livers supplied with [1-^{13}C]-D-glucono-δ-lactone, a convenient water-soluble metabolite that is rapidly taken up by the liver [51]. In that study, HP[1-^{13}C]-D-glucono-δ-lactone was used as a probe to detect flux through the oxidative portion of the pentose phosphate pathway (PPP_{ox}). The appearance of HP $H^{13}CO_3^-$ within seconds after exposure of livers to HP[1-^{13}C]-D-glucono-δ-lactone (Fig. 8.10) demonstrated that this probe rapidly enters hepatocytes, becomes phosphorylated, and enters the PPP_{ox} pathway to produce HP-$H^{13}CO_3^-$ after three enzyme-catalyzed steps (6P-gluconolactonase, 6-phosphogluconate dehydrogenase, and carbonic anhydrase). Livers perfused with octanoate as their sole energy source show no change in the production of HP-$H^{13}CO_3^-$ after exposure to low levels of H_2O_2, while livers perfused with glucose and insulin showed a twofold increase in HP-$H^{13}CO_3^-$ after exposure to

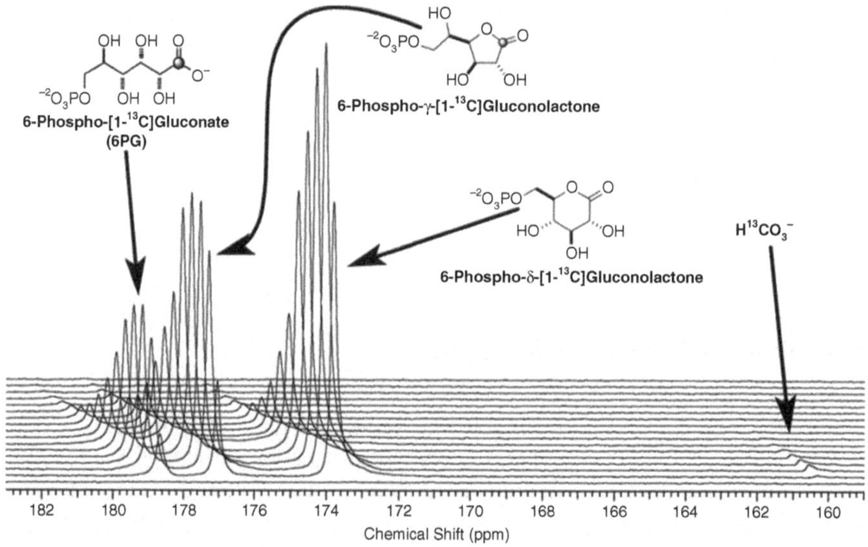

Fig. 8.10 A stacked plot of ^{13}C NMR spectra collected on an isolated mouse liver after the continuous addition of hyperpolarized [1-^{13}C]-D-glucono-δ-lactone. The ^{13}C spectra were collected every 5 s using a 66° pulse. Reproduced with permission from [51]. [NMR in Biomedicine statement: AUTHORS—If you wish to reuse your own article (or an amended version of it) in a new publication of which you are the author, editor, or coeditor, prior permission is not required (with the usual acknowledgments)]

peroxide. This indicates that flux through the PPP_{ox} is stimulated by H_2O_2 in livers perfused with glucose but not in livers perfused with octanoate alone. Subsequent perfusion of livers with nonpolarized [1,2-$^{13}C_2$]glucose followed by ^1H NMR analysis of lactate in the perfusate verified that flux through the PPP_{ox} is indeed low in healthy livers and modestly higher in peroxide-damaged livers. We conclude that hyperpolarized [1-^{13}C]-D-glucono-δ-lactone has the potential to serve as a metabolic imaging probe of this important biological pathway.

8.6 Quantitation: Proceed with Caution

In some respects, the use of hyperpolarized ^{13}C-enriched substrates such as pyruvate, acetate, lactate, α-ketoglutarate, and glutamine is conceptually quite similar to analogous radiotracer experiments. Compared to radiotracer exams, however, HP is distinctive in three respects. *First*, the mass of administered material is massively greater than the comparable radiotracer experiment. Although the exams appear to be well tolerated, investigators should consider that the mass of infused pyruvate or other substrates could have direct metabolic effects on the system. *Second*, the appearance of downstream metabolites will be sensitive to T_1 losses in upstream compartments. For example, the metabolism of HP-[1-^{13}C]pyruvate to HP-bicarbonate and acetyl-CoA occurs in the mitochondria matrix, which has very low water content. The immediate product, HP $^{13}CO_2$, must exit the matrix and react with water via carbonic anhydrase to generate HP-bicarbonate. It is difficult to know the T_1 of intramitochondrial $^{13}CO_2$, yet the T_1 must have an effect on the appearance of HP-bicarbonate. *Third*, the specifics of data acquisition will influence the observed kinetics. Every observe pulse depletes some magnetization so pulses on upstream metabolites such as pyruvate may influence the appearance of downstream metabolites such as bicarbonate. Each of these factors could influence the signal observed in an HP exam and the mathematical models developed for radiotracer kinetics will likely need appropriate modification.

8.7 Summary

Metabolism plays a central role in all body processes, and these processes are disrupted in virtually all of the high-impact diseases of western societies. Because of their complexity, probing these networks in vivo using radiotracers provides only limited insight. The use of ^{13}C as a tracer with detection by NMR is a far richer source of information about tracer distribution but the low sensitivity of ^{13}C for NMR detection excludes meaningful imaging of this nucleus in vivo. Hyperpolarization is an example of new applications of well-known physics that provides a method to probe metabolic events at the cellular, organ, and systemic level. Coupled with metabolomics tools and mathematical analysis of metabolic pathways, new insights in physiopathology are likely. Although applications to biological systems are a relatively new field, hyperpolarization has enabled new insights in some fundamental metabolic pathways. MRI of hyperpolarized ^{13}C-enriched metabolites offers considerable potential as a rapid, noninvasive tool for detecting changes in metabolic fluxes in patients.

Acknowledgments The authors thank the NIH for grant support (P41-EB015908 and R37-HL034557) during the writing of this review.

Problems

1. Production of HP-bicarbonate from HP-[1-^{13}C]pyruvate in the myocardium is thought to reflect the tissue activity of pyruvate dehydrogenase (PDH). Since PDH is a large complex embedded in mitochondria, appearance of HP-bicarbonate might be a good index of mitochondrial integrity. Suppose you perform an HP experiment in vivo and observe very little production of HP-bicarbonate in the heart of an obese rat after injection of HP-[1-^{13}C]pyruvate.
   (a) Does this mean that mitochondria in the heart are defective? Why or why not?
   (b) Suggest a physiological manipulation one might do in vivo to test your conclusion.
2. In the liver, the anaplerotic flux of pyruvate into the TCA cycle via pyruvate carboxylase (PC) is reportedly three- to sixfold higher than the TCA cycle flux itself. Assuming an average of fivefold for a particular metabolic state, calculate the fraction of HP-bicarbonate that would be produced by the entry of HP-[1-^{13}C]pyruvate via PDH versus entry of HP-[1-^{13}C]pyruvate via PC followed by either forward flux through the TCA cycle or PEPCK. State your assumptions. Assume all HP-bicarbonate produced in these pathways is detectable by ^{13}C NMR.
3. Nonhyperpolarized [1,2-^{13}C]glucose has been used as a biomarker of the oxidative pentose phosphate pathway, PPP_{ox}, activity by analysis of the ^{13}C NMR spectrum of lactate produced from this starting material. Suggest an alternative ^{13}C-enriched derivative of glucose that could potentially be used as an HP probe of PPP_{ox} activity. Explain which HP signals one might detect and how these relate to PPP_{ox} activity.

References

1. Collins, R.R.J., Patel, K., Putnam, W.C., Kapur, P., Rakheja, D.: Oncometabolites: a new paradigm for oncology, metabolism, and the clinical laboratory. Clin. Chem. **63**, 1812–1820 (2017)
2. Yang, M., Soga, T., Pollard, P.J.: Oncometabolites: linking altered metabolism with cancer. J. Clin. Invest. **123**, 3652–3658 (2013)

3. Faubert, B., Li, K.Y., Cai, L., Hensley, C.T., Kim, J., Zacharias, L.G., Yang, C., Do, Q.N., Doucette, S., Burguete, D., et al.: Lactate metabolism in human lung tumors. Cell. **171**, 358–371 (2017) e359
4. Hensley, C.T., Faubert, B., Yuan, Q., Lev-Cohain, N., Jin, E., Kim, J., Jiang, L., Ko, B., Skelton, R., Loudat, L., et al.: Metabolic heterogeneity in human lung tumors. Cell. **164**, 681–694 (2016)
5. Maher, E.A., Marin-Valencia, I., Bachoo, R.M., Mashimo, T., Raisanen, J., Hatanpaa, K.J., Jindal, A., Jeffrey, F.M., Choi, C., Madden, C., et al.: Metabolism of [U-13 C]glucose in human brain tumors in vivo. NMR Biomed. **25**, 1234–1244 (2012)
6. Mashimo, T., Pichumani, K., Vemireddy, V., Hatanpaa, K.J., Singh, D.K., Sirasanagandla, S., Nannepaga, S., Piccirillo, S.G., Kovacs, Z., Foong, C., et al.: Acetate is a bioenergetic substrate for human glioblastoma and brain metastases. Cell. **159**, 1603–1614 (2014)
7. Sellers, K., Fox, M.P., Bousamra 2nd, M., Slone, S.P., Higashi, R.M., Miller, D.M., Wang, Y., Yan, J., Yuneva, M.O., Deshpande, R., et al.: Pyruvate carboxylase is critical for non-small-cell lung cancer proliferation. J. Clin. Invest. **125**, 687–698 (2015)
8. Moreno, K.X., Sabelhaus, S.M., Merritt, M.E., Sherry, A.D., Malloy, C.R.: Competition of pyruvate with physiological substrates for oxidation by the heart: implications for studies with hyperpolarized [1-13C]pyruvate. Am. J. Physiol. Heart Circ. Physiol. **298**, H1556–H1564 (2010)
9. Purmal, C., Kucejova, B., Sherry, A.D., Burgess, S.C., Malloy, C.R., Merritt, M.E.: Propionate stimulates pyruvate oxidation in the presence of acetate. Am. J. Physiol. Heart Circ. Physiol. **307**, H1134–H1141 (2014)
10. Jeffrey, F.M., Rajagopal, A., Malloy, C.R., Sherry, A.D.: 13C-NMR: a simple yet comprehensive method for analysis of intermediary metabolism. Trends Biochem. Sci. **16**, 5–10 (1991)
11. Malloy, C.R., Sherry, A.D., Jeffrey, F.M.: Carbon flux through citric acid cycle pathways in perfused heart by 13C NMR spectroscopy. FEBS Lett. **212**, 58–62 (1987)
12. Malloy, C.R., Sherry, A.D., Jeffrey, F.M.: Evaluation of carbon flux and substrate selection through alternate pathways involving the citric acid cycle of the heart by 13C NMR spectroscopy. J. Biol. Chem. **263**, 6964–6971 (1988)
13. Malloy, C.R., Sherry, A.D., Jeffrey, F.M.: Analysis of tricarboxylic acid cycle of the heart using 13C isotope isomers. Am. J. Phys. **259**, H987–H995 (1990)
14. Merritt, M.E., Harrison, C., Storey, C., Jeffrey, F.M., Sherry, A.D., Malloy, C.R.: Hyperpolarized 13C allows a direct measure of flux through a single enzyme-catalyzed step by NMR. Proc. Natl. Acad. Sci. U. S. A. **104**, 19773–19777 (2007)
15. Schroeder, M.A., Cochlin, L.E., Heather, L.C., Clarke, K., Radda, G.K., Tyler, D.J.: In vivo assessment of pyruvate dehydrogenase flux in the heart using hyperpolarized carbon-13 magnetic resonance. Proc. Natl. Acad. Sci. U. S. A. **105**, 12051–12056 (2008)
16. Golman, K., Petersson, J.S., Magnusson, P., Johansson, E., Akeson, P., Chai, C.M., Hansson, G., Mansson, S.: Cardiac metabolism measured noninvasively by hyperpolarized 13C MRI. Magn. Reson. Med. **59**, 1005–1013 (2008)
17. Merritt, M.E., Harrison, C., Storey, C., Sherry, A.D., Malloy, C.R.: Inhibition of carbohydrate oxidation during the first minute of reperfusion after brief ischemia: NMR detection of hyperpolarized 13CO2 and H13CO3. Magn. Reson. Med. **60**, 1029–1036 (2008)
18. Khemtong, C., Carpenter, N.R., Lumata, L.L., Merritt, M.E., Moreno, K.X., Kovacs, Z., Malloy, C.R., Sherry, A.D.: Hyperpolarized 13C NMR detects rapid drug-induced changes in cardiac metabolism. Magn. Reson. Med. **74**, 312–319 (2015)
19. Shulman, R.G., Rothman, D.L.: The "glycogen shunt" in exercising muscle: a role for glycogen in muscle energetics and fatigue. Proc. Natl. Acad. Sci. U. S. A. **98**, 457–461 (2001)
20. Schroeder, M.A., Lau, A.Z., Chen, A.P., Gu, Y., Nagendran, J., Barry, J., Hu, X., Dyck, J.R., Tyler, D.J., Clarke, K., et al.: Hyperpolarized (13)C magnetic resonance reveals early- and late-onset changes to in vivo pyruvate metabolism in the failing heart. Eur. J. Heart Fail. **15**, 130–140 (2013)
21. Schroeder, M.A., Atherton, H.J., Ball, D.R., Cole, M.A., Heather, L.C., Griffin, J.L., Clarke, K., Radda, G.K., Tyler, D.J.: Real-time assessment of Krebs cycle metabolism using hyperpolarized 13C magnetic resonance spectroscopy. FASEB J. **23**, 2529–2538 (2009)

22. Cunningham, C.H., Lau, J.Y., Chen, A.P., Geraghty, B.J., Perks, W.J., Roifman, I., Wright, G.A., Connelly, K.A.: Hyperpolarized 13C metabolic MRI of the human heart: initial experience. Circ. Res. **119**, 1177–1182 (2016)
23. Rider, O.J., Apps, A., Miller, J.J.J.J., Lau, J.Y.C., Lewis, A.J.M., Peterzan, M.A., Dodd, M.S., Lau, A.Z., Trumper, C., Gallagher, F.A., Grist, J.T., Brindle, K.M., Neubauer, S., Tyler, D.J.: Noninvasive In Vivo Assessment of Cardiac Metabolism in the Healthy and Diabetic Human Heart Using Hyperpolarized 13C MRI. Circ. Res. **126**, 725–736 (2020)
24. Park, J.M., Reed, G.D., Liticker, J., Putnam, W.C., Chandra, A., Yaros, K., Afzal, A., MacNamara, J.P., Raza, J., Hall, R.G., Baxter, J., Derner, K., Pena, S., Kallem, R.R., Subramanyan, I., Edpuganti, V., Harrison, C., Muthukumar, A., Lewis, C., Reddy, S., Unni, N., Klemow, D., Syed, S., Li, H.C., Cole, S.M., Froehlich, T., Ayers, C.R., de Lemos, J.A., Malloy, C.R., Haley, B., Zaha, V.G.: Effect of Doxorubicin on Myocardial Bicarbonate Production from Pyruvate Dehydrogenase in Women with Breast Cancer. Circ Res. 2020 Oct 14. https://doi.org/10.1161/CIRCRESAHA.120.317970. Epub ahead of print. PMID: 33054563
25. Marin-Valencia, I., Cho, S.K., Rakheja, D., Hatanpaa, K.J., Kapur, P., Mashimo, T., Jindal, A., Vemireddy, V., Good, L.B., Raisanen, J., et al.: Glucose metabolism via the pentose phosphate pathway, glycolysis and Krebs cycle in an orthotopic mouse model of human brain tumors. NMR Biomed. **25**, 1177–1186 (2012)
26. Marin-Valencia, I., Yang, C., Mashimo, T., Cho, S., Baek, H., Yang, X.L., Rajagopalan, K.N., Maddie, M., Vemireddy, V., Zhao, Z., et al.: Analysis of tumor metabolism reveals mitochondrial glucose oxidation in genetically diverse human glioblastomas in the mouse brain in vivo. Cell Metab. **15**, 827–837 (2012)
27. Yang, C., Harrison, C., Jin, E.S., Chuang, D.T., Sherry, A.D., Malloy, C.R., Merritt, M.E., DeBerardinis, R.J.: Simultaneous steady-state and dynamic 13C NMR can differentiate alternative routes of pyruvate metabolism in living cancer cells. J. Biol. Chem. **289**, 6212–6224 (2014)
28. Burgess, S.C., Hausler, N., Merritt, M., Jeffrey, F.M., Storey, C., Milde, A., Koshy, S., Lindner, J., Magnuson, M.A., Malloy, C.R., et al.: Impaired tricarboxylic acid cycle activity in mouse livers lacking cytosolic phosphoenolpyruvate carboxykinase. J. Biol. Chem. **279**, 48941–48949 (2004)
29. Merritt, M.E., Harrison, C., Sherry, A.D., Malloy, C.R., Burgess, S.C.: Flux through hepatic pyruvate carboxylase and phosphoenolpyruvate carboxykinase detected by hyperpolarized 13C magnetic resonance. Proc. Natl. Acad. Sci. U. S. A. **108**, 19084–19089 (2011)
30. Fernandez, C.A., Des Rosiers, C.: Modeling of liver citric acid cycle and gluconeogenesis based on 13C mass isotopomer distribution analysis of intermediates. J. Biol. Chem. **270**, 10037–10042 (1995)
31. Katz, J.: Determination of gluconeogenesis in vivo with 14C-labeled substrates. Am. J. Phys. **248**, R391–R399 (1985)
32. Williamson, J.R., Walajtys-Rode, E., Coll, K.E.: Effects of branched chain alpha-ketoacids on the metabolism of isolated rat liver cells. I. Regulation of branched chain alpha-ketoacid metabolism. J. Biol. Chem. **254**, 11511–11520 (1979)
33. Moreno, K.X., Moore, C.L., Burgess, S.C., Sherry, A.D., Malloy, C.R., Merritt, M.E.: Production of hyperpolarized (13)CO2 from [1-(13)C]pyruvate in perfused liver does reflect total anaplerosis but is not a reliable biomarker of glucose production. Metabolomics. **11**, 1144–1156 (2015)
34. Lee, P., Leong, W., Tan, T., Lim, M., Han, W., Radda, G.K.: In vivo hyperpolarized carbon-13 magnetic resonance spectroscopy reveals increased pyruvate carboxylase flux in an insulin-resistant mouse model. Hepatology. **57**, 515–524 (2013)
35. Jin, E.S., Moreno, K.X., Wang, J.X., Fidelino, L., Merritt, M.E., Sherry, A.D., Malloy, C.R.: Metabolism of hyperpolarized [1-(13)C]pyruvate through alternate pathways in rat liver. NMR Biomed. **29**, 466–474 (2016)
36. Holyoak, T., Sullivan, S.M., Nowak, T.: Structural insights into the mechanism of PEPCK catalysis. Biochemistry. **45**, 8254–8263 (2006)
37. Sullivan, S.M., Holyoak, T.: Structures of rat cytosolic PEPCK: insight into the mechanism of phosphorylation and decarboxylation of oxaloacetic acid. Biochemistry. **46**, 10078–10088 (2007)

38. Keshari, K.R., Wilson, D.M.: Chemistry and biochemistry of 13C hyperpolarized magnetic resonance using dynamic nuclear polarization. Chem. Soc. Rev. **43**, 1627–1659 (2014)
39. Meier, S., Jensen, P.R., Duus, J.O.: Real-time detection of central carbon metabolism in living Escherichia coli and its response to perturbations. FEBS Lett. **585**, 3133–3138 (2011)
40. Jensen, P.R., Karlsson, M., Lerche, M.H., Meier, S.: Real-time DNP NMR observations of acetic acid uptake, intracellular acidification, and of consequences for glycolysis and alcoholic fermentation in yeast. Chemistry. **19**, 13288–13293 (2013)
41. Meier, S., Karlsson, M., Jensen, P.R., Lerche, M.H., Duus, J.O.: Metabolic pathway visualization in living yeast by DNP-NMR. Mol. BioSyst. **7**, 2834–2836 (2011)
42. Timm, K.N., Hartl, J., Keller, M.A., Hu, D.E., Kettunen, M.I., Rodrigues, T.B., Ralser, M., Brindle, K.M.: Hyperpolarized [U-(2) H, U-(13) C]glucose reports on glycolytic and pentose phosphate pathway activity in EL4 tumors and glycolytic activity in yeast cells. Magn. Reson. Med. **74**, 1543–1547 (2015)
43. Harris, T., Degani, H., Frydman, L.: Hyperpolarized 13C NMR studies of glucose metabolism in living breast cancer cell cultures. NMR Biomed. **26**, 1831–1843 (2013)
44. Rodrigues, T.B., Serrao, E.M., Kennedy, B.W., Hu, D.E., Kettunen, M.I., Brindle, K.M.: Magnetic resonance imaging of tumor glycolysis using hyperpolarized 13C-labeled glucose. Nat. Med. **20**, 93–97 (2014)
45. Teusink, B., Passarge, J., Reijenga, C.A., Esgalhado, E., van der Weijden, C.C., Schepper, M., Walsh, M.C., Bakker, B.M., van Dam, K., Westerhoff, H.V., et al.: Can yeast glycolysis be understood in terms of in vitro kinetics of the constituent enzymes? Testing biochemistry. Eur. J. Biochem. **267**, 5313–5329 (2000)
46. Funk, A.M., Anderson, B.L., Wen, X., Hever, T., Khemtong, C., Kovacs, Z., Sherry, A.D., Malloy, C.R.: The rate of lactate production from glucose in hearts is not altered by perdeuteration of glucose. J. Magn. Reson. **284**, 86–93 (2017)
47. Mishkovsky, M., Anderson, B., Karlsson, M., Lerche, M.H., Sherry, A.D., Gruetter, R., Kovacs, Z., Comment, A.: Measuring glucose cerebral metabolism in the healthy mouse using hyperpolarized (13)C magnetic resonance. Sci. Rep. **7**, 11719 (2017)
48. Park, J.M., Wu, M., Datta, K., Liu, S.C., Castillo, A., Lough, H., Spielman, D.M., Billingsley, K.L.: Hyperpolarized sodium [1-(13)C]-Glycerate as a probe for assessing glycolysis in vivo. J. Am. Chem. Soc. **139**, 6629–6634 (2017)
49. Moreno, K.X., Satapati, S., DeBerardinis, R.J., Burgess, S.C., Malloy, C.R., Merritt, M.E.: Real-time detection of hepatic gluconeogenic and glycogenolytic states using hyperpolarized [2-13C]dihydroxyacetone. J. Biol. Chem. **289**, 35859–35867 (2014)
50. Jin, E.S., Lee, M.H., Murphy, R.E., Malloy, C.R.: Pentose phosphate pathway activity parallels lipogenesis but not antioxidant processes in rat liver. Am. J. Physiol. Endocrinol. Metab. **314**, E543–E551 (2018)
51. Moreno, K.X., Harrison, C.E., Merritt, M.E., Kovacs, Z., Malloy, C.R., Sherry, A.D.: Hyperpolarized delta-[1-(13) C]gluconolactone as a probe of the pentose phosphate pathway. NMR Biomed. **30**, e3713 (2017)

Further Reading

Chaumeil, M.M., Radoul, M., Najac, C., Eriksson, P., Viswanath, P., Blough, M.D., Chesnelong, C., Artee Luchman, H., Gregory Cairncross, J., Ronen, S.M.: Hyperpolarized ^{13}C MR imaging detects no lactate production in mutant IDH1 gliomas: Implications for diagnosis and response monitoring. NeuroImage Clin. **12**, 180–189 (2016)

Comment, A., Merritt, M.E.: Hyperpolarized magnetic resonance as a sensitive detector of metabolic function. Biochemistry. **53**, 7333–7357 (2014)

Hesketh, R.L., Brindle, K.M.: Magnetic resonance imaging of cancer metabolism with hyperpolarized ^{13}C-labeled cell metabolites. Curr. Opin. Chem. Biol. **45**, 187–194 (2018)

Chapter 9
Novel Views on Heart Function from Dynamic Hyperpolarized NMR

Angus Lau, Kerstin Timm, and Damian Tyler

9.1 Introduction

The heart is one of the most metabolically active organs in the body. The contractile function of the human heart begins in the embryo and continues at a rate of around 70 beats/min from birth until death. The heart obtains the energy required to drive mechanical contraction of the actin–myosin machinery of the myofibrils by metabolizing substrates such as fatty acids (60–90%) and glucose (10–30%), which yield adenosine triphosphate (ATP). Impressively, the heart turns over 15 times its own weight in ATP every day [1]. The remarkable robustness of this system results from the heart's ability to modulate the use of fatty acids and glucose as well as other metabolites (lactate, pyruvate, ketone bodies) depending on supply, without a change in contractile function [2]. Oxidation of fatty acids and glucose is reciprocally regulated by the Randle cycle [3] to ensure sparing of one fuel if the other is in plentiful supply. This means that the proportion of fatty acids used for energy generation is increased during fasting, while glucose consumption increases after feeding. The key enzyme governing this metabolic switch is the pyruvate dehydrogenase (PDH) complex, which controls entry of glucose-derived pyruvate into the mitochondria by decarboxylating it to the tricarboxylic acid (TCA)-cycle fuel, acetyl-CoA (coenzyme A).

It is now well understood that myocardial energetics and heart function are compromised in many cardiac diseases [4]. These alterations in ATP production and the

A. Lau · D. Tyler (✉)
Department of Physiology Anatomy and Genetics, University of Oxford, Oxford, UK

Oxford Centre for Clinical Magnetic Resonance Research, University of Oxford, Oxford, UK
e-mail: damian.tyler@dpag.ox.ac.uk

K. Timm
Department of Physiology Anatomy and Genetics, University of Oxford, Oxford, UK

T. Jue, D. Mayer (eds.), *Dynamic Hyperpolarized Nuclear Magnetic Resonance*, Handbook of Modern Biophysics, https://doi.org/10.1007/978-3-030-55043-1_9

molecular pathways leading to energy generation, termed metabolism, are even sometimes implicated as causes, rather than consequences, of heart disease. A better understanding of these metabolic changes will likely lead to novel therapeutic targets for the treatment of a wide array of human cardiac diseases. Hyperpolarized nuclear magnetic resonance (NMR) has the potential to provide real-time, non-invasive readouts of metabolic changes associated with different cardiac disease states.

The organization of this chapter is as follows. First, the basic metabolic processes in the heart will be described, with a view towards the alterations which are induced in heart disease. Next, the utility of hyperpolarized magnetic resonance in the study of heart function will be described and placed in the context of existing metabolic imaging modalities. The chapter will focus on the use of dynamic nuclear polarization (DNP) to elucidate the metabolism of ^{13}C-labeled molecules and the relationships between metabolism and cardiac function. The physiological and disease models of the heart that can be studied using hyperpolarized magnetic resonance (both ex vivo and in vivo) will furthermore be described, as well as the first clinical studies of hyperpolarized MR in the human heart.

9.1.1 Overview of Cardiac Metabolism

The heart is an efficient converter of the chemical energy stored in fuels such as fatty acids and glucose into ATP, and finally into the mechanical energy required for contractile function. This process of metabolism consists of three stages: substrate uptake, utilization and oxidative phosphorylation to generate ATP (Fig. 9.1).

For metabolic substrate utilization, the heart acts as a metabolic omnivore. Cardiomyocytes metabolize many different types of cellular fuel, including long-chain fatty acids, glucose, lactate, pyruvate and ketone bodies. This metabolic flexibility allows the heart to continue functioning under changing conditions by generating ATP from almost any available source. This is a notable feature of the heart because it stands in contrast to other organs such as the brain, which predominantly uses glucose to generate energy to sustain function and cannot use fatty acids for energy generation at all.

The heart has minimal fuel stores in the form of glycogen and fat and thus relies on substrates from the circulation. Under normal oxygen conditions, the adult human heart obtains the majority (60–90%) of ATP from circulating long-chain free fatty acids, which are taken up by cardiomyocytes through specific transporters, transported into the mitochondria via the carnitine shuttle and then oxidized via β-oxidation [2]. The remaining 10–40% comes from the oxidation of pyruvate, which is derived from glucose via glycolysis and circulating lactate in the blood. In extreme but physiological conditions (e.g. starvation), alternate fuels such as ketone bodies become abundant and can be readily oxidized by the heart. These metabolic pathways act as adapters to supply carbon in the form of acetyl-CoA entering the TCA cycle, where a cyclical set of oxidation reactions generate reduced

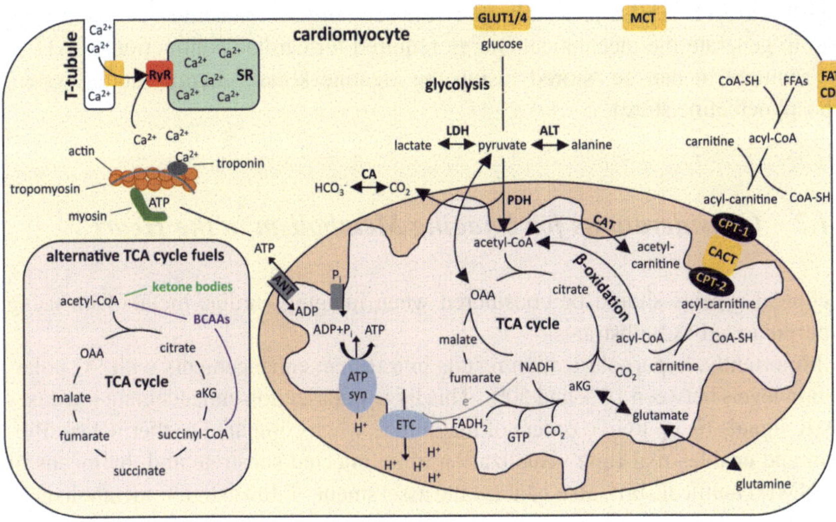

Fig. 9.1 The heart uses predominantly free fatty acids (FFAs) and glucose for energy generation. FFAs are transported into the mitochondria via the carnitine shuttle and then oxidised to acetyl-CoA in successive rounds of β-oxidation. Glucose is oxidised in glycolysis to acetyl-CoA by the gate-keeper enzyme pyruvate dehydrogenase (PDH), yielding carbon dioxide (CO_2) that is in rapid exchange with bicarbonate (HCO_3^-), or alternatively it is anaerobically broken down to lactate in the cytosol. Acetyl-CoA is further oxidised in the tricarboxylic acid (TCA) cycle to yield the electron carriers nicotinamide adenine dinucleotide (NADH) and flavin adenine dinucleotide ($FADH_2$). The TCA cycle can also accept alternative substrates such as ketone bodies and branched-chain amino acids (BCAAs) as shown in the inset. The electron carriers then donate electrons to the electron transport chain (ETC) which yields free energy to pump protons into the mitochondrial intermembrane space. The electrochemical gradient so generated leads to the proton motive force, where protons flow back into the mitochondrial matrix through ATP synthase (ATP syn), which generates ATP. The ATP is then transported into the cytosol in exchange for adenosine diphosphate (ADP) by the adenine nucleotide translocase (ANT). ATP then binds myosin to detach the latter from actin filaments allowing for a new power stroke to be initiated. Action potentials arriving at the cardiomyocytes allow entry of Ca^{2+} ions into the cells through t-tubular Ca^{2+} channels. These Ca^{2+} ions bind onto ryanodine receptors (RyR) on the sarcoplasmic reticulum (SR) mediating Ca^{2+}-induced Ca^{2+} release. Ca^{2+} ions then bind troponin, initiating another power stroke at the sarcomere through ATP hydrolysis at the myosin head. Other abbreviations used in this figure: *aKG* alpha-ketoglutarate, *ALT* alanine aminotransferase, *CA* carbonic anhydrase, *CAT* carnitine acyltransferase, *CACT* carnitine-acylcarnitine translocase, *CPT* carnitine palmitoyltransferase, *FAT* fatty acid transporter, *GLUT* glucose transporter, *LDH* lactate dehydrogenase, *MCT* monocarboxylate transporter, *OAA* oxaloacetate, *PDH* pyruvate dehydrogenase. Figure taken with permission from [14]

nicotinamide adenine dinucleotide (NADH) and flavin adenine dinucleotide ($FADH_2$). These electron carriers are used to drive a series of oxidation–reduction reactions via the electron transport chain in the inner mitochondrial membrane. The energy generated from this process is used to pump protons into the inter membrane space (proton motive force). This proton motive force in turn drives the ATP synthase, which generates ATP, powered by protons flowing back into the mitochondrial matrix, turning the ATP synthase to phosphorylate adenosine diphosphate

(ADP) into ATP. ATP is then utilized by the actin–myosin machinery of the myofibrils to generate the mechanical energy required for cardiac contraction. If ATP is not required, it can be 'stored' using the creatine kinase energy shuttle system (phosphocreatine store).

9.1.2 Considerations for Imaging Metabolism in the Heart

Technical aspects should be considered when imaging cardiac metabolism using hyperpolarized ^{13}C substrates.

Importantly, hyperpolarization results in transient enhancements with ^{13}C polarization levels between 10% and 50%. This large polarization enhancement increases NMR signals by up to five orders of magnitude when compared to thermal equilibrium and enables real-time visualization of an injected substrate and, by means of the NMR chemical shift, also enables the assessment of downstream metabolism of the substrate within the heart.

Despite this great potential, there exist significant technical challenges in efficiently using this large signal increase. The return of the signal to thermal equilibrium is determined by longitudinal T_1 relaxation. It is important to note that for hyperpolarized magnetization, T_1 relaxation is effectively a unidirectional decay process that does not lead to any renewable signal during a repeating pulse sequence. The in vivo ^{13}C T_1 of a carbonyl group is approximately 30 s in the blood, which for practical purposes sets the usable total experiment time window on the order of 1–2 min following injection of the hyperpolarized substrate. For this reason, the efficient utilization of the hyperpolarized magnetization, as well as the optimal choice of acquisition technique, must be carefully considered.

A variety of imaging sequences have been developed for use on both preclinical and clinical MR scanners for visualization of hyperpolarized magnetization in the heart. In addition to the limited T_1 of the hyperpolarized agent, physiological motion must also be accounted for (heartbeat and respiration). This restricts the available imaging time to quiescent periods of both cardiac and respiratory motion. This fundamental difference between imaging of the heart and imaging in other areas of the body (for example, in relatively stationary tumours) leads to an even greater need for efficient acquisition techniques.

To reduce the effect of motion on images, the data acquisition can be synchronized to physiological motion. This enables the combination of data from different cardiac cycles without introducing motional artefacts. In the heart, the beating motion is commonly monitored using electrocardiogram (ECG) triggering or peripheral monitoring using a plethysmograph (blood volume measurement device). Respiratory motion can be reduced by having the patient hold their breath. If this is not feasible, respiratory motion can be monitored using bellows placed on the chest wall, and respiratory triggering can be used to synchronize data acquisition to the same part of the respiratory cycle. The feasibility of these different motion compensation techniques depends on the subject. Rodents typically have much faster heart rates (350 bpm for rats, 600 bpm for mice), while larger subjects such as pigs and humans have slower heart rates (60–120 bpm).

9.1.3 Methods for Imaging Metabolism in the Heart

The simplest acquisition is a pulse-and-acquire free induction decay (FID) acquisition using small flip-angle radiofrequency (RF) pulses. Spectra are then computed by Fourier transformation of the acquired FIDs. This results in dynamic spectra which reflect metabolism over the entire heart. In these sequences, spatial localization to the heart is obtained either by placing a surface receiver coil on the chest wall of the subject or by slice selection. Additional localization is obtained by noting that the appearance of downstream metabolites comes mostly from the myocardium; red blood cells contain no mitochondria and thus no TCA cycle activity. These 'global' acquisitions are useful if the entire heart is of interest, for example, in diabetic or failing hearts where metabolic signatures are not expected to show localized changes.

If images of the spatial distribution of metabolites are desired, as might be the case with focal disease such as myocardial infarction, more sophisticated approaches are required. To resolve multiple metabolites with different resonance frequencies, the *k*-space formalism is especially relevant. The goal is to obtain image data in both space and frequency ('*x*-*f* space') by collecting data in a combined Fourier '*k*-*t* space'. An inverse Fourier transformation over all dimensions will recover images of the spatial distribution of each chemical shift in the object [5].

The most straightforward of these is a chemical shift imaging (CSI) acquisition. In this method, the RF excitation is followed by phase encoding gradient blips that traverse to a specific spatial *k*-space point. A time-domain readout follows and is repeated for each *k*-space point to fill a block in *k*-*t* space. While simple to implement, conventional CSI acquisitions are inefficient with respect to the total number of excitations required to obtain an image. In a 2008 study by Golman et al., a non-cardiac-gated 13.4 s CSI acquisition with 149 excitations was used to reconstruct a single 2-cm thick slice with 7.5 × 7.5 mm^2 nominal in-plane resolution [6].

For cardiac-gated imaging, each excitation is linked to a single cardiac cycle, and thus scan time quickly becomes limiting if a single *k*-space point is obtained per heartbeat. Several methods have been proposed to accelerate spectroscopic acquisitions. On modern MR scanners, gradient strengths are large enough to design *k*-*t* space trajectories by repeatedly tracing over the same *k*-space region during the readout. The inverse of the time between re-acquisition of each *k*-space point gives the spectral bandwidth. This bandwidth needs to be large enough to cover the corresponding ^{13}C spectrum. For example, a 'fly-back' echo-planar (EPI) trajectory can be used to repeatedly trace over a single line of *k*-space [7]. Spiral trajectories can also be used to acquire an entire 2D *k*-space matrix in rapid fashion [8].

To further reduce the number of excitations required, the spectral encoding portion of the scan can be accelerated. Multiple-echo (Dixon) acquisitions were originally developed for the separation of water and fat signals in proton imaging and have been applied to ^{13}C imaging. These methods are useful when the spectrum of interest is known *a priori*, and they accelerate scans by reducing the number of spectral points ('echoes') to be obtained. Provided enough echoes are acquired, the signal equation can be written in matrix form, and the amplitudes of each resonance can be recovered by a matrix inversion [9].

Another method of exploiting the spectral prior knowledge is to directly image each ^{13}C-labeled metabolite in a sequential fashion. This can be accomplished by the design of spectral-spatial RF excitation pulses. These are 2D excitation pulses which excite a single metabolite in a single spatial slice. Placing the spectral encoding portion of the sequence purely into the excitation pulse allows the resulting readout design to become highly simplified. Image reconstruction is straightforward to implement because only a single metabolite is spatially encoded per excitation, using standard spatial Fourier transformation on the *k*-space data. Moreover, this 'interleaved frequency' method is flexible, and any feasible rapid 2D or 3D *k*-space trajectory, such as EPI or spiral, can be used to spatially resolve the metabolic signals [10].

9.2 Probing Physiological Metabolism

The two most commonly used hyperpolarized probes for the measurement of cardiac metabolism are [1-^{13}C]pyruvate and [2-^{13}C]pyruvate (Fig. 9.2). Hyperpolarized [1-^{13}C]pyruvate allows measurements of PDH flux by assessing ^{13}C-label incorporation into CO_2, which rapidly equilibrates with the cellular bicarbonate pool through the carbonic anhydrase enzyme. Hyperpolarized [1-^{13}C]pyruvate furthermore allows assessment of the intracellular lactate pool as the interconversion between pyruvate and lactate is an equilibrium reaction, catalysed by lactate dehydrogenase (LDH). Lastly, hyperpolarized [1-^{13}C]pyruvate also yields [1-^{13}C]alanine, which is derived from the transamination of pyruvate, catalysed by alanine aminotransferase (ALT).

When using hyperpolarized [2-^{13}C]pyruvate the ^{13}C-label is retained in acetyl-CoA after decarboxylation through PDH and thus hyperpolarized [2-^{13}C]pyruvate allows measurements of flux into the TCA cycle, where the label can be seen in [1-^{13}C]citrate and [5-^{13}C]glutamate, which equilibrates with the TCA-cycle intermediate α-ketoglutarate. Hyperpolarized [2-^{13}C]pyruvate furthermore allows measurements of [1-^{13}C]acetylcarnitine, which buffers excess acetyl-CoA and thus reports on carnitine availability [11, 12]. Pyruvate can furthermore be doubly-labelled, and hyperpolarized [1,2-$^{13}C_2$]pyruvate has been used to assess both PDH flux and TCA cycle metabolism simultaneously in the pig heart. This was performed with an interleaved acquisition scheme [13].

Other hyperpolarized probes have been explored for cardiac metabolism, and these have been recently reviewed [14]. Those probes include [1-^{13}C]lactate [15] as an alternative to [1-^{13}C]pyruvate; [1-^{13}C]acetate [16, 17] as a surrogate for

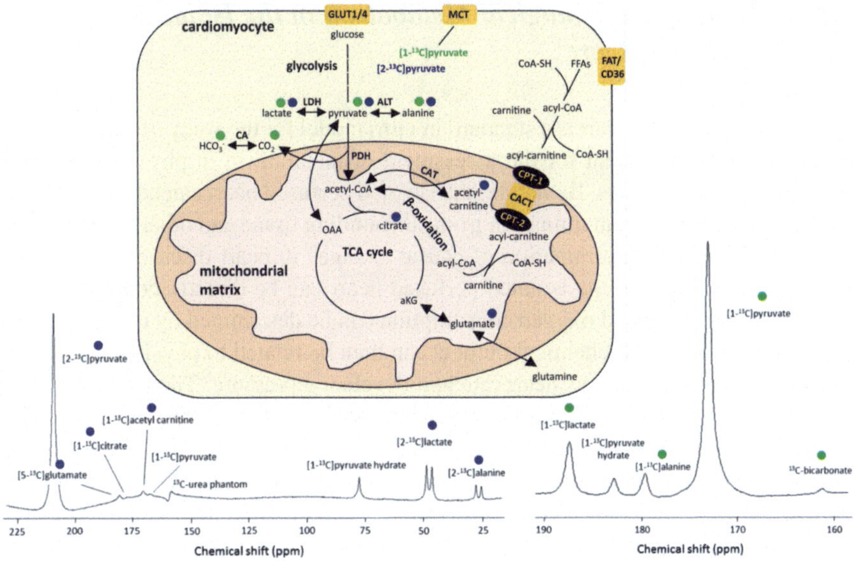

Fig. 9.2 Drawing showing the metabolic fate of hyperpolarized [1-^{13}C]- and [2-^{13}C]pyruvate in the cardiomyocyte (top) and representative MR spectra following injection of hyperpolarized [2-^{13}C]pyruvate (bottom left) and [1-^{13}C]pyruvate (bottom right) during experiments in rat hearts. Hyperpolarized [1-^{13}C]- and [2-^{13}C]pyruvate enter the cardiomyocyte through monocarboxylate transporters (MCT). The ^{13}C label in [1-^{13}C]pyruvate can be exchanged into the existing lactate pool through lactate dehydrogenase (LDH) or alternatively [1-^{13}C]pyruvate can be oxidatively decarboxylated to acetyl-CoA yielding carbon dioxide (CO_2) in the reaction catalysed by pyruvate dehydrogenase (PDH). The CO_2 is in rapid exchange with tissue bicarbonate (HCO_3^-) through carbonic anhydrase (CA). [1-^{13}C]pyruvate can also be transaminated to alanine in the reaction catalysed by alanine aminotransferase (ALT). [2-^{13}C]pyruvate can also yield ^{13}C-labelled lactate and alanine. However, the ^{13}C-label is retained in acetyl-CoA after oxidative decarboxylation by PDH and enters the tricarboxylic acid (TCA) cycle. Acetyl-CoA binds with oxaloacetate (OAA) in the reaction catalysed by citrate synthase to form citrate. The ^{13}C-label in citrate can be transferred into the glutamate pool via several enzymatic steps. The ^{13}C-label in acetyl-CoA can equilibrate with the acetyl-carnitine pool, catalysed by reversible carnitine-acetyltransferase (CAT). Green and blue dots represent [1-^{13}C]- and [2-^{13}C]pyruvate, respectively, and traces the labelled metabolites detected by MR spectroscopy. [1-^{13}C]pyruvate hydrate is a non-metabolically active contaminant arising through the (reversible) heat-induced hydration of hyperpolarized [1-^{13}C]pyruvate during the dissolution process. The peak labelled ^{13}C-urea phantom in the left spectrum is from an 8 M ^{13}C-urea-containing phantom attached to the carbon-13 coil next to the rat to allow for pulse calibration. Other abbreviations used in this figure: *aKG* alpha-ketoglutarate, *CACT* carnitine-acylcarnitine translocase, *CPT* carnitine palmitoyltransferase, *FAT* fatty acid transporter, *FFAs* fatty acids, *GLUT* glucose transporter. Figure taken with permission from [14]

short-chain fatty acid metabolism; [^{13}C]acetoacetate [18, 19] to explore ketone body metabolism and [1-^{13}C]butyrate [20] to assess both short-chain fatty acid metabolism and pseudoketogenesis. Furthermore, simultaneous polarization and injection of [1-^{13}C]pyruvate and [1-^{13}C]butyrate allows the direct investigation of substrate selection in the rodent heart [21]. None of these probes, however, show sufficiently high signal-to-noise ratio (SNR) to allow for in vivo imaging, at least in rodents to date.

9.2.1 Probing Physiological Metabolism in the Isolated Perfused Heart

The isolated perfused heart is a standard in vitro model for the study of cardiac function. It is an ideal system for the assessment of metabolism in physiological and pathophysiological states. By nature, the isolated perfused heart can be studied in an environment without contamination from surrounding tissue and other in vivo influences, enabling the determination of causal changes in heart disease. Delivery of metabolic substrates to the isolated perfused heart can be directly controlled, and metabolic utilization and oxygen consumption can be determined by examining the output buffer. These biochemical outputs can then be related to physiological heart function parameters such as heart rate and developed pressure. The perfused heart thus provides an attractive setting in which hyperpolarized ^{13}C measurements can be validated. The information gained from such experiments can guide the further development and validation of experiments in the in vivo setting. One note of caution, however, is that many studies involving isolated perfused hearts do not account for hormonal influences, such as from insulin/glucagon, which are rarely added to the perfusate. Results from metabolic studies in perfused hearts therefore must be interpreted with this in mind.

From the viewpoint of hyperpolarized NMR, the isolated perfused heart provides several key advantages over corresponding in vivo experiments: physiological interventions, such as the induction of myocardial ischemia, can be performed in situ within the MR environment, enabling subsequent or concurrent assessment of rapid metabolic changes following intervention. The delivery of substrates is performed by addition of the hyperpolarized ^{13}C substrate to the buffer, as opposed to intravenous infusion. This can reduce sharply the transit time between injection and arrival in the heart. Furthermore, perfused heart experiments are typically performed in vertical bore systems, for which higher field strengths and narrower bore magnets result in better spectral separation and narrower linewidths of the metabolites of interest. High signal-to-noise ratios and improved signal transmit and receive homogeneity can be obtained due to the small heart size and proximity of the RF coil to the heart.

9.2.1.1 Measurement of Flux Through a Single Enzymatic Step

The isolated perfused heart system has been used to validate and determine the source of signal in the hyperpolarized ^{13}C experiment. As described above, the in vitro setting allows substrate concentrations to be easily controlled. Thus, PDH flux can be modulated by the availability of fatty acids within the buffer. Fatty acid oxidation yields acetyl-CoA, which is oxidized in the TCA cycle to generate reduced NADH and $FADH_2$ for subsequent oxidation in the electron transport chain. Acetyl-CoA further acts as an allosteric enzyme modulator and inhibits PDH via the Randle cycle [3] to ensure inhibition of glucose oxidation if fatty acids are plentiful.

Short-chain odd-numbered fatty acids, such as propionate (a three-carbon fatty acid), act as potent activators of PDH [22], as these enter the TCA cycle at the level of succinyl-CoA and then increase the pyruvate pool through the malic enzyme [23]. On the other hand, longer even-numbered fatty acids, such as octanoate (an eight-carbon fatty acids), strongly inhibit PDH as they yield increased levels of acetyl-CoA, inhibiting PDH. By supplying hyperpolarized [1-^{13}C]pyruvate in the presence and absence of these competing substrates, the hyperpolarized signals can be correlated with gold-standard steady state ^{13}C NMR isotopomer analyses.

The study by Merritt et al. [24] demonstrated that the decarboxylation of the ^{13}C label from [1-^{13}C]pyruvate directly reflects metabolic flux through PDH. Rapid pulse-and-acquire spectra were obtained at 1 s temporal resolution following delivery of hyperpolarized [1-^{13}C]pyruvate to the perfused heart (Fig. 9.3). The levels of hyperpolarized $H^{13}CO_3^-$ and $^{13}CO_2$ were measured in hearts supplied with either hyperpolarized [1-^{13}C]pyruvate alone or hyperpolarized [1-^{13}C]pyruvate plus both octanoate and propionate. The fatty acid competition dramatically reduced

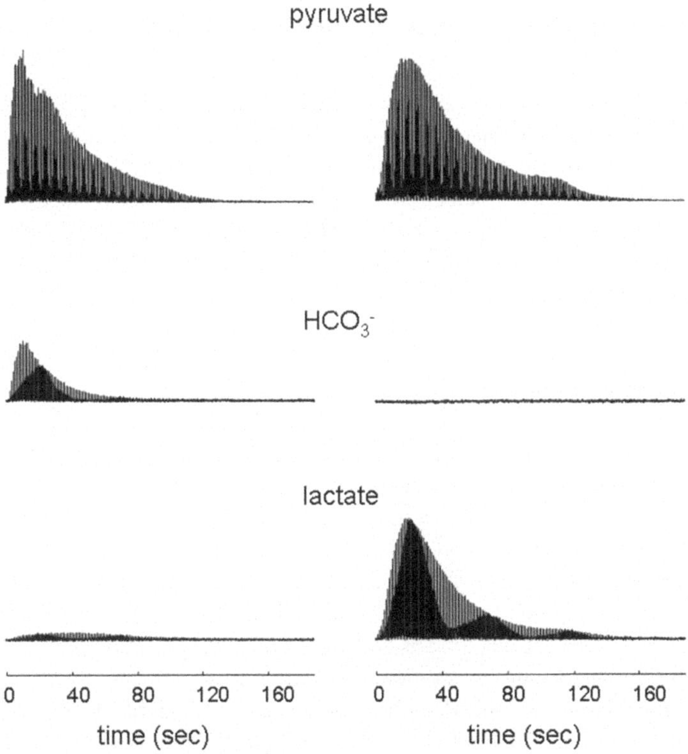

Fig. 9.3 Time-course plots of ^{13}C MR spectra from perfused hearts supplied with [1-^{13}C]pyruvate. Each point is a single scan from the chemical shift ± ≈ 10 Hz from the region assigned to pyruvate C1, $H^{13}CO_3^-$, or lactate C1. The left column is from a heart supplied with [1-^{13}C]pyruvate. The right column is from a heart supplied with [1-^{13}C]pyruvate plus unlabelled octanoate. Figure taken with permission from [24]

$H^{13}CO_3^-$ and $^{13}CO_2$ appearance, even in the presence of propionate. To measure the contribution to TCA cycle flux from these different substrates, steady-state 'isotopomer' experiments were then performed. Perfused hearts were supplied with uniformly labelled [U-$^{13}C_3$]pyruvate, or [2,4,6,8-$^{13}C_4$]octanoate plus [U-^{13}C]pyruvate plus unlabelled propionate. The different labelling patterns resulted in different steady-state labelling of acetyl-CoA and easily distinguished ^{13}C glutamate multiplets, which arise from either pyruvate or β-oxidation. This analysis reveals that TCA cycle flux is equally active in the presence of all substrate combinations. (Refer to reference 24 for details on methodology and labelling pattern).

Apart from PDH modulation in perfused hearts, lactate levels can also be modulated leading to increased [1-^{13}C]lactate measurements from hyperpolarized [1-^{13}C]pyruvate. This was performed by infusing isoproterenol, a β-adrenergic agonist, in the isolated perfused heart, which led to a rapid increase in labelled lactate [25].

9.2.2 Probing Physiological Metabolism In vivo

While the perfused heart provides a controlled system in which to validate hyperpolarized ^{13}C MR measurements, the true utility of the technique lies in its non-invasive nature and ability to inform on the same biochemical pathways in vivo. The detection of changes in substrate selection and PDH activity in cardiac disease indicates the potential utility of hyperpolarized ^{13}C MR in the clinical diagnosis of heart disease. The following section will focus on global measurements of in vivo cardiac metabolism using hyperpolarized ^{13}C spectroscopic acquisitions.

9.2.2.1 Rodent Studies

Hyperpolarized [1-^{13}C]pyruvate MRS of the heart is sensitive to physiological perturbations of PDH flux. Hyperpolarized pyruvate is commonly delivered via the intravenous route by tail vein injection. Rapid pulse-and-acquire spectra with cardiac gating can then be obtained, typically with 1 s temporal resolution following injection of hyperpolarized [1-^{13}C]pyruvate. The signals are localized to the heart by placement of a surface coil over the chest or slice-selection. It is certain that contamination from ventricular blood as well as neighbouring organs, such as liver and skeletal muscle, contribute to the detected signal, especially in [1-^{13}C]lactate and [1-^{13}C]alanine. However, the decarboxylation of [1-^{13}C]pyruvate to $H^{13}CO_3^-$ is dominated by cardiac PDH activity, based on both the kinetics of an intravenous injection, as well as the relative PDH flux in the heart when compared to the liver and skeletal muscle. In a first in vivo demonstration in a rodent model, Schroeder et al. [26] showed that PDH flux, quantified using the appearance of hyperpolarized $H^{13}CO_3^-$, is sensitive to overnight fasting, where increased levels of circulating fatty acids and increased cardiac fatty acid oxidation inhibit PDH via the Randle cycle, thus decreasing glucose oxidation. This decrease in PDH flux measured by hyperpolarized [1-^{13}C]pyruvate was subsequently shown to correlate with PDH activity in tissue extracts [27].

Most cardiac hyperpolarized MR studies are performed in rats. However, mice are in many ways more attractive model systems, as a wide range of well characterized transgenic models exist in which alterations in cardiac metabolism could be directly matched to gene function. Nevertheless, mouse heart physiology (e.g. heart rate) differs dramatically from humans (and rats), and caution is therefore required in the interpretation of results. Notwithstanding the clear benefit of genetic mouse models, performing hyperpolarized ^{13}C MR is challenging in mice, due to the much smaller size of the heart (1/10th of a rat) and faster heart rate (50% higher). Nevertheless, measurements of PDH flux with hyperpolarized [1-^{13}C]pyruvate have been successfully performed. The first study by Dodd et al. [28] showed that like measurements in rats, PDH flux is decreased in the mouse heart after fasting. It was furthermore shown that by using the potent activator of PDH, dichloroacetate (DCA), PDH flux could be increased in the mouse heart. DCA inhibits pyruvate dehydrogenase kinase (PDK), which phosphorylates and inhibits PDH. Thus, inhibiting PDK shifts PDH towards the active state and stimulates PDH activity.

9.2.2.2 Large Animal Studies

Physiological cardiac metabolism can also be assessed in larger animal models, such as the pig. The porcine heart more closely resembles the human heart in size, anatomy and in physiology (similar heart rate) than the rodent heart and is therefore an ideal model system to study cardiac metabolism. To this effect, initial spectroscopic experiments were undertaken by Menichetti et al. [29] where they investigated the impact of increased workload on cardiac metabolism using a slice-selective spectroscopy sequence. They observed that a doubling of the rate pressure product following infusion of the β-adrenergic agonist, dobutamine, led to a two-fold increase in the incorporation of the injected hyperpolarized [1-^{13}C]pyruvate into both bicarbonate and lactate.

Additionally, hyperpolarized [1,2-$^{13}C_2$]pyruvate MRS has proved useful to assess PDH flux and TCA cycle metabolism in the porcine heart during the same experiment, by implementing an interleaved MRS sequence [13]. The same authors subsequently found that upon dobutamine infusion, cytosolic metabolic rates increased, while glutamate-labelling was maintained. The authors conclude that this reflects the rate-limiting step in NADH-transport across the inner mitochondrial membrane, namely the malate-aspartate shuttle [30]. NADH is required for glutamate dehydrogenase, which equilibrates the mitochondrial α-ketoglutarate pool (TCA cycle intermediate) with the cytosolic glutamate pool (observable as [5-^{13}C]glutamate downstream of [2-^{13}C]pyruvate).

9.2.3 *Imaging In vivo Cardiac Metabolism Using Hyperpolarized MRI*

The previous sections have described hyperpolarized ^{13}C methods for assessing global cardiac function and metabolism. This is sufficient for homogeneous cardiac diseases, such as diabetic cardiomyopathy or heart failure. However, focal diseases, such as ischemia or myocardial infarction result in localized changes to cardiac metabolism. For this reason, methods to spatially localize metabolic changes in the heart are important.

9.2.3.1 Large Animal Studies

The most promising pre-clinical work demonstrating feasibility has been performed in porcine models of cardiac disease. The porcine heart is of comparable size to the human heart, which allows for testing of directly translatable imaging techniques for clinical use. The cardiac cycle is similar to the human cardiac cycle, with a relatively long (200–300 ms) quiescent diastolic window for imaging.

As described in Sect. 9.1.2 (Considerations for Imaging Metabolism in the Heart), acceleration and optimization of the scan pulse sequence must be undertaken in order to improve spatial and temporal resolution, and to minimize motion artefacts. If each excitation is linked to a single cardiac cycle the total scan time quickly becomes limiting when a single *k*-space point is obtained per heartbeat. Several methods have been proposed to accelerate spectroscopic acquisitions.

Lau et al. presented cardiac-gated CSI data showing the spatial distribution of pyruvate and bicarbonate with 6×6 mm^2 in-plane resolution in a single 10-mm thick axial slice through the pig myocardium [8]. The images they obtained showed pyruvate localized mainly to the left ventricle, but also to the right ventricle and the internal thoracic arteries, while bicarbonate was localized to the myocardium. The SNR of the images generated dropped off near the posterior myocardium, which was related to a reduction in B_1 due to the surface transmit/receive coil used for imaging. They also presented a spectrum, which was a sum over all magnitude spectra from all voxels, demonstrating the uniformity of the shim over the heart. The linewidth of both pyruvate and bicarbonate was $\sim$1 ppm full width half maximum (FWHM). The slightly broader bicarbonate resonance was presumably due to susceptibility differences from the myocardium being closer to the air-tissue interface with the lung. No lactate resonance appeared in the sum spectrum at its expected chemical shift.

Acceleration via spectral-spatial excitation of single resonances has found its greatest success in cardiac imaging of large animals. Based on previously acquired spectra following hyperpolarized [1-^{13}C]pyruvate injection, Lau et al. designed a spectral–spatial excitation pulse with a duration of 13 ms and sub-lobes of 1 ms giving spectral pass-bands with FWHM of 200 Hz separated by a 1-kHz-wide stopband and a minimum slice thickness of 6 mm [8].

To cope with the challenging motion and flow conditions encountered in cardiac imaging, a flow-insensitive spiral readout gradient was selected. The spiral gradient was designed for a field of view (FOV) of 24 cm and a nominal in-plane resolution of 3.7 mm.

The above methods result in metabolic maps at a single time point. However, the selective excitation approach provides inherent flexibility by enabling imaging of only certain metabolites at certain time points. This allows for interrogation of the injected pyruvate during the arrival phase (the 'input function') at low flip angles, while preserving magnetization for conversion into its downstream metabolites. At later times, a higher flip angle can be used to image [1-^{13}C]lactate and $H^{13}CO_3^-$, producing high quality metabolic maps indicating the presence or absence of tissue viability and ischemia within the heart [31].

In more recent work, Hansen et al. [32] used a multiple-echo IDEAL spiral imaging sequence to image the localization of the hyperpolarized [1-^{13}C]pyruvate signal and downstream metabolites (lactate, alanine, bicarbonate) in Danish domestic pig hearts. They showed that PDH flux was increased after an infusion of glucose-insulin-potassium (GIK). The alanine and lactate signals from hyperpolarized [1-^{13}C]pyruvate were also increased after GIK. This was associated with decreased plasma fatty acid levels and increased plasma insulin levels.

9.2.3.2 Rodent Studies

Rodent models of heart disease have gained widespread use in basic research. Small animals can be easily handled, enabling rapid throughput studies. They have rapid generation time, and it is possible to perform genetic manipulation, especially in murine models.

From an imaging standpoint, some interesting challenges arise with the imaging of rats and mice. The rat heart is approximately 1 cm in diameter, and the myocardial wall is usually 1–2 mm thick. The typical resting heart rate is 300–400 beats/min. Average cardiac cycles are 150 ms in duration, with a diastolic window of 70 ms. These physiological parameters lead to imaging requirements of simultaneously very high spatial and temporal resolution. Fast spectroscopic imaging methods are required to dynamically image the spatial distribution of pyruvate metabolism in this very small, constantly moving organ.

The principles described in the previous section for imaging of hyperpolarized ^{13}C pyruvate in large animals apply to small animals, but the spatial and temporal scales involved restrict the feasibility of some of these techniques. However, despite these restrictions, multiple groups have demonstrated cardiac metabolic imaging in rodents by using various approaches to reduce the demands that such high spatial/temporal resolution imaging places on the gradient hardware of the MRI scanner or through acceleration techniques to reduce the coverage of *k-t* space required.

For example, Miller et al. demonstrated physiological changes in cardiac metabolism (fed vs. fasted vs. DCA) in the in vivo rat heart using a high spatial resolution ($2 \times 2 \times 3.8$ mm) three-dimensional (3D) echo planar imaging approach [33]. In this work, a 3D approach was taken as this reduces the demands placed on the gradient system by spectral-spatial excitation pulses when trying to obtain the thin slices required by 2D approaches. Sigfridsson et al. took a different approach to reducing gradient demands by utilizing a multi-echo, multi-band excitation approach that allowed for a longer duration RF pulse that was designed to illicit a low flip angle excitation on pyruvate and a higher flip angle excitation on the downstream metabolites [34]. In other work, the same group utilized spatiotemporal correlations to accelerate their acquisition five-fold and achieve a spatial resolution of 1 × 1 mm at a temporal resolution of 3 s in the in vivo rat heart [35]. However, the use of a 2D spectral-spatial excitation limited the minimum slice thickness to 8 mm.

9.3 Non-metabolic Imaging

For substances which are not metabolized (e.g. ^{13}C-urea), standard fast imaging techniques can be used to provide rapid visualization of the injected agent. The lack of any signal background can be exploited to provide high quality images of the hyperpolarized compound. Because little or no metabolism of the injected agent occurs, the signal lifetime is typically longer than compounds such as ^{13}C pyruvate. Furthermore, the magnetization does not need to be preserved to observe downstream metabolites; rather, the sequence should be optimized to observe the initial delivery of the agent to the tissue. The potential cardiac applications of non-metabolic imaging include angiography, perfusion mapping, as well as interventional applications for catheter tracking. Furthermore, tissue redox state and pH, which are linked to cellular metabolism, can also be assessed with hyperpolarized NMR.

9.3.1 Angiography

Due to the lack of signal background (only 1.1% natural abundance of ^{13}C nuclei with low thermal polarization), any signal in the images arises purely from the injected substance. Projection imaging through the entire subject reveals only vessels containing the hyperpolarized agent. A rapid balanced steady-state free precession (SSFP) sequence allows one to exploit the long T_1 and T_2 relaxation times of the ^{13}C nucleus, the high initial signal amplitude and large flip angles, to obtain images with high spatial and temporal resolution. The balanced gradients ensure that the resulting transverse magnetization constructively adds, leading to images with high SNR. This method of angiography was first used by Olssen et al. in healthy pigs [36].

9.3.2 Perfusion

The high contrast-to-noise provided by the hyperpolarized imaging approaches described above can also be applied to the assessment of cardiac perfusion. However, when trying to isolate the signal contribution of the hyperpolarized agent perfusing the myocardial tissue from the large signal contribution presented by the hyperpolarized agent in the myocardial chambers, partial volume artefacts can become prohibitively large. To overcome this, it is necessary to employ velocity selective [37] or flow sensitizing [38] techniques to minimize the signal from the fast-flowing spins within the myocardial chambers and to limit the signal to that from the myocardial tissue. Using these approaches and appropriate modelling of the signals obtained, Fuetter et al. have shown that quantitative assessment of myocardial perfusion can be achieved in the porcine heart [39]. Figure 9.4 depicts dynamic hyperpolarized perfusion images showing the arrival of the perfusion probe in the right ventricle and left ventricle followed by perfusion of the coronary arteries in just 6 s. In addition to the quantitative assessment of perfusion, the ability to hyperpolarize multiple compounds at the same time can enable the simultaneous assessment of cardiac perfusion and metabolism in a single scan through the use of co-polarized ^{13}C-urea (to assess perfusion) and [1-^{13}C]pyruvate (to assess metabolism) [40].

9.3.3 Assessment of Cellular Redox State

A seminal paper from Hans Krebs in 1967 [41] showed that the cytoplasmic and mitochondrial redox states (NAD:NADH ratio) can be assessed by measuring the metabolite pairs lactate:pyruvate and β-hydroxybutyrate:acetoacetate, respectively. Measuring metabolite pairs is desirable over measuring the reduced and oxidized electron carrier, as these degrade rapidly upon cell/tissue extraction. Measurement of the metabolites usually involves metabolite extraction and spectrophotometric or mass spectrometric analysis of the extracts. However, cytosolic and mitochondrial redox state can now also be assessed using hyperpolarized MR. For cytosolic

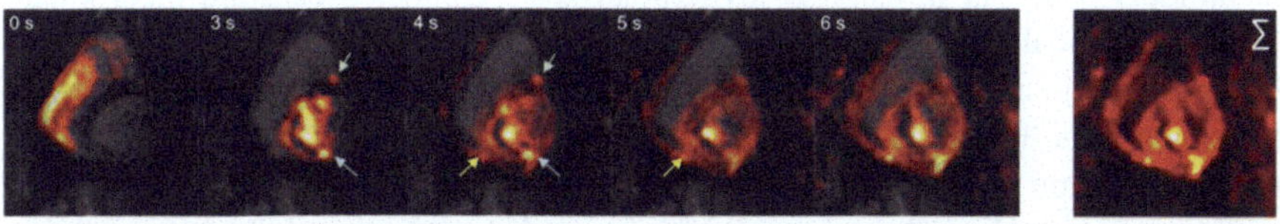

Fig. 9.4 Representative dynamic perfusion image series acquired in vivo (left). Myocardial bolus peak occurs between 4 and 6 s after start of acquisition. Arrows in left panel indicate the left anterior descending artery (LAD) (green, top right), left circumflex (LCx) (blue, bottom right) and right coronary artery (RCA) (yellow, bottom left). Myocardial bolus peak image summed over two frames (right panel). Enhanced signal in the lateral inferior segment is caused by signal leakage from the LCx and signal pile-up stemming from geometric distortions in the EPI image due to local off-resonances. Figure taken with permission from [39]

cardiac redox state hyperpolarized ^{13}C-labelled pyruvate [42] has been utilized in vivo, while mitochondrial redox state can be assessed with hyperpolarized ^{13}C-labelled acetoacetate [43].

9.3.4 Assessment of Intracellular pH

Myocardial ischemia results in the rapid onset of acidosis. In the poorly perfused, ischemic heart, increased anaerobic glycolysis produces lactic acid that accumulates in the intracellular and extracellular spaces. Transient acidosis may be beneficial to the heart due to reduced contractility and conservation of ATP for ion transport. However, sustained, severe acidosis decreases Na^+/K^+ ATPase activity, which increases myocardial Na^+ levels. This results in elevated myocardial Ca^{2+} levels via the Na^+/Ca^{2+} exchanger, damaging the myocardium. In ^{31}P-MRS, the inorganic phosphate (P_i) resonance exhibits a pH-dependent chemical shift. This feature of phosphorus spectra has been used as a gold standard for intracellular pH (pH_i) measurement in the isolated, perfused heart. However, non-invasive measurement of pH_i using phosphorus spectroscopy in vivo has remained challenging due to the contamination of the myocardial P_i resonance with the nearby 2,3-diphosphoglycerate (2,3-DPG), found in the ventricular blood.

Hyperpolarized ^{13}C studies allow for non-invasive measurement of pH by taking advantage of the pH-dependent equilibrium between bicarbonate (HCO_3^-) and carbon dioxide (CO_2). The rapid equilibration of these two species takes place within the intracellular, extracellular and blood pools, and is catalysed by different isozymes of the carbonic anhydrase enzyme. Assuming the two species are at equilibrium, pH can be computed by taking the ratio between the HCO_3^- and CO_2 signal intensities and using the Henderson-Hasselbalch equation,

$$pH = pK_a + \log\left(\left[HCO_3^-\right]/\left[CO_2\right]\right).$$

Where $pK_a = -\log_{10}K_a$.

An initial proof of concept measuring pH in vivo in this way was demonstrated in tumours using an infusion of hyperpolarized ^{13}C-labeled caesium bicarbonate [44]. A chemical shift imaging sequence was used to image bicarbonate and CO_2. This resulted in a map of the extracellular pH within the hypoxic tumour microenvironment. The liquid state polarization of hyperpolarized $H^{13}CO_3^-$ was 16% and the T_1 relaxation time was 10 s in this study.

Intracellular pH can be assessed using an analogous approach by taking advantage of the PDH-catalysed cleavage of the ^{13}C label from [1-^{13}C]pyruvate. This mitochondrial reaction produces $^{13}CO_2$ and $H^{13}CO_3^-$, which are rapidly equilibrated by mitochondrial carbonic anhydrase. The success of this strategy depends strongly on whether there is sufficient PDH flux and TCA cycle activity within the tissue of interest. In the normoxic fed state, the heart is typically highly aerobic. Assessment

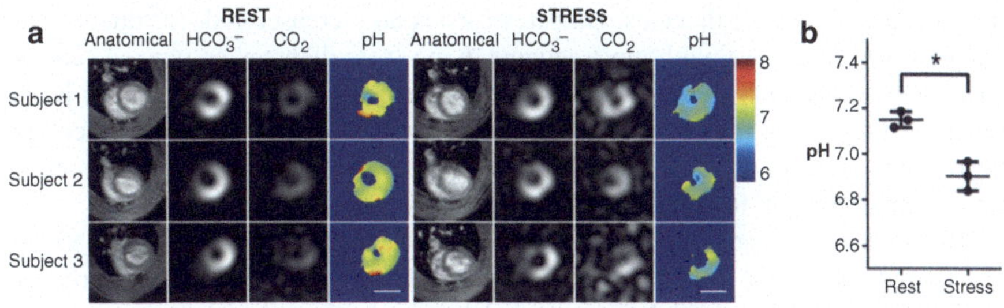

Fig. 9.5 Anatomical, HCO_3, CO_2, and pH images and pH determination from resting and stressed hearts. Panel (**a**) shows intracellular pH measurements obtained using the ratio between HCO_3^- and CO_2 from a mid-ventricular slice. Images are cropped to a 30 × 30 mm^2 FOV. The scale bar indicates 1 cm. The ^{13}C images are taken from a single time point, 20 s after the start of injection. Panel (**b**) shows that increased cardiac workload results in decreased apparent pH in the healthy heart. Figure taken with permission from [46]

of intracellular pH via [1-^{13}C]pyruvate has been demonstrated in the isolated perfused heart [45], in vivo rodent heart [46], and in the porcine heart [13], though in the latter dually-labelled [1,2-$^{13}C_2$]pyruvate was used, allowing for simultaneous assessment of PDH flux, TCA cycle metabolism and pH. Typical liquid state polarization and T_1 values of hyperpolarized [1-^{13}C]pyruvate are 30% and 40 s, respectively, making pH measurements in this way more sensitive. Figure 9.5 shows a false-colour map of myocardial intracellular pH in the rat before and after dobutamine stress, assessed with hyperpolarized [1-^{13}C]pyruvate.

9.4 Applications in Preclinical Disease Models

9.4.1 Metabolic Assessment in the Diabetic Heart

Diabetes is categorized into multiple sub-forms, the two most common being insulin-dependent, or type I, and insulin-independent, or type II diabetes. Type I diabetes is caused by autoimmune-destruction of the insulin-producing pancreatic β-cells (cause still unknown) and always requires treatment with insulin-injections. Type II diabetes, however, is a multifactorial disease involving some genetic components as well as risk factors such as obesity, resulting in insulin-resistance, initially with a compensatory increase in insulin-secretion followed by decompensated hypo-insulinemia. Type II diabetes can be treated with diet and exercise in its earliest form, with oral medications such as metformin (which decreases hepatic gluconeogenesis) used if diet and exercise are unsuccessful and subsequently with insulin (and combined oral-antidiabetics) in more severe forms. Historically, type I diabetes was a disease of the young, with a median onset of 14 years of age, whereas type II diabetes was late onset, in adult life. However, with the rise of obesity in the general population, even in early childhood, young children are now also routinely diagnosed with type II diabetes. Diabetes carries an increased risk for micro- and macro-vascular disease, with most patients dying from diabetes-associated cardiovascular disease, such as myocardial infarction or heart failure. The diabetic heart relies on fatty acids for energy-generation even more so than the normal healthy heart, and while fatty acids generate more energy per gram than glucose, they require more oxygen for complete oxidation to CO_2.

This decrease in cardiac glucose oxidation can be measured with hyperpolarized [1-^{13}C]pyruvate. In the first in vivo paper of cardiac hyperpolarized MRS, Schroeder et al. [26] showed in a model of type I diabetes (rats injected with streptozotocin) that PDH flux was decreased compared to healthy control rats. This experiment confirmed that hyperpolarized MRS could be used to detect the well-known decrease in glucose oxidation in the diabetic heart. A model of untreated type I diabetes is, however, not a great model of diabetic cardiomyopathy, as all type I diabetic patients are treated with insulin. Furthermore, type II diabetes is much more common and not all patients with type II diabetes are treated with insulin or treated at all (if undiagnosed). Measurements of PDH flux were thus also performed in the heart of type II diabetic rats (high fat diet plus low dose streptozotocin injections [47]). Similar to the type I diabetic heart, a study by LePage et al. [48] not only showed decreased cardiac PDH flux in type II diabetic rats, but also that this could be rescued with DCA treatment. In addition to reversing glucose oxidation capacity, DCA treatment in this rat model of type II diabetes also reversed diastolic dysfunction.

Metabolism in the diabetic heart has also been assessed with other hyperpolarized probes. Abdurrachim et al. [18] used hyperpolarized [3-^{13}C]acetoacetate to explore cardiac ketone body metabolism in the Goto-Kakazaki (GK) non-obese rat model of type II diabetes. The authors found that incorporation of the ^{13}C-label into TCA-cycle intermediates was increased in the hearts of GK rats compared to wild-type (WT) Wistar rats, suggesting increased usage of acetoacetate as an energetic fuel in the mitochondria. Interestingly, labelling of β-hydroxybutyrate was decreased in this model, suggesting an altered mitochondrial redox state [41] in the GK rat heart.

9.4.2 Metabolic Assessment in the Ischemic Heart

Cardiac ischemia changes metabolic fuel choice and as oxygen levels in the tissue and oxidative metabolism decreases, anaerobic glycolysis increases in compensation to ensure energy supply by substrate level phosphorylation. This switch is coordinated by differential translocation of fatty acid and glucose transporters in the sarcolemma [49]. However, anaerobic glycolysis is less efficient at generating energy, yielding only two molecules of ATP per molecule of glucose compared to 32 ATP molecules if glucose is fully oxidized. This leads to energetic impairment of the myocardium, with a successive onset of heart failure if ischemia is not reversed. Ischemia can occur slowly over time with coronary artery disease (Angina), or can happen rapidly in myocardial infarction (MI). In preclinical models, ischemia is either achieved by global low-flow ischemia of isolated perfused hearts (reduced buffer flow through the coronary arteries) or surgically in animal models. Here, both ligation of the left anterior descending artery (LAD) and cryo-infarction with a liquid nitrogen-cooled probe placed on the myocardium can be used to produce an MI in species such as mice, rats and pigs. Since hyperpolarized MRS/MRI can visualize real-time metabolism, it is ideally placed to evaluate the rapid changes that occur in cardiac metabolism due to both the onset of ischemia as well as the effects of reperfusion.

In the perfused heart, low-flow ischemia leads to reduced PDH flux using hyperpolarized [1-^{13}C]pyruvate as a probe, while reperfusion leads to an increased lactate signal [50]. Increased ^{13}C-label flux into lactate was also shown immediately post reperfusion in a rodent model of closed-chest ischemia-reperfusion (intra-arterial balloon-inflation), and this increase in anaerobic metabolism normalized to baseline levels by 1 week after reperfusion, assessed using cardiac hyperpolarized [1-^{13}C]pyruvate CSI [51].

In large animals, the study by Golman et al. [6] used hyperpolarized ^{13}C MR imaging of [1-^{13}C]pyruvate to demonstrate the detection of myocardial viability. In this study, pigs were subjected to either 15 or 45 min occlusion of the left circumflex artery, followed by 2 h of reperfusion. Hyperpolarized [1-^{13}C]pyruvate was then injected intravenously, and ^{13}C metabolic images were acquired with a non-cardiac-gated 13.4 s CSI acquisition of a single 2-cm thick slice with 7.5 × 7.5 mm^2 nominal in-plane resolution. Cardiac function was also assessed using gold standard cardiac MR cine (wall motion) and late gadolinium enhancement (LGE) techniques. The hyperpolarized ^{13}C MR images allow one to distinguish between the 15-min occlusion and the 45-min occlusion, and classify the tissue state as either viable, necrotic, or akinetic but with viable myocardium. This important classification, based on the metabolic signatures of each condition, indicates potentially salvageable tissue which might be rescued by reperfusion.

Hyperpolarized MRI presents a powerful tool to assess myocardial ischemia. Current clinical decision-making regarding revascularization of narrowed coronary arteries relies on perfusion defects or fractional flow reserve measurements. However, the COURAGE trial showed that revascularization of narrowed arteries assessed by these methods only yields clinical benefit if the underlying myocardium is ischemic [52]. Hyperpolarized MRI is thus uniquely placed to assess the viability of the myocardium by assessing cellular metabolism at the site of the perfusion defect, which may inform clinical decision making in the future and minimize the use of percutaneous coronary intervention.

Myocardial infarction as an acute ischemic event has also been assessed in rodent models in vivo. In a rat model with LAD occlusion, CSI of hyperpolarized [1-^{13}C]pyruvate showed that in the chronically infarcted rat (4 weeks post LAD occlusion) both PDH and LDH flux are decreased [53]. In the same study, ex vivo LAD occlusion showed an acute increase in LDH flux and decrease in PDH flux using the same imaging approach. This shows how the myocardium reacts to acute ischemia by increasing anaerobic glycolysis, while all metabolic flux is decreased in the chronically infarcted heart. Hyperpolarized MRI after LAD-occlusion in

Sprague-Dawley rats also showed regional changes in metabolism, where the whole myocardium displayed decreased PDH flux while lactate and alanine labelling from hyperpolarized [1-^{13}C]pyruvate were only decreased in the apex [54]. The ischemic region of the heart in this model covered half of the anterior wall of the left ventricle as well as the apex.

Interestingly, a study by Lewis et al. has shown that the acute increase in lactate-labelling from hyperpolarized [1-^{13}C]pyruvate in a rat model of cryo-infarction was due to infiltrating macrophages rather than myocardial LDH flux [55]. This was confirmed by depleting cardiac monocytes/macrophages in vivo prior to myocardial infarction (Fig. 9.6) and by assessing LDH flux in macrophages in solution with hyperpolarized [1-^{13}C]pyruvate. This highlights that hyperpolarized MRI can be a powerful imaging modality that can not only assess real-time metabolic fluxes in vivo, but that these can be correlated to ex vivo analysis, providing mechanistic insight into disease pathogenesis. Another study by Miller et al. [56] showed in the same rodent model of cryo-infarction that hyperpolarized [1,4-$^{13}C_2$]fumarate, which was first used to image cellular necrosis in a cancer model [57], can display necrotic changes to the rodent myocardium post cryo-MI. Thus, hyperpolarized MRI can be used to explore different aspects of not just cellular metabolism but also molecular function and viability by using different carbon-13 labelled probes.

9.4.3 Metabolic Assessment in the Hypertrophied/ Failing Heart

Cardiac hypertrophy can occur for example due to increased afterload (hypertension), genetic disorders (hypertrophic cardiomyopathy) or hyperthyroidism. The increased myocardial mass counteracts pressure changes to ensure adequate cardiac output for circulation. However, the heart can only compensate for a period of time before the onset of decompensated heart failure. Changes in myocardial metabolism have been proposed to be a cause rather than a consequence of this onset of heart failure [4], and substrate selection changes with disease progression. Initially, glucose oxidation increases with hypertrophy, though both fatty acid and glucose oxidation decrease with overt heart failure. Heart failure can furthermore occur as a consequence of ischemic heart disease and myocardial infarction as well as treatment with cardiotoxic chemotherapeutic agents.

In a rat model of spontaneous hypertension, increased glucose oxidation was shown through an increase in PDH flux using hyperpolarized [1-^{13}C]pyruvate MRS [58]. Heart failure due to increased afterload can furthermore be achieved by abdominal aortic banding in animal models. Using this technique, Seymour et al. explored cardiac metabolism with hyperpolarized [1-^{13}C]pyruvate MRS and found that in this disease setting, PDH flux was not increased [59]. However, the authors report increased anaerobic glycolysis through ^{13}C-labelling of the lactate pool.

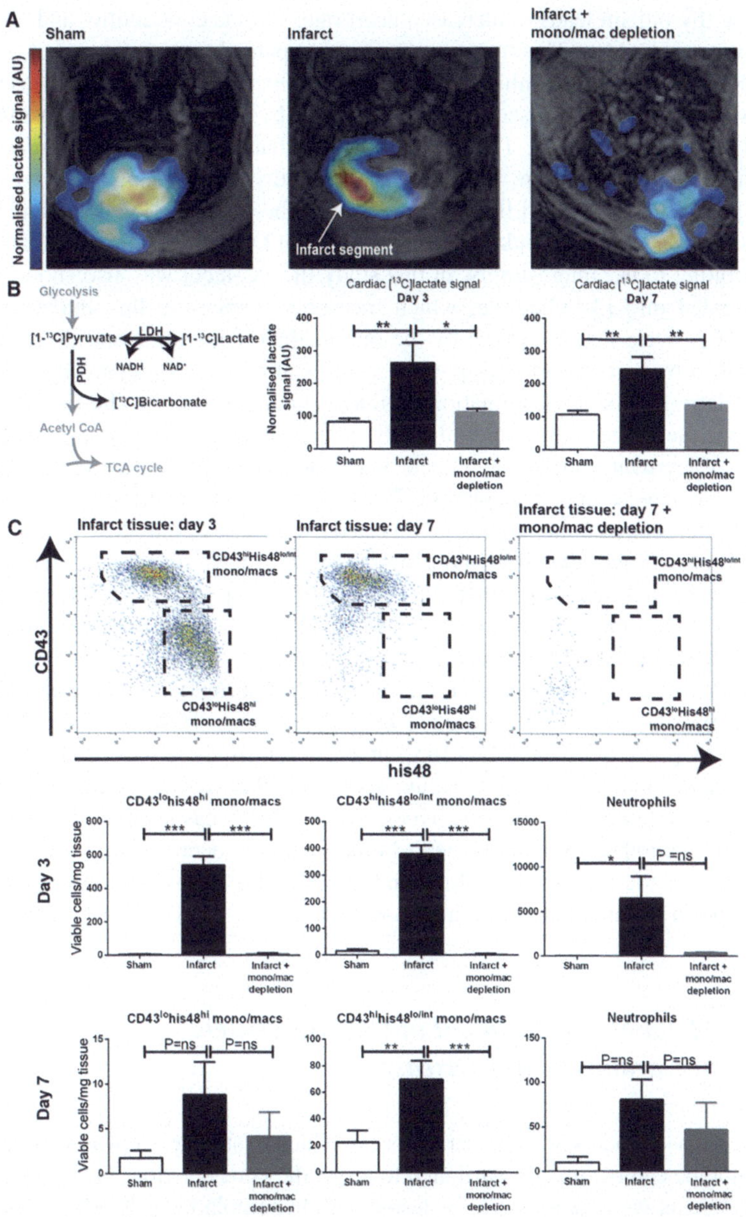

Fig. 9.6 Hyperpolarized lactate generation in cryo-infarction experiments in rodent hearts under (**a**) sham, (**b**) infarct, and (**c**) infarct + macrophage depletion. Hyperpolarized MRI demonstrated intense [1-^{13}C]lactate signal at both days 3 and 7 post-experimental myocardial infarction, which was normalized in rats undergoing pharmacological macrophage depletion. Higher lactate signal is detected from the anterior portion of the heart because of some inhomogeneity from the ^{13}C surface coil; lactate signal detected from the chest wall may reflect inflammation in the healing incision. Figure taken with permission from [55]

Hyperthyroidism leads to increased heart rate, cardiac contractility and, eventually, hypertrophy [60]. This hypertrophy is accompanied with an increased activity of PDK, resulting in PDH inhibition and reduced glucose oxidation [61]. This metabolic switch has been assessed in real-time in vivo by Atherton et al. [62] using a rat model of hyperthyroidism. The authors showed that PDH flux was decreased by 59% after only 7 daily i.p. injections of triiodothyronine, using hyperpolarized [1-^{13}C]pyruvate. This could be reversed by both short- (single i.v. dose) and long-term (daily for 7 days in drinking water) delivery of DCA, which restored PDH flux by inhibiting PDK. Interestingly, in this study the ^{13}C-label was also shown to be incorporated into [3-^{13}C]citrate, which measures anaplerotic flux into the TCA cycle. TCA cycle flux is limited by the intermediate pool sizes, and these can be replenished by carboxylation of pyruvate into oxaloacetate, catalysed by pyruvate carboxylase (PC) or by deamination of glucogenic amino acids. The [3-^{13}C]citrate resonance measured in this study directly reflects anaplerotic entry of hyperpolarized [1-^{13}C]pyruvate into the TCA cycle, via oxaloacetate. Hyperthyroid rats showed a four-fold increase in this anaplerotic flux, which could be reduced by a third with DCA. Hyperpolarized metabolites in this study were assessed by MRS using a surface coil placed over the chest and a simple pulse-acquire spectroscopy sequence. Spectra were analysed using kinetic fits, however, since the [3-^{13}C]citrate resonance was very small, only data from 10 s of summed spectra were used here to derive metabolite ratios of [3-^{13}C]citrate/[1-^{13}C]pyruvate.

A study by Dodd et al. showed in a rat model of chronic ischemic heart failure that decreased PDH flux is preceded by a decrease in TCA cycle metabolism, assessed by [2-^{13}C]pyruvate MRS [63]. In a different study in pigs, dilated cardiomyopathy was induced by rapid pacing, TCA-cycle flux was also shown to precede changes in PDH flux [64]. Lastly, hyperpolarized MRS has been used in a rodent model of chemotherapy-induced heart failure, where a decrease in PDH flux was shown to precede functional decline, and this was due to loss and dysfunction of mitochondria (Timm et al., Proc. Intl. Soc. Mag. Reson. Med. 272019, 0786).

9.4.4 *Metabolic Assessment in Genetic Models of Cardiovascular Disease*

Genetic mouse models are a powerful tool in which to study cardiovascular disease as phenotype can be directly attributed to a specific gene deletion. This also allows genetic effects on metabolism to be tested with hyperpolarized ^{13}C MRS. The first paper showing hyperpolarized [1-^{13}C]pyruvate MRS in the mouse heart was described above by Dodd et al. [28]. In this paper, cardiac metabolism was assessed with hyperpolarized [1-^{13}C]pyruvate MRS and compared in three different mouse strains; C57BL/6, Balb/c and 129 Sv/Ev. This revealed strain-related differences in cardiac metabolism, where PDH flux was highest in 129 Sv/Ev mice and LDH flux was lowest in Balb/c mice. Future studies with genetically engineered mice should take these strain differences into consideration in the experimental design.

Following on from this work, several genetically-modified mouse models have now been trialled in conjunction with hyperpolarized cardiac MR. Bakermans et al. [65] assessed the metabolism of hyperpolarized [1-^{13}C]pyruvate in the heart of the long-chain acyl-CoA dehydrogenase knockout (LCAD KO) mice (C57BL/6 background). While there was no difference in cardiac PDH flux in these mice in the fed state, increased PDH flux was observed in LCAD KO mice compared to WT controls in the fasted state. Since these mice have impaired capacity to oxidize long-chain fatty acids, they rely more on glucose oxidation during fasting. However, in this study, this was associated with decreased cardiac energetics and function, measured by ^{31}P MRS and CINE-MRI respectively. This study highlights the importance of fatty acids as an energy-rich fuel for cardiac metabolism in the fasted state. Other genetic models that have been studied include mouse models of Chuvash polycythemia [66], muscular dystrophy [67] and neonatal diabetes [68].

Further genetically modified mouse models, such as the peroxisome proliferator-activated receptor α (PPARα) knock-out mouse [69], would be interesting models to study cardiac metabolism. PPARα is a fatty-acid target and an important transcriptional regulator of fatty acid metabolism. These mice display severe cardiac lipotoxicity [70] as they are unable to oxidize the fatty acids entering the heart, especially during fasting.

9.5 Human Studies

While the application of hyperpolarized MRI in pre-clinical cardiac studies has generated interesting insights into the aetiology and treatment of a range of cardiovascular diseases, its true potential lies in its clinical application. As a fundamental approach, it offers the potential to provide a real-time readout of cardiac metabolism without the ionizing radiation dose delivered during positron emission tomography (PET) scans, allowing for repeat monitoring of the same patient throughout the course of a disease or its treatment. The potential clinical cardiovascular applications of hyperpolarized MRI have been well reviewed elsewhere [71, 72] and are starting to be explored by several groups around the world.

Following the initial 'first-in-man' demonstration of the use of hyperpolarized MRI by Nelson et al. in patients with prostate cancer [73], initial clinical cardiovascular studies have been undertaken. Rider et al. have demonstrated both physiological and pathological alterations in cardiac metabolism by exploring healthy control subjects and patients with type II diabetes in the fed and fasted state [74]. The work combined hyperpolarized MRS with other MR-based techniques (e.g. CINE MRI, ^{31}P MRS and ^{1}H MRS) to characterize the metabolic alterations seen in the diabetic heart (Fig. 9.7). They have shown that hyperpolarized MRI can detect the expected Randle cycle induced alterations in metabolism that occur between the fed and fasted state by scanning subjects after an overnight fast and then again following the consumption of 70 g of glucose. They have further shown significant reductions in

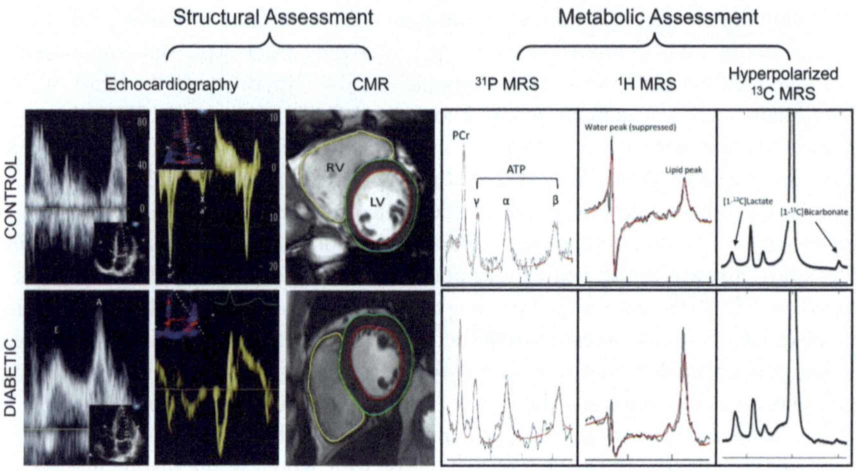

Fig. 9.7 Example data collected during the study from a recruited control (top row) and a subject with type 2 diabetes (bottom row). Left panel shows the structural assessment, while the right panel shows the corresponding metabolic assessment. In characterizing the subjects both structurally (CMR/Echo) and metabolically (^{31}P MRS, ^{1}H MRS, hyperpolarized ^{13}C MRS), the study collates the most comprehensive MR based study of the diabetic cardiac phenotype to date. Figure taken with permission from [74]

PDH flux in the type II diabetic heart, in agreement with the results observed in many animal models.

In further work, Cunningham et al. have demonstrated the potential for high spatial resolution imaging (1 cm think slices, 8.8 × 8.8 mm in plane resolution) in the healthy human heart [75]. Following injection of hyperpolarized [1-^{13}C]pyruvate, it is possible to generate exquisite images visualizing the arrival of hyperpolarized pyruvate in the chambers of the heart and its subsequent metabolic conversion into lactate and bicarbonate within the myocardial tissue.

9.6 Conclusion

The chapter has shown how hyperpolarized metabolic probes can be used to perform both MRS and MRI in ex vivo perfused hearts, in vivo in animal models and in the human heart. This imaging technique is non-radioactive, non-toxic and suffers from virtually no background signal. It is thus uniquely placed to interrogate cardiac cellular metabolism non-invasively in living systems. Alongside metabolic fluxes, hyperpolarized MRI can also be used to assess tissue pH, redox state, perfusion and necrosis, allowing for multi-parametric assessment of tissue integrity and function. First in-man trials have been concluded, demonstrating clinical feasibility and several trials are underway to assess physiological and pathological cardiac metabolism. While human imaging is the ultimate goal of this technique to inform clinical decision making (personalized medicine), hyperpolarized MRS/MRI is and will remain an extremely powerful tool with which to explore disease onset, progression and mechanism in preclinical models.

Acknowledgments The authors would also like to acknowledge financial support provided by the British Heart Foundation (BHF) in the form of a BHF Immediate Research Fellowship and BHF Senior Research Fellowships respectively (KT: FS/16/7/31843, DJT: FS/14/17/30634 and FS/19/18/34252). All authors would also like to acknowledge the support provided by the OXFORD-BHF Centre for Research Excellence (grant no. RE/13/1/30181) and the National Institute for Health Research Oxford Biomedical Research Centre programme.

Problems

1. At a normal physiological intracellular pH of 7.2 and assuming a pKa of 6.17 [44] for the carbonic anhydrase reaction, what is the expected ratio of hyperpolarized bicarbonate and CO_2 that you would expect to see following injection of hyperpolarized [1-^{13}C]pyruvate?
2. What are the challenges faced when trying to use hyperpolarized agents to assess perfusion in the in vivo heart?
3. Why does care need to be taken when interpreting the results of hyperpolarized [2-^{13}C]pyruvate data in relation to incorporation of the ^{13}C label into the TCA cycle in situations where PDH flux is impaired (e.g. in the diabetic heart)?

References

1. Ingwall: ATP and the heart. Kluwer Academic, Boston (2002)
2. Lopaschuk, G.D., Ussher, J.R., Folmes, C.D.L., Jaswal, J.S., Stanley, W.C.: Myocardial fatty acid metabolism in health and disease. Physiol. Rev. **90**, 207–258 (2010). https://doi.org/10.1152/physrev.00015.2009
3. Hue, L., Taegtmeyer, H.: The Randle cycle revisited: a new head for an old hat. Am. J. Physiol. Metab. **297**, E578 (2009). https://doi.org/10.1152/ajpendo.00093.2009
4. Neubauer, S.: The failing heart—an engine out of fuel. N. Engl. J. Med. **356**, 1140 (2007)
5. Brown, T.R., Kincaid, B.M., Ugurbil, K.: NMR chemical shift imaging in three dimensions. Proc. Natl. Acad. Sci. U. S. A. **79**, 3523 (1982). https://doi.org/10.1073/pnas.79.11.3523
6. Golman, K., Petersson, J.S., Magnusson, P., Johansson, E., Åkeson, P., Chai, C.M., Hansson, G., Månsson, S.: Cardiac metabolism measured noninvasively by hyperpolarized 13C MRI. Magn. Reson. Med. **59**, 1005 (2008). https://doi.org/10.1002/mrm.21460
7. Cunningham, C.H., Chen, A.P., Albers, M.J., Kurhanewicz, J., Hurd, R.E., Yen, Y.F., Pauly, J.M., Nelson, S.J., Vigneron, D.B.: Double spin-echo sequence for rapid spectroscopic imaging of hyperpolarized 13C. J. Magn. Reson. **187**, 357 (2007). https://doi.org/10.1016/j.jmr.2007.05.014
8. Lau, A.Z., Chen, A.P., Ghugre, N.R., Ramanan, V., Lam, W.W., Connelly, K.A., Wright, G.A., Cunningham, C.H.: Rapid multislice imaging of hyperpolarized 13C pyruvate and bicarbonate in the heart. Magn. Reson. Med. **64**, 1323 (2010). https://doi.org/10.1002/mrm.22525

9. Wiesinger, F., Weidl, E., Menzel, M.I., Janich, M.A., Khegai, O., Glaser, S.J., Haase, A., Schwaiger, M., Schulte, R.F.: IDEAL spiral CSI for dynamic metabolic MR imaging of hyperpolarized [1-13C]pyruvate. Magn. Reson. Med. **68**, 8 (2012). https://doi.org/10.1002/mrm.23212
10. Lau, A.Z., Chen, A.P., Hurd, R.E., Cunningham, C.H.: Spectral-spatial excitation for rapid imaging of DNP compounds. NMR Biomed. **24**, 988 (2011). https://doi.org/10.1002/nbm.1743
11. Schroeder, M.A., Atherton, H.J., Ball, D.R., Cole, M.A., Heather, L.C., Griffin, J.L., Clarke, K., Radda, G.K., Tyler, D.J.: Real-time assessment of Krebs cycle metabolism using hyperpolarized 13C magnetic resonance spectroscopy. FASEB J. **23**, 2529 (2009). https://doi.org/10.1096/fj.09-129171
12. Schroeder, M.A., Atherton, H.J., Dodd, M.S., Lee, P., Cochlin, L.E., Radda, G.K., Clarke, K., Tyler, D.J.: The cycling of acetyl-coenzyme a through acetylcarnitine buffers cardiac substrate supply: a hyperpolarized 13C magnetic resonance study. Circ. Cardiovasc. Imaging. **5**, 201 (2012). https://doi.org/10.1161/CIRCIMAGING.111.969451
13. Chen, A.P., Hurd, R.E., Schroeder, M.A., Lau, A.Z., Gu, Y.P., Lam, W.W., Barry, J., Tropp, J., Cunningham, C.H.: Simultaneous investigation of cardiac pyruvate dehydrogenase flux, Krebs cycle metabolism and pH, using hyperpolarized [1,2-13C 2]pyruvate in vivo. NMR Biomed. **25**, 305–311 (2012). https://doi.org/10.1002/nbm.1749
14. Timm, K.N., Miller, J.J., Henry, J.A., Tyler, D.J.: Cardiac applications of hyperpolarised magnetic resonance. Prog. Nucl. Magn. Reson. Spectrosc. **106–107**, 66 (2018). https://doi.org/10.1016/j.pnmrs.2018.05.002
15. Chen, A.P., Lau, J.Y.C., Alvares, R.D.A., Cunningham, C.H.: Using [1-13C]lactic acid for hyperpolarized 13C MR cardiac studies. Magn. Reson. Med. **73**, 2087–2093 (2015). https://doi.org/10.1002/mrm.25354
16. Koellisch, U., Gringeri, C.V., Rancan, G., Farell, E.V., Menzel, M.I., Haase, A., Schwaiger, M., Schulte, R.F.: Metabolic imaging of hyperpolarized [1-13C]acetate and [1-13C]acetylcarnitine—investigation of the influence of dobutamine induced stress. Magn. Reson. Med. **74**, 1011–1018 (2015). https://doi.org/10.1002/mrm.25485
17. Bastiaansen, J.A.M., Cheng, T., Lei, H., Gruetter, R., Comment, A.: Direct noninvasive estimation of myocardial tricarboxylic acid cycle flux in vivo using hyperpolarized 13C magnetic resonance. J. Mol. Cell. Cardiol. **87**, 129–137 (2015). https://doi.org/10.1016/j.yjmcc.2015.08.012
18. Abdurrachim, D., Woo, C.C., Teo, X.Q., Chan, W.X., Radda, G.K., Lee, P.T.H.: A new hyperpolarized 13 C ketone body probe reveals an increase in acetoacetate utilization in the diabetic rat heart. Sci. Rep. **9**, 5532 (2019). https://doi.org/10.1038/s41598-019-39378-w
19. Miller, J.J., Ball, D.R., Lau, A.Z., Tyler, D.J.: Hyperpolarized ketone body metabolism in the rat heart. NMR Biomed. **31**, e3912 (2018). https://doi.org/10.1002/nbm.3912
20. Ball, D.R., Rowlands, B., Dodd, M.S., Le Page, L., Ball, V., Carr, C.A., Clarke, K., Tyler, D.J.: Hyperpolarized butyrate: a metabolic probe of short chain fatty acid metabolism in the heart. Magn. Reson. Med. **71**, 1663 (2014). https://doi.org/10.1002/mrm.24849
21. Bastiaansen, J.A.M., Merritt, M.E., Comment, A.: Measuring changes in substrate utilization in the myocardium in response to fasting using hyperpolarized [1-13C]butyrate and [1-13C]pyruvate. Sci. Rep. **6**, 25573 (2016). https://doi.org/10.1038/srep25573
22. Latipää, P.M., Peuhkurinen, K.J., Hiltunen, J.K., Hassinen, I.E.: Regulation of pyruvate dehydrogenase during infusion of fatty acids of varying chain lengths in the perfused rat heart. J. Mol. Cell. Cardiol. **17**, 1161 (1985). https://doi.org/10.1016/S0022-2828(85)80112-7
23. Sherry, A.D., Malloy, C.R., Roby, R.E., Rajagopal, A., Jeffrey, F.M.H.: Propionate metabolism in the rat heart by 13C n. m.r. spectroscopy. Biochem. J. **254**(2), 593–598 (1988). https://doi.org/10.1042/bj2540593
24. Merritt, M.E., Harrison, C., Storey, C., Jeffrey, F.M., Sherry, A.D., Malloy, C.R.: Hyperpolarized 13C allows a direct measure of flux through a single enzyme-catalyzed step by NMR. Proc. Natl. Acad. Sci. U. S. A. **104**, 19773–19777 (2007). https://doi.org/10.1073/pnas.0706235104

25. Khemtong, C., Carpenter, N.R., Lumata, L.L., Merritt, M.E., Moreno, K.X., Kovacs, Z., Malloy, C.R., Sherry, A.D.: Hyperpolarized 13C NMR detects rapid drug-induced changes in cardiac metabolism. Magn. Reson. Med. **74**, 312 (2015). https://doi.org/10.1002/mrm.25419
26. Schroeder, M.A., Cochlin, L.E., Heather, L.C., Clarke, K., Radda, G.K., Tyler, D.J.: In vivo assessment of pyruvate dehydrogenase flux in the heart using hyperpolarized carbon-13 magnetic resonance. Proc. Natl. Acad. Sci. **105**, 12051–12056 (2008). https://doi.org/10.1073/pnas.0805953105
27. Atherton, H.J., Schroeder, M.A., Dodd, M.S., Heather, L.C., Carter, E.E., Cochlin, L.E., Nagel, S., Sibson, N.R., Radda, G.K., Clarke, K., Tyler, D.J.: Validation of the in vivo assessment of pyruvate dehydrogenase activity using hyperpolarised 13C MRS. NMR Biomed. **24**, 201 (2011). https://doi.org/10.1002/nbm.1573
28. Dodd, M.S., Ball, V., Bray, R., Ashrafian, H., Watkins, H., Clarke, K., Tyler, D.J.: In vivo mouse cardiac hyperpolarized magnetic resonance spectroscopy. J. Cardiovasc. Magn. Reson. **15**, 19 (2013). https://doi.org/10.1186/1532-429X-15-19
29. Menichetti, L., Frijia, F., Flori, A., Wiesinger, F., Lionetti, V., Giovannetti, G., Aquaro, G.D., Recchia, F.A., Ardenkjaer-Larsen, J.H., Santarelli, M.F., Lombardi, M.: Assessment of real-time myocardial uptake and enzymatic conversion of hyperpolarized [1-13C]pyruvate in pigs using slice selective magnetic resonance spectroscopy. Contrast Media Mol. Imaging. **7**, 85 (2012). https://doi.org/10.1002/cmmi.480
30. Chen, A.P., Lau, A.Z., Gu, Y.P., Schroeder, M.A., Barry, J., Cunningham, C.H.: Probing the cardiac malate–aspartate shuttle non-invasively using hyperpolarized [1,2-13C2]pyruvate. NMR Biomed. **31**, e3845 (2018). https://doi.org/10.1002/nbm.3845
31. Lau, A.Z., Chen, A.P., Barry, J., Graham, J.J., Dominguez-Viqueira, W., Ghugre, N.R., Wright, G.A., Cunningham, C.H.: Reproducibility study for free-breathing measurements of pyruvate metabolism using hyperpolarized 13C in the heart. Magn. Reson. Med. **69**, 1063 (2013). https://doi.org/10.1002/mrm.24342
32. Hansen, E.S.S., Tougaard, R.S., Nørlinger, T.S., Mikkelsen, E., Nielsen, P.M., Bertelsen, L.B., Bøtker, H.E., Jørgensen, H.S., Laustsen, C.: Imaging porcine cardiac substrate selection modulations by glucose, insulin and potassium intervention: a hyperpolarized [1-13C]pyruvate study. NMR Biomed. **30** (2017). https://doi.org/10.1002/nbm.3702
33. Miller, J.J., Lau, A.Z., Teh, I., Schneider, J.E., Kinchesh, P., Smar, T.S., Ball, V., Sibson, N.R., Tyler, D.J.: Robust and high resolution hyperpolarized metabolic imaging of the rat heart at 7 t with 3d spectral-spatial EPI. Magn. Reson. Med. **75**(4), 1515–1524 (2016). https://doi.org/10.1002/mrm.25730
34. Sigfridsson, A., Weiss, K., Wissmann, L., Busch, J., Krajewski, M., Batel, M., Batsios, G., Ernst, M., Kozerke, S.: Hybrid multiband excitation multiecho acquisition for hyperpolarized 13C spectroscopic imaging. Magn. Reson. Med. **73**, 1713 (2015). https://doi.org/10.1002/mrm.25294
35. Weiss, K., Sigfridsson, A., Wissmann, L., Busch, J., Batel, M., Krajewski, M., Ernst, M., Kozerke, S.: Accelerating hyperpolarized metabolic imaging of the heart by exploiting spatiotemporal correlations. NMR Biomed. **26**, 1380 (2013). https://doi.org/10.1002/nbm.2963
36. Olsson, L.E., Chai, C.M., Axelsson, O., Karlsson, M., Golman, K., Petersson, J.S.: MR coronary angiography in pigs with intraarterial injections of a hyperpolarized 13C substance. Magn. Reson. Med. **55**, 731 (2006). https://doi.org/10.1002/mrm.20847
37. Fuetterer, M., Busch, J., Peereboom, S.M., Von Deuster, C., Wissmann, L., Lipiski, M., Fleischmann, T., Cesarovic, N., Stoeck, C.T., Kozerke, S.: Hyperpolarized 13C urea myocardial first-pass perfusion imaging using velocity-selective excitation. J. Cardiovasc. Magn. Reson. **19**, 46 (2017). https://doi.org/10.1186/s12968-017-0364-4
38. Lau, A.Z., Miller, J.J., Robson, M.D., Tyler, D.J.: Cardiac perfusion imaging using hyperpolarized 13c urea using flow sensitizing gradients. Magn. Reson. Med. **75**, 1474 (2016). https://doi.org/10.1002/mrm.25713
39. Fuetterer, M., Busch, J., Traechtler, J., Wespi, P., Peereboom, S.M., Sauer, M., Lipiski, M., Fleischmann, T., Cesarovic, N., Stoeck, C.T., Kozerke, S.: Quantitative myocardial first-pass

cardiovascular magnetic resonance perfusion imaging using hyperpolarized [1-13 C] pyruvate. J. Cardiovasc. Magn. Reson. **20**, 73 (2018). https://doi.org/10.1186/s12968-018-0495-2
40. Lau, A.Z., Miller, J.J., Robson, M.D., Tyler, D.J.: Simultaneous assessment of cardiac metabolism and perfusion using copolarized [1-13C]pyruvate and 13C-urea. Magn. Reson. Med. **77**, 151 (2017). https://doi.org/10.1002/mrm.26106
41. Williamson, D.H., Lund, P., Krebs, H.A.: The redox state of free nicotinamide-adenine dinucleotide in the cytoplasm and mitochondria of rat liver. Biochem. J. **103**, 514 (1967). https://doi.org/10.1042/bj1030514
42. Lewis, A.J.M., Miller, J.J.J., McCallum, C., Rider, O.J., Neubauer, S., Heather, L.C., Tyler, D.J.: Assessment of metformin-induced changes in cardiac and hepatic redox state using hyperpolarized [1-13C]pyruvate. Diabetes. **65**, 3544 (2016). https://doi.org/10.2337/db16-0804
43. Chen, W., Sharma, G., Jiang, W., Maptue, N.R., Malloy, C.R., Sherry, A.D., Khemtong, C.: Metabolism of hyperpolarized 13C-acetoacetate to β-hydroxybutyrate detects real-time mitochondrial redox state and dysfunction in heart tissue. NMR Biomed. **32**, e4091 (2019). https://doi.org/10.1002/nbm.4091
44. Gallagher, F.A., Kettunen, M.I., Day, S.E., Hu, D.E., Rdenkjaer-Larsen, J.H., in't Zandt, R., Jensen, P.R., Karlsson, M., Golman, K., Lerche, M.H., Brindle, K.M.: Magnetic resonance imaging of pH in vivo using hyperpolarized C-13-labelled bicarbonate. Nature. **453**, 940-U73 (2008)
45. Schroeder, M.A., Swietach, P., Atherton, H.J., Gallagher, F.A., Lee, P., Radda, G.K., Clarke, K., Tyler, D.J.: Measuring intracellular pH in the heart using hyperpolarized carbon dioxide and bicarbonate: a 13C and 31P magnetic resonance spectroscopy study. Cardiovasc. Res. **86**, 82–91 (2010). https://doi.org/10.1093/cvr/cvp396
46. Lau, A.Z., Miller, J.J., Tyler, D.J.: Mapping of intracellular pH in the in vivo rodent heart using hyperpolarized [1-13C]pyruvate. Magn. Reson. Med. **77**, 1810–1817 (2017). https://doi.org/10.1002/mrm.26260
47. Mansor, L.S., Gonzalez, E.R., Cole, M.A., Tyler, D.J., Beeson, J.H., Clarke, K., Carr, C.A., Heather, L.C.: Cardiac metabolism in a new rat model of type 2 diabetes using high-fat diet with low dose streptozotocin. Cardiovasc. Diabetol. **12**, 136 (2013). https://doi.org/10.1186/1475-2840-12-136
48. Le Page, L.M., Rider, O.J., Lewis, A.J., Ball, V., Clarke, K., Johansson, E., Carr, C.A., Heather, L.C., Tyler, D.J.: Increasing pyruvate dehydrogenase flux as a treatment for diabetic cardiomyopathy: a combined 13C hyperpolarized magnetic resonance and echocardiography study. Diabetes. **64**, 2735–2743 (2015). https://doi.org/10.2337/db14-1560
49. Heather, L.C., Pates, K.M., Atherton, H.J., Cole, M.A., Ball, D.R., Evans, R.D., Glatz, J.F., Luiken, J.J., Griffin, J.L., Clarke, K.: Differential translocation of the fatty acid transporter, FAT/CD36, and the glucose transporter, GLUT4, coordinates changes in cardiac substrate metabolism during ischemia and reperfusion. Circ. Heart Fail. **6**, 1058–1066 (2013). https://doi.org/10.1161/CIRCHEARTFAILURE.112.000342
50. Merritt, M.E., Harrison, C., Storey, C., Sherry, A.D., Malloy, C.R.: Inhibition of carbohydrate oxidation during the first minute of reperfusion after brief ischemia: NMR detection of hyperpolarized 13CO2 and H13CO3. Magn. Reson. Med. **60**, 1029–1036 (2008). https://doi.org/10.1002/mrm.21760
51. Oh-Ici, D., Wespi, P., Busch, J., Wissmann, L., Krajewski, M., Weiss, K., Sigfridsson, A., Messroghli, D., Kozerke, S.: Hyperpolarized metabolic MR imaging of acute myocardial changes and recovery after ischemia-reperfusion in a small-animal model. Radiology. **278**, 742–751 (2016). https://doi.org/10.1148/radiol.2015151332
52. Shaw, L.J., Berman, D.S., Maron, D.J., Mancini, G.B.J., Hayes, S.W., Hartigan, P.M., Weintraub, W.S., O'Rourke, R.A., Dada, M., Spertus, J.A., Chaitman, B.R., Friedman, J., Slomka, P., Heller, G.V., Germano, G., Gosselin, G., Berger, P., Kostuk, W.J., Schwartz, R.G., Knudtson, M., Veledar, E., Bates, E.R., McCallister, B., Teo, K.K., Boden, W.E.: Optimal medical therapy with or without percutaneous coronary intervention to reduce ischemic burden: results from the Clinical Outcomes Utilizing Revascularization and Aggressive Drug

Evaluation (COURAGE) trial nuclear substudy. Circulation. **117**, 1283–1291 (2008). https://doi.org/10.1161/CIRCULATIONAHA.107.743963
53. Ball, D.R., Cruickshank, R., Carr, C.A., Stuckey, D.J., Lee, P., Clarke, K., Tyler, D.J.: Metabolic imaging of acute and chronic infarction in the perfused rat heart using hyperpolarised [1-13C]pyruvate. NMR Biomed. **26**, 1441–1450 (2013). https://doi.org/10.1002/nbm.2972
54. Lauritzen, M.H., Magnusson, P., Laustsen, C., Butt, S.A., Ardenkjær-Larsen, J.H., Sogaard, L.V., Paulson, O.B., Akeson, P.: Imaging regional metabolic changes in the ischemic rat heart in vivo using hyperpolarized [1-13 C]pyruvate. Tomography. **3**, 123 (2017). https://doi.org/10.18383/j.tom.2017.00008
55. Lewis, A.J., Miller, J.J., Lau, A.Z., Curtis, M.K., Rider, O.J., Choudhury, R.P., Neubauer, S., Cunningham, C.H., Carr, C.A., Tyler, D.J.: Non-invasive immuno-metabolic cardiac inflammation imaging using hyperpolarized magnetic resonance. Circ. Res. **122**(8), 1084–1093 (2018). https://doi.org/10.1161/CIRCRESAHA.117.312535
56. Miller, J.J., Lau, A.Z., Nielsen, P.M., McMullen-Klein, G., Lewis, A.J., Jespersen, N.R., Ball, V., Gallagher, F.A., Carr, C.A., Laustsen, C., Bøtker, H.E., Tyler, D.J., Schroeder, M.A.: Hyperpolarized [1,4-13C2]fumarate enables magnetic resonance-based imaging of myocardial necrosis. JACC Cardiovasc. Imaging. **11**(11), 1594–1606 (2018)
57. Gallagher, F.A., Kettunen, M.I., Hu, D.-E., Jensen, P.R., Zandt, R., Karlsson, M., Gisselsson, A., Nelson, S.K., Witney, T.H., Bohndiek, S.E., Hansson, G., Peitersen, T., Lerche, M.H., Brindle, K.M.: Production of hyperpolarized [1,4-13C2]malate from [1,4-13C2]fumarate is a marker of cell necrosis and treatment response in tumors. Proc. Natl. Acad. Sci. **106**, 19801–19806 (2009). https://doi.org/10.1073/pnas.0911447106
58. Dodd, M.S., Ball, D.R., Schroeder, M.A., Le Page, L.M., Atherton, H.J., Heather, L.C., Seymour, A.M., Ashrafian, H., Watkins, H., Clarke, K., Tyler, D.J.: In vivo alterations in cardiac metabolism and function in the spontaneously hypertensive rat heart. Cardiovasc. Res. **95**, 69–76 (2012). https://doi.org/10.1093/cvr/cvs164
59. Seymour, A.-M.L., Giles, L., Ball, V., Miller, J.J., Clarke, K., Carr, C.A., Tyler, D.J.: In vivo assessment of cardiac metabolism and function in the abdominal aortic banding model of compensated cardiac hypertrophy. Cardiovasc. Res. **106**, 249–260 (2015). https://doi.org/10.1093/cvr/cvv101
60. Siehl, D., Chua, B.H.L., Lautensack-Belser, N., Morgan, H.E.: Faster protein and ribosome synthesis in thyroxine-induced hypertrophy of rat heart. Am. J. Phys. Cell Phys. **248**, C309 (1985). https://doi.org/10.1152/ajpcell.1985.248.3.c309
61. Orfali, K.A., Fryer, L.G.D., Holness, M.J., Sugden, M.C.: Interactive effects of insulin and triiodothyronine on pyruvate dehydrogenase kinase activity in cardiac myocytes. J. Mol. Cell. Cardiol. **27**, 901 (1995). https://doi.org/10.1016/0022-2828(95)90040-3
62. Atherton, H.J., Dodd, M.S., Heather, L.C., Schroeder, M.A., Griffin, J.L., Radda, G.K., Clarke, K., Tyler, D.J.: Role of pyruvate dehydrogenase inhibition in the development of hypertrophy in the hyperthyroid rat heart: a combined magnetic resonance imaging and hyperpolarized magnetic resonance spectroscopy study. Circulation. **123**, 2552–2561 (2011). https://doi.org/10.1161/CIRCULATIONAHA.110.011387
63. Dodd, M.S., Atherton, H.J., Carr, C.A., Stuckey, D.J., West, J.A., Griffin, J.L., Radda, G.K., Clarke, K., Heather, L.C., Tyler, D.J.: Impaired in vivo mitochondrial Krebs cycle activity after myocardial infarction assessed using hyperpolarized magnetic resonance spectroscopy. Circ. Cardiovasc. Imaging. **7**, 895–904 (2014). https://doi.org/10.1161/CIRCIMAGING.114.001857
64. Schroeder, M.A., Lau, A.Z., Chen, A.P., Gu, Y., Nagendran, J., Barry, J., Hu, X., Dyck, J.R.B., Tyler, D.J., Clarke, K., Connelly, K.A., Wright, G.A., Cunningham, C.H.: Hyperpolarized 13C magnetic resonance reveals early- and late-onset changes to in vivo pyruvate metabolism in the failing heart. Eur. J. Heart Fail. **15**, 130–140 (2013). https://doi.org/10.1093/eurjhf/hfs192
65. Bakermans, A.J., Dodd, M.S., Nicolay, K., Prompers, J.J., Tyler, D.J., Houten, S.M.: Myocardial energy shortage and unmet anaplerotic needs in the fasted long-chain acyl-co a dehydrogenase knockout mouse. Cardiovasc. Res. **100**, 441 (2013). https://doi.org/10.1093/cvr/cvt212

66. Slingo, M., Cole, M., Carr, C., Curtis, M., Dodd, M., Giles, L., Heather, L., Tyler, D., Clarke, K., Robbins, P.A.: The von hippel-Lindau Chuvash mutation in mice alters cardiac substrate and high-energy phosphate metabolism. Am. J. Physiol. Heart Circ. Physiol. **311**, H759 (2016). https://doi.org/10.1152/ajpheart.00912.2015
67. Betts, C.A., McClorey, G., Healicon, R., Hammond, S.M., Manzano, R., Muses, S., Ball, V., Godfrey, C., Merritt, T.M., Van Westering, T., O'Donovan, L., Wells, K.E., Gait, M.J., Wells, D.J., Tyler, D., Wood, M.J.: Cmah-dystrophin deficient mdx mice display an accelerated cardiac phenotype that is improved following peptide-PMO exon skipping treatment. Hum. Mol. Genet. **28**(3), 396–406 (2019). https://doi.org/10.1093/hmg/ddy346
68. Rohm, M., Savic, D., Ball, V., Curtis, M.K., Bonham, S., Fischer, R., Legrave, N., MacRae, J.I., Tyler, D.J., Ashcroft, F.M.: Cardiac dysfunction and metabolic inflexibility in a mouse model of diabetes without dyslipidemia. Diabetes. **67**, 1057 (2018). https://doi.org/10.2337/db17-1195
69. Lee, S.S., Pineau, T., Drago, J., Lee, E.J., Owens, J.W., Kroetz, D.L., Fernandez-Salguero, P.M., Westphal, H., Gonzalez, F.J.: Targeted disruption of the alpha isoform of the peroxisome proliferator-activated receptor gene in mice results in abolishment of the pleiotropic effects of peroxisome proliferators. Mol. Cell. Biol. **15**, 3012 (1995). https://doi.org/10.1128/mcb.15.6.3012
70. Djouadi, F., Weinheimer, C.J., Saffitz, J.E., Pitchford, C., Bastin, J., Gonzalez, F.J., Kelly, D.P.: A gender-related defect in lipid metabolism and glucose homeostasis in peroxisome proliferator-activated receptor α-deficient mice. J. Clin. Invest. **102**, 1083 (1998). https://doi.org/10.1172/JCI3949
71. Schroeder, M.A., Clarke, K., Neubauer, S., Tyler, D.J.: Hyperpolarized magnetic resonance: a novel technique for the in vivo assessment of cardiovascular disease. Circulation. **124**, 1580 (2011). https://doi.org/10.1161/CIRCULATIONAHA.111.024919
72. Rider, O.J., Tyler, D.J.: Clinical implications of cardiac hyperpolarized magnetic resonance imaging. J. Cardiovasc. Magn. Reson. **15**(1), 93 (2013)
73. Nelson, S.J., Kurhanewicz, J., Vigneron, D.B., Larson, P.E.Z., Harzstark, A.L., Ferrone, M., Van Criekinge, M., Chang, J.W., Bok, R., Park, I., Reed, G., Carvajal, L., Small, E.J., Munster, P., Weinberg, V.K., Ardenkjaer-Larsen, J.H., Chen, A.P., Hurd, R.E., Odegardstuen, L.I., Robb, F.J., Tropp, J., Murray, J.A.: Metabolic imaging of patients with prostate cancer using hyperpolarized [1–13C]pyruvate. Sci. Transl. Med. **5**, 198ra108 (2013)
74. Rider, O.J., Apps, A., Miller, J.J., Lau, J.Y., Lewis, A.J., Peterzan, M.A., Dodd, M.S., Lau, A.Z., Trumper, C., Gallagher, F., Grist, J.T., Brindle, K., Neubauer, S., Tyler, D.J.: Non-invasive in vivo assessment of cardiac metabolism in the healthy and diabetic human heart using hyperpolarized ^{13} C MRI. Circ. Res. **126**, 725 (2020). https://doi.org/10.1161/CIRCRESAHA.119.316260
75. Cunningham, C.H., Lau, J.Y.C., Chen, A.P., Geraghty, B.J., Perks, W.J., Roifman, I., Wright, G.A., Connelly, K.A.: Hyperpolarized 13C metabolic MRI of the human heart: initial experience. Circ. Res. **119**, 1177 (2016). https://doi.org/10.1161/CIRCRESAHA.116.309769

Further Reading

Cunningham, C.H., Lau, J.Y.C., Chen, A.P., Geraghty, B.J., Perks, W.J., Roifman, I., Wright, G.A., Connelly, K.A.: Hyperpolarized 13C metabolic MRI of the human heart: initial experience. Circ. Res. **119**, 1177 (2016). https://doi.org/10.1161/CIRCRESAHA.116.309769

Fuetterer, M., Busch, J., Traechtler, J., Wespi, P., Peereboom, S.M., Sauer, M., Lipiski, M., Fleischmann, T., Cesarovic, N., Stoeck, C.T., Kozerke, S.: Quantitative myocardial first-pass cardiovascular magnetic resonance perfusion imaging using hyperpolarized [1-13 C] pyruvate. J. Cardiovasc. Magn. Reson. **20**, 73 (2018). https://doi.org/10.1186/s12968-018-0495-2

Ingwall: ATP and the heart. Kluwer Academic, Boston (2002)

Lewis, A.J., Miller, J.J., Lau, A.Z., Curtis, M.K., Rider, O.J., Choudhury, R.P., Neubauer, S., Cunningham, C.H., Carr, C.A., Tyler, D.J.: Non-invasive immuno-metabolic cardiac inflammation imaging using hyperpolarized magnetic resonance. Circ. Res. **122**(8), 1084–1093 (2018). https://doi.org/10.1161/CIRCRESAHA.117.312535

Rider, O.J., Apps, A., Miller, J.J., Lau, J.Y., Lewis, A.J., Peterzan, M.A., Dodd, M.S., Lau, A.Z., Trumper, C., Gallagher, F., Grist, J.T., Brindle, K., Neubauer, S., Tyler, D.J.: Non-invasive in vivo assessment of cardiac metabolism in the healthy and diabetic human heart using hyperpolarized 13C MRI. Circ. Res. **126**, 725 (2020). https://doi.org/10.1161/CIRCRESAHA.119.316260

Schroeder, M.A., Cochlin, L.E., Heather, L.C., Clarke, K., Radda, G.K., Tyler, D.J.: In vivo assessment of pyruvate dehydrogenase flux in the heart using hyperpolarized carbon-13 magnetic resonance. Proc. Natl. Acad. Sci. **105**, 12051–12056 (2008). https://doi.org/10.1073/pnas.0805953105

Chapter 10
Insights on Lactate Metabolism in Skeletal Muscle Based on ^{13}C Dynamic Nuclear Polarization Studies

Jae Mo Park and Thomas Jue

10.1 Introduction

Conventional biochemistry often views lactate (lac) as a glycolytic end-product, whose appearance signals O_2 insufficiency. Glycolysis can still produce adenosine triphosphate (ATP) during anaerobiosis. As a result, many researchers mark a rise in lac as a hypoxia threshold [1]. Because orthodox physiology does not envision lac as an immediate fuel source in muscle, the body must export it to liver for clearance and recycling as part of the Cori cycle [2]. The paradigm states that muscle consumes glucose to produce lac, while liver uses lac to produce glucose via gluconeogenesis.

However, studies have begun to challenge a monolithic Cori cycle. Researchers have observed lac forming in muscle even under well-oxygenated conditions [3]. A rise in lac does not always indicate the presence of anaerobiosis [4–6]. Indeed, radioactive tracer experiments have detected muscle consuming lac and generating CO_2. The observation of CO_2 formation indicates that lac serves as a precursor for the pyruvate dehydrogenase (PDH) reaction and tricarboxylic acid (TCA) cycle [7–9]. ^{13}C nuclear magnetic resonance (NMR) studies have corroborated the finding by observing the conversion of [3-^{13}C]lac to ^{13}C labeled glutamate (glu) in a perfused hindquarter model [10, 11]. Contrary to the prevailing paradigm, many studies support a different view: lac can serve as an immediate metabolic precursor in muscle and can shuttle from glycolytic fibers to oxidative fibers. In fact, the

J. M. Park
Advanced Imaging Research Center, University of Texas Southwestern Medical Center, Dallas, TX, USA

T. Jue (✉)
Biochemistry and Molecular Medicine, University of California Davis, Davis, CA, USA
e-mail: tjue@ucdavis.edu

T. Jue, D. Mayer (eds.), *Dynamic Hyperpolarized Nuclear Magnetic Resonance*, Handbook of Modern Biophysics, https://doi.org/10.1007/978-3-030-55043-1_10

detection of lactate dehydrogenase (LDH) in the mitochondria has expanded the notion to include a lac shuttle between the cytosol and the mitochondria [12–16].

Nevertheless, the biochemical mechanism underpinning lac accumulation under aerobic condition remains unclear. In 2001, a novel hypothesis rationalized aerobic lac formation as part of a dynamic glycogen flux during a muscle twitch [17]. Since NMR has detected a large energy fluctuation during each contraction, the cell must dynamically restore glycogen in order to sustain any extended period of contractions [18]. Even though phosphocreatine (PCr) and glycogen provide the immediate energy source during muscle contractions lasting for milliseconds, a rapid glycogen resynthesis using glucose and/or lac during the extended relaxation phase must replenish the glycogen pool to sustain a long period of muscle contractions. The accumulated lac serves then as a precursor for oxidative ATP production, which can fuel the resynthesis of glycogen [19]. In essence, the cell uses the accumulated lac during aerobic muscle contraction to buffer energy utilization.

Testing the tenability of these provocative models, however, has posed many technical challenges, because innovative methodological approaches must track in seconds the lac kinetics muscle in situ. Both the glycogen shunt and intracellular lac shuttle models require such rapid mobilization and utilization of lac. If lac cannot mobilize rapidly, then the results invalidate immediately the hypotheses in both models.

Dynamic nuclear polarization (DNP) NMR experiments can then play a unique, pivotal role to clarify and validate these models by investigating the kinetics of [1-^{13}C]lac and [2-^{13}C]pyr in skeletal muscle [20]. With the increased signal enhancement in the DNP approach, ^{13}C DNP NMR can follow in seconds the dynamics of [1-^{13}C]lac and [2-^{13}C]pyr in real time and can confirm or refute the tenability of both the glycogen shunt and intracellular lac shuttle models in muscle.

10.2 Biodistribution

To analyze properly the DNP data requires an estimate of the endogenous metabolite pool in muscle. Resting muscle cells and normoxic rat blood contain about 50 μM pyruvate (pyr) and 700 μM lac [21]. Recent experiments have corroborated the finding and have determined 60 μM pyr in normoxic rat blood [22]. After a tail vein injection of 80 mM of pyr, a standard amount used in DNP experiments, pyr reaches 250 μM in blood after 1 min. It remains well above 150 μM even after 10 min. Despite the elevated levels of pyr, other metabolite levels, such as glucose, insulin, lac, triacylglyceride (TAG), and non-esterified fatty acid (NEFA), do not change [22]. Physiological functions exhibit no impairment. A similar amount of injected lac would presumably reach a similar concentration.

The injected pyr and lac distribute first in the vasculature and then in the interstitial volume, before they enter the cell. At high concentration (>10 mM), they can also enter the cell by free diffusion [23, 24].

Volumetric chemical shift imaging (CSI) has confirmed that metabolites in the vasculature do not contribute significantly to the observed DNP signal, consistent with blood volume constituting <5% of the total muscle volume. The interstitial and intracellular spaces, however, constitute respectively 20% and 35% of the total muscle volume [25]. Consequently, the DNP signals of pyr, lac, and the metabolite product reflect predominantly signals from both the interstitial and intracellular spaces. Both [1-^{13}C]lac and [1-^{13}C]pyr will decay with an apparent spin lattice relaxation time T_1 of ~40 s [26].

Monocarboxylate transporter 1 (MCT1) and monocarboxylate transporter 4 (MCT4) convey lac and pyr into the cell with a K_m of 13–40 mM for L-lac and >50 mM for pyr. MCT4 distributes evenly in all muscle fibers, while MCT1 localizes preferentially in oxidative fibers. With endurance training, MCT1 expression increases. Lac transport also increases 30–100% [6, 27, 28]. The increased MCT1 expression and lac transport in oxidative fibers support the model of lac shuttle between glycolytic and oxidative muscle fibers [6, 27–29].

MCTs utilize an ordered sequential mechanism in which a proton binds first and then the lac anion binds [30, 31]. Both lac and proton translocate across the membrane followed by a sequential release. Because the cycling of the free carrier across the membrane constitutes the rate-limiting step, monocarboxylate exchange proceeds substantially faster than net transport [24, 32].

Given the approximately 0.1-mM pyr and 1-mM lac in resting myocyte, the sudden arrival of 0.25 mM hyperpolarized pyr or lac, and the cycling of the monocarboxylate as the rate-limiting step in MCT transport, the observed hyperpolarized ^{13}C pyr and lac signals in muscle in situ will have significant contributions from both exchange and net transport.

10.3 Lac in the Cell

In muscle the near-equilibrium enzyme, LDH, catalyzes usually the conversion of pyr to lac to regenerate the requisite oxidizing equivalent (NAD^+) to sustain glycolysis or glycogenolysis during contraction. According to the Cori cycle, lac exports to liver for clearance and recycling. From this vantage, exogenous lac should not convert significantly to pyr. Yet, when muscle receives a surge of lac, the DNP experiments detect clearly the reverse flow from lac to pyr [33, 34]. Figure 10.1 compares the 2-min time-averaged ^{13}C signals in rat leg muscle following an injection of hyperpolarized [1-^{13}C]lac with and without the addition of dichloroacetate (DCA), which stimulates PDH activity.

In the DNP experiment, the incoming [1-^{13}C]lac represents the largest fraction of the total ^{13}C (tC) signals, lac/tC = 0.805. But clearly, the ^{13}C label distributes to pyr (0.034), alanine (ala: 0.151), and bicarbonate (bic: 0.011). After DCA injection, which stimulates PDH activity, the lac/tC signal does not change significantly (0.795). However, the ^{13}C label distribution has decreased to pyr but has increased to ala: pyr (0.014), ala (0.073). Bicarbonate (bic, HCO_3^-) level has increased even more dramatically, from 0.011 to 0.120, almost 11 times (Table 10.1).

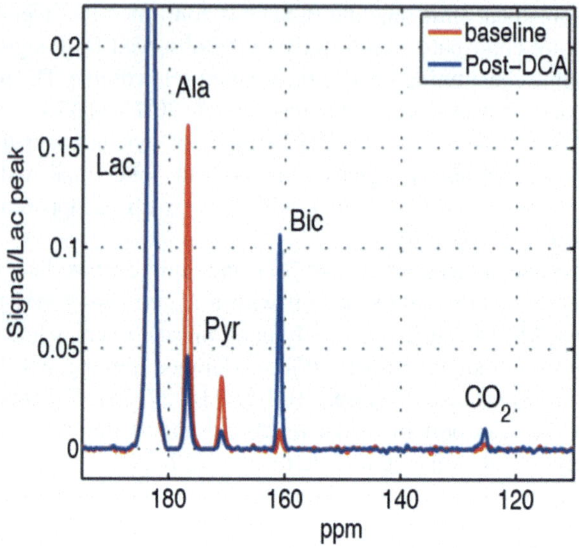

Fig. 10.1 Time-averaged (0–2 min) ^{13}C signals from rat skeletal muscle after injection of 40-mM hyperpolarized [1-^{13}C]lac: Control (red); 1 h after dichloroacetate (DCA) infusion (blue). Each spectrum is normalized by the corresponding [1-^{13}C]lac peak intensity. Relative to the signal intensity of [1-^{13}C]lac, the DCA stimulation of pyruvate dehydrogenase (PDH) produces a marked change in ^{13}C spectra, as noted in intensity changes of the alanine (ala), pyruvate (pyr), bicarbonate (bic), and CO_2 signals. Scaling all signals to the initial lac signal intensity corrects for variation in ^{13}C polarization. The peak assignments are as follows: [1-^{13}C]lac (185 ppm), [1-^{13}C]ala (179 ppm), [1-^{13}C]pyr (173 ppm), [^{13}C]bic (163 ppm), and $^{13}CO_2$ (127 ppm)

Table 10.1 ^{13}C biodistribution from [1-^{13}C]lactate

Metabolite/tC	Control	Post-DCA (1 h)
Lac/tC	0.805	0.795
Pyr/tC	0.034	0.014
Ala/tC	0.151	0.073
Bic/tC	0.011	0.120

Skeletal muscle metabolite values reflect the average of all transients between 0 and 2 min normalized to the total integrated ^{13}C carbon signals between 0 and 2 min after a bolus injection of hyperpolarized [1-^{13}C]lac [34]

Ala alanine, *Bic* bicarbonate, *DCA* dichloroacetate, *Lac* lactate, *Pyr* pyruvate, *tC* total ^{13}C

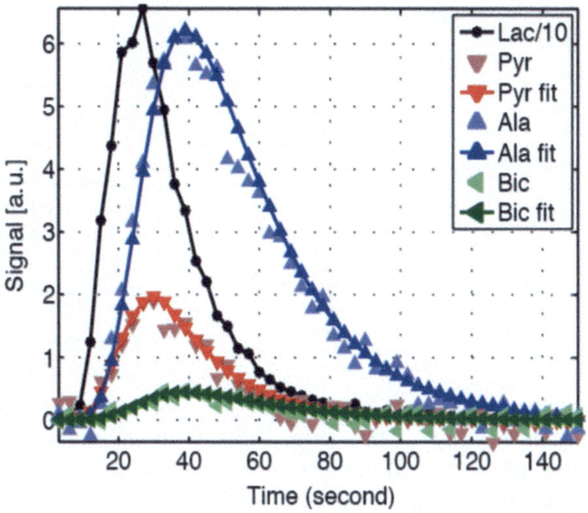

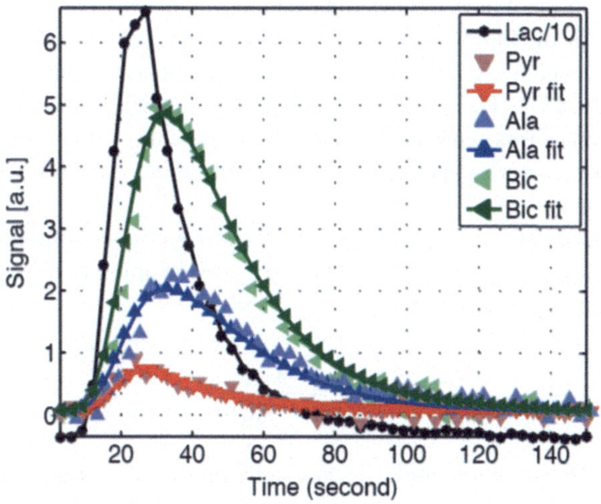

Fig. 10.2 Dynamic changes (3-s temporal resolution) in the ^{13}C signals from rat skeletal muscle after injection of 40-mM hyperpolarized [1-^{13}C]lac: (**a**) Control; (**b**) 1 h after dichloroacetate (DCA) infusion. Each time course is normalized by the peak value of the corresponding [1-^{13}C]lac curve. Relative to the signal intensity of [1-^{13}C]lac, the DCA stimulation of pyruvate dehydrogenase (PDH) produces a rapid time-dependent change in ^{13}C spectra, as noted in dynamics of the alanine (ala), pyruvate (pyr), and bicarbonate (bic) signals

The conversion of lac to pyr occurs very quickly. Figure 10.2 shows the corresponding kinetics after injection of hyperpolarized [1-^{13}C]lac in control rat leg muscle. Without DCA, the composite metabolism and exchange rates ($r_{x \to y}$) relative to the time-dependent change of the lac signal are as follows: $r_{lac \to pyr} = 0.023$, $r_{lac \to ala} = 0.041$, $r_{lac \to bic} = 0.002$. DCA alters sharply these rates (Table 10.2) [34]. The lac to pyr, ala, and bic rates change most significantly: $r_{lac \to pyr} = 0.012$,

Table 10.2 Relative conversion rate from [1-^{13}C]lactate

	Control	Post-DCA (1 h)
$r_{lac \rightarrow pyr}$	0.023	0.012
$r_{lac \rightarrow ala}$	0.041	0.036
$r_{lac \rightarrow bic}$	0.002	0.039

The metabolite conversion rate (r) reflects the temporal change in the metabolite signal intensity relative to the kinetics of the lac signal [34]

ala alanine, *bic* bicarbonate, *DCA* dichloroacetate, *lac* lactate, *pyr* pyruvate

$r_{lac \rightarrow ala}$ = 0.036, and $r_{lac \rightarrow bic}$ = 0.039 (Fig. 10.2). With DCA, $r_{lac \rightarrow bic}$ increases by almost a factor of 10, $r_{lac \rightarrow pyr}$ decreases by a factor of 2, and $r_{lac \rightarrow ala}$ decreases slightly. The higher rate of lac to ala vs lac to pyr conversion reflects the higher endogenous ala pool, which implies the exchange contribution.

The line fit for the kinetics analysis estimates apparent conversion rates (known as the metrics), which are less sensitive to experiment parameters as compared to metabolite ratios [35]. The apparent conversion rate constants were calculated by fitting the processed time-curves of metabolites to the following multi-site exchange models, which are more generalized forms of the previously proposed three-site exchange model.

Radio frequency (RF) sampling was corrected from the measured [1-^{13}C]lactate time-curve ($\text{lac}_{\text{measured},n}$) using Eq. (10.1) to estimate the amount of ^{13}C-lactate in the muscle at n^{th} time-point ($\text{lac}_{\text{RFcor},n}$),

$$\text{lac}_{\text{RFcor},n} = \frac{\text{lac}_{\text{measured},n}}{\sin\theta} \tag{10.1}$$

where θ is the RF excitation angle. Then locally produced pyruvate, alanine, bicarbonate (pyr_n, ala_n, bic_n) at each time-point were estimated from a set of $T_{1,\text{pyr}}$, $T_{1,\text{ala}}$, $T_{1,\text{bic}}$, k_{LP} ($k_{\text{lac} \rightarrow \text{pyr}}$), k_{PA} ($k_{\text{pyr} \rightarrow \text{ala}}$), and k_{PB} ($k_{\text{pyr} \rightarrow \text{bic}}$) using Eq. (10.2a–c), where k values are the apparent conversion rate constants and T_1's are longitudinal relaxation times.

$$\text{TR} \cdot k_{\text{LP}} = \frac{\text{ala}_n + \text{pyr}_n + \text{bic}_n}{\text{lac}_{\text{RFcor},n} + \text{ala}_n + \text{pyr}_n + \text{bic}_n} \tag{10.2a}$$

$$\text{TR} \cdot k_{\text{PA}} = \frac{\text{ala}_n}{\text{ala}_n + \text{pyr}_n + \text{bic}_n} \tag{10.2b}$$

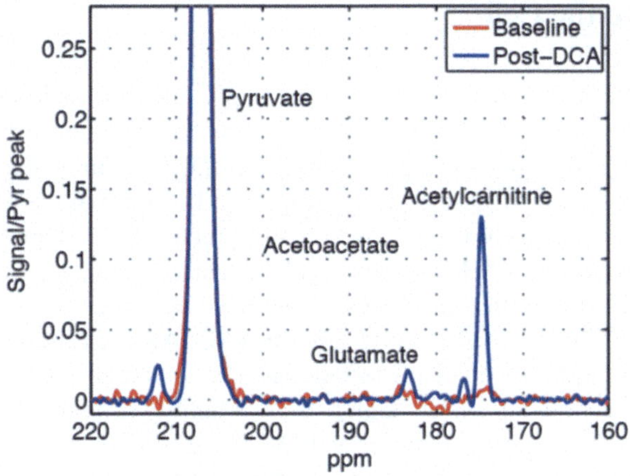

Fig. 10.3 Time-averaged (0–2 min) ^{13}C signals from rat skeletal muscle after injection of 80-mM hyperpolarized [2-^{13}C]pyr: Control (Red); (Blue) 1 h after dichloroacetate (DCA) infusion. Each spectrum is normalized by the corresponding [2-^{13}C]pyr peak intensity. Relative to the signal intensity of [2-^{13}C]pyr, the DCA stimulation of pyruvate dehydrogenase (PDH) produces a marked change in ^{13}C spectra, as noted in intensity changes of the acetyl carnitine (ALCAR), acetoacetate (AcAc), and glutamate (glu) signals. Scaling all signals to the initial pyruvate (pyr) signal intensity corrects for variation in ^{13}C polarization

$$\mathrm{TR}\cdot k_{\mathrm{PB}} = \frac{\mathrm{bic}_n}{\mathrm{ala}_n + \mathrm{pyr}_n + \mathrm{bic}_n} \tag{10.2c}$$

Finally, accumulated metabolite products ($\mathrm{pyr}_{\mathrm{acm},n}$, $\mathrm{ala}_{\mathrm{acm},n}$, $\mathrm{bic}_{\mathrm{acm},n}$) were calculated using Eq. (10.3a–c) to take the residual signals from previous time-points into account. The T_1 and k values that generate the accumulated signal curves that fit the most to the detected pyr, ala, bic time-curves were chosen as the estimates.

$$\mathrm{pyr}_{\mathrm{acm},n} = \sum_{i=0}^{n-1} \mathrm{pyr}_{n-i} \cdot \mathrm{e}^{-i\cdot \mathrm{TR}/T_{1,\mathrm{pyr}}} \cdot \cos^i \theta \tag{10.3a}$$

$$\mathrm{ala}_{\mathrm{acm},n} = \sum_{i=0}^{n-1} \mathrm{ala}_{n-i} \cdot \mathrm{e}^{-i\cdot \mathrm{TR}/T_{1,\mathrm{ala}}} \cdot \cos^i \theta \tag{10.3b}$$

$$\mathrm{bic}_{\mathrm{acm},n} = \sum_{i=0}^{n-1} \mathrm{bic}_{n-i} \cdot \mathrm{e}^{-i\cdot \mathrm{TR}/T_{1,\mathrm{bic}}} \cdot \cos^i \theta \tag{10.3c}$$

The model-fitted analysis yields the apparent conversion rate constants at baseline as follows: $k_{\mathrm{LP}} = 0.011 \pm 0.001\ \mathrm{s}^{-1}$, $k_{\mathrm{PA}} = 0.245 \pm 0.015\ \mathrm{s}^{-1}$, $k_{\mathrm{PB}} = 0.012 \pm 0.004\ \mathrm{s}^{-1}$ (Figs. 10.2 and 10.3). Similar to the metabolite ratio analysis, the apparent conversion rate constants after DCA infusion are changed to $k_{\mathrm{LP}} = 0.011 \pm 0.001\ \mathrm{s}^{-1}$, $k_{\mathrm{PA}} = 0.132 \pm 0.04\ \mathrm{s}^{-1}$, $k_{\mathrm{PB}} = 0.158 \pm 0.061\ \mathrm{s}^{-1}$. The change in lac conversion to pyr due to DCA is not significant unlike the changes in pyr-to-ala and pyr-to-bic conversions or metabolite ratios [36].

10.4 Pyr in the Cell

The capacity for muscle to convert lac to pyr implies that activated PDH can compete effectively for pyr, which appears to question the rigid perception of a rapid glycolysis with a slow oxidative phosphorylation. DNP confirms that supposition. Figure 10.3 shows a typical time-averaged spectra (0–2 min) of the metabolite signals in muscle generated from injecting the precursor [2-^{13}C]pyr (208 ppm). [2-^{13}C]pyr converts to [2-^{13}C]lac (70 ppm), [2-^{13}C]ala (52 ppm), and [2-^{13}C]pyruvate hydrate (pyh) (96 ppm). In control muscle, the DNP experiments cannot detect any conversion of [2-^{13}C]pyr to glu, acetyl carnitine (ALCAR), or acetoacetate (AcAc). However, with DCA, the metabolism rate has increased sufficiently, so that DNP can detect the conversion of [2-^{13}C]pyr to [5-^{13}C]glu (184 ppm), [1-^{13}C]ALCAR (175 ppm), [1-^{13}C]AcAc (177 ppm), and [3-^{13}C]AcAc (213 ppm). The dynamic metabolic response to injected hyperpolarized [2-^{13}C]pyr in the presence of DCA shows a rapid production of ALCAR at $r_{\text{pyr} \to \text{alcar}} = 0.026$, Tables 10.3 and 10.4. Because ALCAR equilibrates with the acetyl coenzyme A (CoA) pool, the observation implies a rapid increase in acetyl CoA.

10.5 Bicarbonate in the Mitochondria

Because the formation bic reflects the unidirectional flux from pyr to bic via PDH, exchange does not play a significant role [26, 37]. In the presence of DCA, PDH activity increases markedly and shifts the metabolism toward PDH and oxidative metabolism. Since both DCA and muscle contraction stimulate PDH, the increased unidirectional flux from pyr to bic suggests that muscle can adapt quickly and convert pyr to acetyl CoA, when energy demand increases suddenly [38]. But bic/tC increases 11 times over the control level. Moreover, the increased bic/tC accompanies an increased lac/pyr ratio from 24.6 to 59.0, which arises from a decreasing pyr/tC. The results imply that activated PDH can compete immediately and effectively with the near-equilibrium LDH and ALT for pyr.

10.6 NAD$^+$/NADH

The LDH enzymatic rate depends upon the turnover of a smaller NAD$^+$ and NADH nucleotide pool. In the cell, cytosol comprises 90% of the total volume, while mitochondria represent only 10% volume. Even though the cytosol comprises a larger fraction of the cell, most of the NAD$^+$ and NADH reside in the mitochondria. In skeletal muscle cell comprising of 450 and 50 μmol/kg cell wet wt, the cytosol contains 150 and 0.28 μmol/kg cytosolic wet wt of NAD$^+$ and NADH. In contrast, mitochondria contain 3150 and 500 μmol/kg mitochondrial wet wt [39, 40]. However, the cytosol and mitochondria redox states stand at a different poise. The

Table 10.3 ^{13}C biodistribution from [2-^{13}C]pyruvate

Metabolite/tC	Control	Post-DCA (1 h)
Lac/tC	0.044	0.048
Pyr/tC	0.960	0.882
Pyh/tC	0.082	0.080
Ala/tC	0.024	0.011
ALCAR/tC	ND	0.051
AcAc/tC	ND	0.021
Glu/tC	ND	0.012

Skeletal muscle metabolite values reflect the average of all transients between 0 and 2 min normalized to the total integrated ^{13}C carbon signals between 0 and 2 min after a bolus injection of hyperpolarized [2-^{13}C]pyr [34]

AcAc acetoacetate, *Ala* alanine, *ALCAR* acetyl carnitine, *DCA* dichloroacetate, *Glu* glutamate, *Lac* lactate, *ND* not determined, *Pyh* pyruvate hydrate, *Pyr* pyruvate, *tC* total ^{13}C

Table 10.4 Relative conversion rate from [2-^{13}C]pyruvate

	Control	Post-DCA (1 h)
$r_{pyr \rightarrow lac}$	0.053	0.048
$r_{pyr \rightarrow pyh}$	0.070	0.062
$r_{pyr \rightarrow alcar}$	ND	0.026

The metabolite conversion rate (*r*) reflects the temporal change in the metabolite signal intensity relative to the kinetics of the pyr signal [34]

alcar acetyl carnitine, *DCA* dichloroacetate, *lac* lactate, *ND* not determined, *pyh* pyruvate hydrate, *pyr* pyruvate

~540 NAD^+/NADH ratio stands at a higher oxidization poise than the 6.3 NAD^+/NADH ratio in the mitochondria. The nucleus has a very small NAD^+ and NADH pool and comprises only about 1% of muscle cell volume.

The observed chemical exchange/metabolism of lac and pyr in muscle indicates that NAD^+/NADH turnover occurs rapidly. In skeletal muscle, LDH has no apparent limitation in catalyzing either the forward direction (pyr to lac) or the backward direction (lac to pyr), even though the lac to pyr reaction produces NADH, which would limit glycolysis and glycogenolysis.

The rapid conversion of lac to pyr and quick NAD^+/NADH turnover support key tenets of the intracellular lac shuttle model.

10.7 Pyruvate Dehydrogenase

Relative to the reaction from lac to pyr ($r_{lac \to pyr}$), DCA increases by 19.5 times the reaction from lac to bic ($r_{lac \to bic}$). Previous studies have demonstrated an excellent linear correlation between bic/pyr, bic/ala, and $r_{pyr \to bic}$, with PDH activity determined by conventional biochemical assays [22]. Indeed, the $r_{lac \to bic}/r_{lac \to pyr}$ ratio in the presence of DCA shows the PDH activity (PDHa) increasing by 37.1 times from 0.096 to 3.559, consistent with human muscle biopsy assays 30 s after the onset of contraction [38, 41]. The DNP experiments imply a large capacity for PDH to activate and can shift the metabolic flux toward oxidative phosphorylation. As envisioned by the glycogen shunt theory (vide infra), the accumulated lac during aerobic muscle contraction can buffer energy loss by reacting immediately with PDH during the resting phase to form acetyl CoA, a precursor for the TCA cycle and oxidative phosphorylation. Steady state, ^{13}C NMR, and tracer studies have confirmed that lac can serve as the precursor for the PDH and TCA [8–11].

10.8 Metabolic Inertia, PDH, and VO_2

Whether the activated PDH controlling acetyl CoA formation controls respiration remains an open question. However, some researchers have posited that muscle contraction requires a stimulated PDHa to increase oxidative phosphorylation and reduce lac formation [42]. In essence, they envision the conversion of pyr to acetyl CoA via PDH as the rate-limiting step in controlling respiration.

Consequently, at the onset of muscle contraction, the sudden surge in energy demand requires an activated PDH [38, 41]. Nevertheless, researchers have proposed that a metabolic inertia exists in activating PDH in order to explain the observed delay between muscle O_2 uptake and consumption [43, 44].

However, studies have shown that VO_2 measured by arterial-venous blood oxygen content and thermodilution do not differ in control vs. DCA-treated muscle, despite the difference in the PDH activation kinetics [41, 45]. Moreover, the arterial-venous based measurements do not consider the role of myoglobin (Mb). Indeed, ^1H NMR measurements of intracellular Mb show an immediate desaturation with a $t_{1/2}$ = 24 s at the onset of contraction [46, 47]. Mb contributes immediately its O_2 store to respiration, and the intracellular VO_2 as determined by the Mb desaturation kinetics shows no sign of any delay.

The DNP experiments with [2-^{13}C]pyr also do not detect any kinetic delay in the step from pyr to acetyl CoA upon DCA activation. ALCAR forms immediately upon DCA activation. Human muscle studies also show that DCA infusion increases within 30 s ALCAR level 5.2 times from 2500 to 13,000 μmol/kg dry muscle, while the acetyl CoA level rises correspondingly 4.1 times from 11 to 45 μmol/kg dry muscle [38, 41].

10.9 Lactate Shuttle and Glycogen Shunt

Conventionally, glycogenolytic ATP generation during the transient mismatch between O_2 dependent energy supply and demand rationalizes the increased production of lac observed during muscle contraction. Brooks et al. have shown that muscle produces lac in the presence of abundant O_2 and have proposed a lac shuttle from glycolytic white fibers to oxidative red fibers to redistribute the energy supply. The detection of a mitochondrial LDH has expanded the idea to include a potential shuttle from the cytosol to the mitochondria [48].

A biochemical mechanism underlying the lac production under aerobic conditions has also emerged. Predicated on triggered ^{31}P NMR studies that show a rapid and significant millisecond consumption of ATP during a contraction cycle, Shulman and Rothman have proposed a glycogen shunt, which posits glycogen supplying the ATP needs in milliseconds and serving as a temporal energy buffer [17, 18]. The limited supply of glycogen in the cell and the significant energy demand during a millisecond contraction cycle require a dynamic glycogen replenishment during the relaxation phase, which can last for seconds. Otherwise, glycogen would deplete completely, and no sustained period of contraction can occur. But the glycogen shunt introduces an energy inefficiency even under aerobic conditions. In essence, lac accumulates during muscle contraction until the lac export equals net production. The accumulated lac can then serve as a precursor for acetyl CoA and oxidative ATP generation and also the restoration of the glycogen pool [19]. In muscle, lac can serve as a carbon precursor for glyconeogenesis [49–52].

Both the intracellular lac shuttle and glycogen shunt models require that muscle must mobilize and utilize lac rapidly and can convert it quickly to acetyl CoA for oxidative metabolism in the mitochondria. Indeed, DNP studies have established that muscle can instantly mobilize and use lac. PDH poses no metabolic inertia and can compete readily with LDH to divert accumulated lac to acetyl CoA. Consequently, the DNP experiments support the critical underpinning of both the glycogen shunt model and the intracellular lac shuttle hypothesis.

10.10 Summary

The present DNP NMR experiments have tested the premises of the intracellular lac shuttle and glycogen shunt hypotheses. Both require a rapid mobilization and metabolism of lac. If experiments cannot detect a rapid lac dynamics, they would falsify the premise. These hypotheses would fail immediately.

But the DNP experiment data appear consistent with the hypothesis that lac can serve as a transient energy buffer and as a precursor for mitochondrial LDH in muscle. Given the rapid kinetics of lac in muscle, further studies must now investigate the role in regulating metabolism. Indeed, the findings caution against an overly simplistic interpretation of lac and other glycolytic end-products as merely hypoxia biomarkers [53, 54].

Acknowledgments We acknowledge funding support from NIH (P41 EB015891 and R01 NS107409-01A1), the Welch Foundation (I-2009-20190330), and France Berkeley Fund, California Department of Public Health 18-10923.

Problems

1. How does the glycogen shunt hypothesis explain the formation of lactate during exercise, when the cell has not crossed any hypoxia threshold?
2. Do you expect any changes of metabolic products in skeletal muscle after exercise when hyperpolarized either [1-^{13}C]lactate or [2-^{13}C]pyruvate is injected?
3. From hyperpolarized [2-^{13}C]pyruvate, no TCA intermediates are observed in the spectra although glutamate was detected. Can you explain why? On the other hand, a large acetyl-carnitine peak was shown. What does it imply?

References

1. Kreutzer, U., Jue, T.: Critical intracellular oxygen in the myocardium as determined with the 1H NMR signal of myoglobin. Am. J. Phys. **268**, H1675–H1681 (1995)
2. Brooks, G.A., Fahey, T.D., White, T.P., Baldwin, K.M.: Exercise physiology: human bioenergetics and its application. Mayfield Publishing, Mountain View (2000)
3. Brooks, G.A.: Lactate production under fully aerobic conditions: the lactate shuttle during rest and exercise. Fed. Proc. **45**(13), 2924–2929 (1986)
4. Wasserman, K.: Determinants and detection of anaerobic threshold and consequences of exercise above it. Circulation. **76**(6 Pt 2), VI29–VI39 (1987)
5. Brooks, G.A.: The lactate shuttle during exercise and recovery. Med. Sci. Sports Exerc. **18**(3), 360–368 (1986)
6. Gladden, L.B.: Lactate metabolism: a new paradigm for the third millennium. J. Physiol. **558**(Pt 1), 5–30 (2004). https://doi.org/10.1113/jphysiol.2003.058701
7. Stanley, W.C., Brooks, G.A.: Measuring lactate production. Am. J. Phys. **253**(4 Pt 1), E472–E473 (1987)
8. Stanley, W.C., Gertz, E.W., Wisneski, J.A., Morris, D.L., Neese, R.A., Brooks, G.A.: Systemic lactate kinetics during graded exercise in man. Am. J. Phys. **249**(6 Pt 1), E595–E602 (1985)
9. Stanley, W.C., Gertz, E.W., Wisneski, J.A., Neese, R.A., Morris, D.L., Brooks, G.A.: Lactate extraction during net lactate release in legs of humans during exercise. J. Appl. Physiol. **60**(4), 1116–1120 (1986)
10. Bertocci, L.A., Jones, J.G., Malloy, C.R., Victor, R.G., Thomas, G.D.: Oxidation of lactate and acetate in rat skeletal muscle: analysis by 13C-nuclear magnetic resonance spectroscopy. J. Appl. Physiol. **83**(1), 32–39 (1997)

11. Bertocci, L.A., Lujan, B.F.: Incorporation and utilization of [3-13C]lactate and [1,2-13C]acetate by rat skeletal muscle. J. Appl. Physiol. **86**(6), 2077–2089 (1999)
12. Brooks, G.A.: Intra- and extra-cellular lactate shuttles. Med. Sci. Sports Exerc. **32**(4), 790–799 (2000)
13. Brooks, G.A., Hashimoto, T.: Investigation of the lactate shuttle in skeletal muscle mitochondria. J. Physiol. **584**(Pt 2), 705–706 (2007).;author reply 707–708). https://doi.org/10.1113/jphysiol.2007.142992
14. Sahlin, K., Fernstrom, M., Svensson, M., Tonkonogi, M.: No evidence of an intracellular lactate shuttle in rat skeletal muscle. J. Physiol. **541**(Pt 2), 569–574 (2002)
15. Yoshida, Y., Holloway, G.P., Ljubicic, V., Hatta, H., Spriet, L.L., Hood, D.A., Bonen, A.: Negligible direct lactate oxidation in subsarcolemmal and intermyofibrillar mitochondria obtained from red and white rat skeletal muscle. J. Physiol. **582**(Pt 3), 1317–1335 (2007). https://doi.org/10.1113/jphysiol.2007.135095
16. Passarella, S., Paventi, G., Pizzuto, R.: The mitochondrial L-lactate dehydrogenase affair. Front. Neurosci. **8**, 407 (2014). https://doi.org/10.3389/fnins.2014.00407
17. Shulman, R.G., Rothman, D.L.: The "glycogen shunt" in exercising muscle: a role for glycogen in muscle energetics and fatigue. Proc. Natl. Acad. Sci. U. S. A. **98**(2), 457–461 (2001)
18. Chung, Y., Sharman, R., Carlsen, R., Unger, S.W., Larson, D., Jue, T.: Metabolic fluctuation during a muscle contraction cycle. Am. J. Phys. Cell Physiol. **274**, C846–C852 (1998)
19. Shulman, R.G.: Glycogen turnover forms lactate during exercise. Exerc. Sport Sci. Rev. **33**(4), 157–162 (2005)
20. Ardenkjaer-Larsen, J.H., Fridlund, B., Gram, A., Hansson, G., Hansson, L., Lerche, M.H., Servin, R., Thaning, M., Golman, K.: Increase in signal-to-noise ratio of >10,000 times in liquid-state NMR. Proc. Natl. Acad. Sci. U. S. A. **100**(18), 10158–10163 (2003)
21. Goodman, M.N., Ruderman, N.B., Aoki, T.T.: Glucose and amino acid metabolism in perfused skeletal muscle. Effect of dichloroacetate. Diabetes. **27**(11), 1065–1074 (1978)
22. Atherton, H.J., Schroeder, M.A., Dodd, M.S., Heather, L.C., Carter, E.E., Cochlin, L.E., Nagel, S., Sibson, N.R., Radda, G.K., Clarke, K., Tyler, D.J.: Validation of the in vivo assessment of pyruvate dehydrogenase activity using hyperpolarised 13C MRS. NMR Biomed. **24**(2), 201–208 (2011). https://doi.org/10.1002/nbm.1573
23. Juel, C.: Muscle lactate transport studied in sarcolemmal giant vesicles. Biochim. Biophys. Acta. **1065**(1), 15–20 (1991)
24. Poole, R.C., Halestrap, A.P.: Transport of lactate and other monocarboxylates across mammalian plasma membranes. Am. J. Phys. **264**(4 Pt 1), C761–C782 (1993)
25. Boron, W.F., Boulpaep, E.: Medical physiology: a cellular and molecular approach, 1st ed. Elsevier, Philadelphia (2003)
26. Day, S.E., Kettunen, M.I., Gallagher, F.A., Hu, D.E., Lerche, M., Wolber, J., Golman, K., Ardenkjaer-Larsen, J.H., Brindle, K.M.: Detecting tumor response to treatment using hyperpolarized 13C magnetic resonance imaging and spectroscopy. Nat. Med. **13**(11), 1382–1387 (2007)
27. McDermott, J.C., Bonen, A.: Lactate transport by skeletal muscle sarcolemmal vesicles. Mol. Cell. Biochem. **122**(2), 113–121 (1993)
28. Pilegaard, H., Juel, C., Wibrand, F.: Lactate transport studied in sarcolemmal giant vesicles from rats: effect of training. Am. J. Phys. **264**(2 Pt 1), E156–E160 (1993)
29. Juel, C., Halestrap, A.P.: Lactate transport in skeletal muscle - role and regulation of the monocarboxylate transporter. J. Physiol. **517**(Pt 3), 633–642 (1999)
30. Roth, D.A., Brooks, G.A.: Lactate and pyruvate transport is dominated by a pH gradient-sensitive carrier in rat skeletal muscle sarcolemmal vesicles. Arch. Biochem. Biophys. **279**(2), 386–394 (1990)
31. Roth, D.A., Brooks, G.A.: Lactate transport is mediated by a membrane-bound carrier in rat skeletal muscle sarcolemmal vesicles. Arch. Biochem. Biophys. **279**(2), 377–385 (1990)
32. Carpenter, L., Halestrap, A.P.: The kinetics, substrate and inhibitor specificity of the lactate transporter of Ehrlich-lettre tumour cells studied with the intracellular pH indicator BCECF. Biochem. J. **304**(Pt 3), 751–760 (1994)

33. Bastiaansen, J.A.M., Yoshihara, H.A.L., Takado, Y., Gruetter, R., Comment, A.: Hyperpolarized ^{13}C lactate as a substrate for in vivo metabolic studies in skeletal muscle. Metabolomics. **10**, 986–994 (2014). https://doi.org/10.1007/s11306-014-0630-5
34. Park, J.M., Josan, S., Mayer, D., Hurd, R.E., Chung, Y., Bendahan, D., Spielman, D.M., Jue, T.: Hyperpolarized 13C NMR observation of lactate kinetics in skeletal muscle. J. Exp. Biol. **218**(Pt 20), 3308–3318 (2015). https://doi.org/10.1242/jeb.123141
35. Park, J.M., Josan, S., Jang, T., Merchant, M., Yen, Y.F., Hurd, R.E., Recht, L., Spielman, D.M., Mayer, D.: Metabolite kinetics in C6 rat glioma model using magnetic resonance spectroscopic imaging of hyperpolarized [1-(13)C]pyruvate. Magn. Reson. Med. **68**(6), 1886–1893 (2012). https://doi.org/10.1002/mrm.24181
36. Chen, J., Hackett, E.P., Kovacs, Z., Malloy, C.R., Park, J.M.: Assessment of hepatic pyruvate carboxylase activity using hyperpolarized [1-13C]-l-lactate. Magn. Reson. Med. **00**, 1–8 (2020). https://doi.org/10.1002/mrm.28489
37. Kettunen, M.I., Hu, D.E., Witney, T.H., McLaughlin, R., Gallagher, F.A., Bohndiek, S.E., Day, S.E., Brindle, K.M.: Magnetization transfer measurements of exchange between hyperpolarized [1-13C]pyruvate and [1-13C]lactate in a murine lymphoma. Magn. Reson. Med. **63**(4), 872–880 (2010). https://doi.org/10.1002/mrm.22276
38. Howlett, R.A., Heigenhauser, G.J., Hultman, E., Hollidge-Horvat, M.G., Spriet, L.L.: Effects of dichloroacetate infusion on human skeletal muscle metabolism at the onset of exercise. Am. J. Phys. **277**(1), E18–E25 (1999)
39. Li, Y., Dash, R.K., Kim, J., Saidel, G.M., Cabrera, M.E.: Role of NADH/NAD+ transport activity and glycogen store on skeletal muscle energy metabolism during exercise: in silico studies. Am. J. Phys. Cell Physiol. **296**(1), C25–C46 (2009). https://doi.org/10.1152/ajpcell.00094.2008
40. White, A.T., Schenk, S.: NAD(+)/NADH and skeletal muscle mitochondrial adaptations to exercise. Am. J. Physiol. Endocrinol. Metab. **303**(3), E308–E321 (2012). https://doi.org/10.1152/ajpendo.00054.2012
41. Bangsbo, J., Gibala, M.J., Krustrup, P., Gonzalez-Alonso, J., Saltin, B.: Enhanced pyruvate dehydrogenase activity does not affect muscle O_2 uptake at onset of intense exercise in humans. Am. J. Phys. **282**(1), R273–R280 (2002)
42. Howlett, R.A., Hogan, M.C.: Dichloroacetate accelerates the fall in intracellular PO_2 at onset of contractions in xenopus single muscle fibers. Am. J. Phys. **284**, R481–R485 (2002)
43. Tschakovsy, M.E., Hughson, R.L.: Interaction of factors determining oxygen uptake at the onset of exercise. J. Appl. Physiol. **86**(4), 1101–1113 (1999)
44. Grassi, B., Gladden, L.B., Samaja, M., Stary, C.M., Hogan, M.C.: Faster adjustment of O2 delivery does not affect V(O2) on-kinetics in isolated in situ canine muscle. J. Appl. Physiol. **85**(4), 1394–1403 (1998)
45. Bangsbo, J., Gibala, M.J., Howarth, K.R., Krustrup, P.: Tricarboxylic acid cycle intermediates accumulate at the onset of intense exercise in man but are not essential for the increase in muscle oxygen uptake. Pflugers Arch. - Eur. J. Physiol. **452**(6), 737–743 (2006). https://doi.org/10.1007/s00424-006-0075-4
46. Whipp, B.J.: The slow component of O2 uptake kinetics during heavy exercise. Med. Sci. Sports Exerc. **26**(11), 1319–1326 (1994)
47. Chung, Y., Mole, P.A., Sailasuta, N., Tran, T.K., Hurd, R., Jue, T.: Control of respiration and bioenergetics during muscle contraction. Am. J. Phys. Cell Physiol. **288**(3), C730–C738 (2005)
48. Brooks, G.A.: Cell-cell and intracellular lactate shuttles. J. Physiol. **587**(Pt 23), 5591–5600 (2009). https://doi.org/10.1113/jphysiol.2009.178350
49. Bonen, A., McDermott, J.C., Tan, M.H.: Glycogenesis and glyconeogenesis in skeletal muscle: effects of pH and hormones. Am. J. Phys. **258**(4 Pt 1), E693–E700 (1990)
50. McDermott, J.C., Bonen, A.: Glyconeogenic and oxidative lactate utilization in skeletal muscle. Can. J. Physiol. Pharmacol. **70**(1), 142–149 (1992)
51. McLane, J.A., Holloszy, J.O.: Glycogen synthesis from lactate in the three types of skeletal muscle. J. Biol. Chem. **254**(14), 6548–6553 (1979)

52. Johnson, J.L., Bagby, G.J.: Gluconeogenic pathway in liver and muscle glycogen synthesis after exercise. J. Appl. Physiol. **64**(4), 1591–1599 (1988)
53. Grieshaber, M.K., Hardewig, I., Kreutzer, U., Portner, H.O.: Physiological and metabolic responses to hypoxia in invertebrates. Rev. Physiol. Biochem. Pharmacol. **125**, 43–147 (1994)
54. Kreutzer, U., Jue, T.: Metabolic response in Arenicola marina to limiting oxygen as reflected in the 1H-NMR oxymyoglobin signal. Eur. J. Biochem. **243**, 233–239 (1997)

Further Reading

Brooks, G.A.: Cell-cell and intracellular lactate shuttles. J. Physiol. **587**, 5591–5600 (2009)

Chung, Y., Sharman, R., Carlsen, R., Unger, S.W., Larson, D., Jue, T.: Metabolic fluctuation during a muscle contraction cycle. Am. J. Phys. Cell Physiol. **274**, C846–CC52 (1998)

Park, J.M., Josan, S., Jang, T., Merchant, M., Yen, Y.F., et al.: Metabolite kinetics in C6 rat glioma model using magnetic resonance spectroscopic imaging of hyperpolarized [1-(13)C]pyruvate. Magn. Reson. Med. **68**, 1886–1893 (2012)

Shulman, R.G., Rothman, D.L.: The "glycogen shunt" in exercising muscle: a role for glycogen in muscle energetics and fatigue. Proc. Natl. Acad. Sci. U. S. A. **98**, 457–461 (2001)

Chapter 1

Solutions

1. 15 mmol/L (mM) equals 15 μmol/cm^3, since 1 L is 1000 cm^3. 1 mol equals $6.022 \cdot 10^{23}$ (Avogadro's number). Multiplication of the two numbers gives $9.0 \cdot 10^{18}$ spins/cm^3.
   Similarly, 40 mol/L would give $2.4 \cdot 10^{19}$ spins/cm^3.
2. Taking the cubic root of the spin densities from the solution of problem 1, gives $\sqrt[-3]{9.0 \cdot 10^{18} \text{spins}/\text{cm}^3} = 4.8 \cdot 10^{-7}\ \text{cm} = 4.8\ \text{nm}$ for 15 mmol/L EPA and

   $\sqrt[-3]{2.4 \cdot 10^{19} \text{spins}/\text{cm}^3} = 3.5 \cdot 10^{-7}\ \text{cm} = 3.5\ \text{nm} = 35\ \mathring{A}$ for 40 mmol/L EPA.
3. $C = \frac{\mu_0}{4\pi}\frac{\hbar\gamma_e^2}{2\pi} \approx 5.20 \cdot 10^{-20}\ \text{Hz} \cdot \text{m}^3$ for an electron spin. Dividing by 10 Å = 10^{-9} m cubed gives $5.20 \cdot 10^7$ Hz = 52 MHz. Thus, at minimal distance of approach, the electron-electron dipolar interaction is much stronger than the average, and of the order of the nuclear Larmor frequency depending on nucleus and magnetic field strength.
4. The ^{13}C magnetic moment is 2,700 smaller than the electron magnetic moment. The factor, C, would thus in this case equal $\frac{\mu_0}{4\pi}\frac{\hbar\gamma_e\gamma_{13C}}{2\pi} \approx 1.93 \cdot 10^{-23}\ \text{Hz} \cdot m^3$. Division by 5 Å = $5 \cdot 10^{-9}$ m cubed gives 154 kHz. This is a relatively strong interaction compared to the ^{13}C line width, thus shifting the resonance frequency far into the wings of the ^{13}C resonance, and possibly preventing ^{13}C–^{13}C spin diffusion.
5. Neat pyruvic acid is 14 mol/L. The ^{13}C natural abundance is 1.1%. Pyruvic acid has three carbon atoms. Thus, the carbon concentration is $14 \cdot 1.1\% \cdot 3 = 0.46$ mol/L. In the simplistic lattice model, this leads to a spin-spin distance of 15 Å $\left(\sqrt[-3]{1000cN_A}\right)$. Equation (1.13) then gives a diffusion constant of 3.8 10^{-19} m^2/s. The average spin diffusion distance is $\sqrt{6D_{SD}t}$ and should reach as far as 35 and 48 Å (problem 2 answers). The average time to reach this far would be 5.4 and 10 s, respectively.

T. Jue, D. Mayer (eds.), *Dynamic Hyperpolarized Nuclear Magnetic Resonance*, Handbook of Modern Biophysics, https://doi.org/10.1007/978-3-030-55043-1

Chapter 2

Solutions

Solution Problem 1

From Eq. (2.2), we can calculate the room-temperature thermal ^{13}C polarization at 3 T:

$$P_{\text{thermal}}\left(3\,\text{T},300\,\text{K}\right)=\frac{\frac{h}{2\pi}\gamma B_0}{2k_{\text{B}}T}=\frac{1.05\cdot 10^{-34}\cdot 6.73\cdot 10^{7}\cdot 3}{2\cdot 1.38\cdot 10^{-23}\cdot 300}=2.56\cdot 10^{-6}$$

The ^{13}C polarization at the time of the hyperpolarized ^{13}C MR measurement inside the MR scanner was therefore:

$$P\left(t=60\,\text{s}\right)=\varepsilon\cdot P_{\text{thermal}}\left(3\,\text{T},300\,\text{K}\right)$$

When the solution was first introduced inside the MR scanner, the ^{13}C polarization was:

$$P\left(t=40\text{s}\right)=\exp\left(\frac{20}{65}\right)\cdot P\left(t=60\,\text{s}\right),$$

and the ^{13}C polarization at the time of dissolution, prior to transport to the MR scanner, was:

$$P\left(t=0\right)=\exp\left(\frac{40}{50}\right)\cdot P\left(t=40\,\text{s}\right)=\exp\left(\frac{40}{50}\right)\cdot\exp\left(\frac{20}{65}\right)\cdot P\left(t=60\,\text{s}\right).$$

T. Jue, D. Mayer (eds.), *Dynamic Hyperpolarized Nuclear Magnetic Resonance*, Handbook of Modern Biophysics, https://doi.org/10.1007/978-3-030-55043-1

Hence,

$$P(t=0)=\exp\left(\frac{40}{50}\right)\cdot\exp\left(\frac{20}{65}\right)\cdot\varepsilon\cdot P_{\text{thermal}}(3\,\mathrm{T},300\,\mathrm{K})=42.7\%.$$

From the slope of the linear regression presented in Fig. 2.2,

$$P(t=0)=8.56\cdot\frac{B_0}{T}.$$

The operating temperature of the hyperpolarizer was therefore:

$$T=8.56\cdot\frac{B_0}{P(t=0)}=8.56\cdot\frac{7}{42.7}=\underline{1.4\,\mathrm{K}}.$$

Solution Problem 2

From Eq. (2.3), we can deduce the field B_0 at which the polarizer should operate to have the optimal $T_{1,e}$ value:

$$\frac{1.5\cdot B_0^4}{1.2\cdot(3.35)^4}=\frac{1}{0.3},$$

which leads to

$$B_0=\sqrt[4]{\frac{1.2\cdot(3.35)^4}{1.5\cdot 0.3}}=\underline{4.28\,\mathrm{T}}.$$

The model could therefore be tested using the described polarizer.

The largest ^{13}C polarization that can be expected at this field and temperature can then be estimated from the slope of the linear regression presented in Fig. 2.2:

$$P=8.56\cdot\frac{B_0}{T}=8.56\cdot\frac{4.28}{1.5}=\underline{24.4\%}.$$

Chapter 3

Solutions

1. Following from Eq. (3.2) and Fig. 3.1 there is a linear relationship between the frequency profile $f(\nu)$ of an RF pulse and its corresponding slice profile $f(z)$: $f(z)=\frac{2\pi}{\gamma G}f(\nu)$.
   Therefore, an offset $\Delta\nu$ for two chemical shift species leads to an offset in slice locations of
   $$\Delta z = \frac{2\pi}{\gamma}\frac{\Delta \nu}{G}.$$
   Hence, the CSDA is minimized by maximizing the amplitude of the slice selection gradient. However, at the same time, the same linear relationship exists between the slice thickness FWHM_z and the bandwidth of the RF pulse. Increasing the amplitude of the slice selection gradient requires an increased bandwidth for a fixed slice thickness. This can be achieved by reducing the duration of the pulse while simultaneously increasing the B_1 amplitude by the same amount to achieve the same flip angle. The shortest duration of the designed RF pulse for a 90° excitation that can be applied on the scanner is
   $$\tau_{\mathrm{min}} = \frac{B_{1,\mathrm{nom}}}{B_{1,\mathrm{max}}}\tau_{\mathrm{nom}} = 0.9\,\mathrm{ms}.$$
   At this length, the pulse has a bandwidth of 4.4 kHz and the gradient amplitude for a 1-cm slice is
   $$G_z = \frac{2\pi}{\gamma}\frac{\mathrm{FWHM}_\nu}{\mathrm{FWHM}_z} = 4.11\,\mathrm{G/cm}.$$
   At 3 T, the chemical shift difference of 12.2 ppm corresponds to a frequency offset of
   $$\nu = \frac{\gamma}{2\pi}B_0\,\Delta CS = 392\,\mathrm{Hz}.$$

T. Jue, D. Mayer (eds.), *Dynamic Hyperpolarized Nuclear Magnetic Resonance*, Handbook of Modern Biophysics, https://doi.org/10.1007/978-3-030-55043-1

Therefore, the offset in slice location between [1-^{13}C]pyruvate and [1-^{13}C]lactate is 0.09 mm.

2. (a) The sum of the duration of the two gradient lobes corresponds to the dwell time in the temporal sampling direction, that is, the inverse of the SW. At the same time, the spatial resolution Δx is the inverse of the total extend in k-space k_{max} covered during the duration Δt_1 of the readout gradient lobe divided by 2π. The same extend in k-space in the opposite direction, hence, the negative gradient amplitude, has to be covered during the refocusing gradient lobe of the waveform in order to return to the same position in k-space as at the beginning of the readout lobe:

$$\frac{1}{\Delta x} = \frac{k_{max}}{2\pi} = \frac{\gamma}{2\pi} G_{read}\,\Delta t_1 = \frac{\gamma}{2\pi}\left|G_{refocus}\right|\Delta t_2.$$

From this follows

$$\Delta t_1 = \frac{1}{SW} - \frac{2\pi}{\gamma G_{max}\Delta x} = 0.626\,\text{ms}, G_{read} = 2.982\,\text{G/cm}.$$

An FOV of 8 cm with 0.5-cm nominal resolution requires sampling $N_x = 16$ data points per readout gradient, corresponding to

$$\text{BW}_{read} = \frac{N_x}{\Delta t_1} = 25.541\,\text{kHz}.$$

(b) As the corresponding sampling interval Δt_s of 0.039 ms cannot be realized on the MRI scanner you need to reduce Δt_s to 0.036 ms. Increasing it to 0.040 ms while maintaining a SW of 1 kHz is not possible due gradient strength limitations as this would require shortening the duration of the refocusing gradient, and hence, increasing its amplitude. Therefore, the gradient durations and amplitudes are:

$$\Delta t_1 = 0.576\,\text{ms}, G_{read} = 3.243\,\text{G/cm}$$

$$\Delta t_2 = 0.424\,\text{ms}, G_{refocus} = 4.405\,\text{G/cm}.$$

Chapter 4

Solutions

1. At 3 T ^{13}C resonates at 32.115 MHz. Therefore the 137 ppm difference between [2-^{13}C]pyruvate and [2-^{13}C]lactate is 4400 Hz or 4.4 × the slice width. [2-^{13}C] pyruvate is detected in a 5 mm slice 22 mm away from the slice where [2-^{13}C] lactate is detected. Not very practical if you want to measure a metabolic process in a single location.
2. It can be a disadvantage to measure the injected hyperpolarized pyruvate if a ratio of the products is the goal. The measurement of hyperpolarized pyruvate comes at a cost in the pool of hyperpolarized spins in the products.

T. Jue, D. Mayer (eds.), *Dynamic Hyperpolarized Nuclear Magnetic Resonance*, Handbook of Modern Biophysics, https://doi.org/10.1007/978-3-030-55043-1

Chapter 5

Solutions

A1. No. The rate constant calculated from hyperpolarized [1-^{13}C]pyruvate using the given model still contains both information of metabolic flux and isotopic exchange. To calculate the net flux of pyruvate to lactate, more sophisticated kinetic models and well-planned data acquisition scheme are needed. For example, Kettunen et al. proposed to use saturation and inversion transfer experiments to estimate the exchange rate. Witney et al. also showed how lactate dehydrogenase activity can be determined from the isotope exchange measurements of hyperpolarized pyruvate and lactate.

M. I. Kettunen et al., *"Magnetization transfer measurements of exchange between hyperpolarized [1- 13C]pyruvate and [1- 13C]lactate in a murine lymphoma," Magn. Reson. Med., vol. 63, no. 4, pp. 872–880, May 2010.*

T. H. Witney et al., *"Kinetic modeling of hyperpolarized 13C label exchange between pyruvate and lactate in tumor cells.," J. Biol. Chem., vol. 286, no. 28, pp. 24572–24580, Jul. 2011.*

A2. The negligible increase of ^{13}C-alanine production is likely due to the predominant pyruvate to alanine flux over isotopic exchange. Several factors potentially contribute this. For example, high metabolic demand of alanine or imbalance in the co-factors for pyruvate transamination to alanine (e.g., α-amino acid or α-keto acid) can drive the preferential conversion of pyruvate to alanine. Alternatively, concentration of the co-injected ^{12}C-alanine could be much smaller than the intrinsic ^{12}C-alanine pool.

T. Jue, D. Mayer (eds.), *Dynamic Hyperpolarized Nuclear Magnetic Resonance*, Handbook of Modern Biophysics, https://doi.org/10.1007/978-3-030-55043-1

Chapter 6

Solutions

Solution Problem 1

To determine the reaction kinetics of hyperpolarized [1-^{13}C]-DHA conversion to ascorbic acid (AA), all that is required for this nonenzymatic conversion is a reducing agent, namely glutathione (GSH) in excess. For example if hyperpolarized [1-^{13}C]-DHA (2 mL) was injected into a 10 mm NMR tube containing a solution of GSH (2 mL, 100 mM) in phosphate buffer (100 mM, pH 7.4) with EDTA (0.3 mM), to yield final concentrations of 5 mM [1-^{13}C]-DHA and 50 mM GSH at pH 7.3. To obtain the rate kinetics, this can be repeated with varying concentrations of GSH such as 1, 5, 10, and 25 mM. The rate of [1-^{13}C]-AA production, can be determined by fitting the signals to the modified Bloch equations (Chap. 6) for two-site exchange with the back-flux of the reaction set to zero [63, 64].

Solution Problem 2

Of the two peaks of lactate the downfield lactate corresponds to extracellular and the upfield signal denotes the intracellular compartmentation of lactate. This measure of instantaneous lactate efflux using hyperpolarized MR has been validated by the works of Sriram et al. [65] and Breukels et al. [65, 66]. The three modulations: (a) increase in cell density would result in increased intracellular signal without altering the extracellular lacate signal (when the alginate bead loading and flow rate are not altered); (b) Increasing the bioreactor flow rate would result in a decreased extracellular lactate signal without much change in the intracellular signal due to a reduced residence time of the lactate molecule in the MR sensitive region of the tube; (c) With MCT4 inhibition, the transporter through which lactate predominantly is exported out of the cell, the extracellular signal would be diminished.

T. Jue, D. Mayer (eds.), *Dynamic Hyperpolarized Nuclear Magnetic Resonance*, Handbook of Modern Biophysics, https://doi.org/10.1007/978-3-030-55043-1

This might also result in the intracellular initially increasing but may be reduced eventually due to product inhibition of the LDH enzyme.

Solution Problem 3

Some of the biological challenges associated with hyperpolarized imaging of live animals are as listed in Sect. 6.4.1. The key element is keeping a patent vascular catheter to inject the probe, followed by repetitive similar depth of anesthesia across animals. If prior in vitro characterizations of the hyperpolarized probes are not performed, known chemical shift tables are used to preliminarily identify the products of metabolism. These are often then followed by modulations to the product to alter the signal via, small molecule inhibition or genetic alterations as in the works of Sriram et al. [66, 67].

Chapter 7

Solutions

1. **The answer should cover the following points:**
   [1-^{13}C]pyruvic acid is a self-glassing, water-soluble small molecule that is liquid at room temperature.
   It has a high concentration (14 M), therefore a small volume sample can be polarized and dissoluted easily.
   [1-^{13}C]pyruvate polarizes well with trityl radicals and has a relatively long T_1 (~60 s at 3 T). This gives ample time for administration and observation of metabolism.
   It has the potential to provide its own radical for hyperpolarization by photogeneration with UV light which would avoid the requirement for filtration of the radical prior to administration into patients.
   The sample requires pH neutralization prior to injection. In patients the pH has to be confirmed prior to injection which delays injection and reduces the available signal as the hyperpolarization decays.
   The major metabolic products that can be observed from [1-^{13}C] pyruvate are [1-^{13}C] lactate, [1-^{13}C]alanine and [^{13}C]bicarbonate.
   Pyruvate is the product of glycolysis. In normal cells, the major metabolic flux is from glycolysis into the tricarboxylic acid (TCA) cycle. However, cancer cells typically have increased rates of glycolysis and reduce pyruvate to lactate, even in the presence of oxygen, a process called the Warburg effect. Increased lactate production is also seen in a variety of inflammatory and ischemic pathologies.
   The exchange of pyruvate to lactate is catalyzed by lactate dehydrogenase. Following injection of hyperpolarized [1-^{13}C]pyruvate rapid exchange with the lactate pool results in [1-^{13}C]lactate labelling which is detected by a change in chemical shift. Increased [1-^{13}C]lactate labelling can be observed in areas of pathology such as cancer.

T. Jue, D. Mayer (eds.), *Dynamic Hyperpolarized Nuclear Magnetic Resonance*,
Handbook of Modern Biophysics, https://doi.org/10.1007/978-3-030-55043-1

A wide range of potential clinical applications have been proposed for hyperpolarized [1-^{13}C]pyruvate particularly in cancer. These include aiding diagnosis, providing prognostic information, stratifying patients for treatment and monitoring response to treatment. Many of these potential uses have been demonstrated in preclinical studies.

[1-^{13}C]pyruvate was the first hyperpolarized imaging agent to be used in patients. It has been demonstrated as safe and increased [1-^{13}C]lactate production has been detected in tumors.

Limitations of [1-^{13}C]pyruvate:

It may not be possible to discriminate between inflammatory and tumor cells with this imaging agent.

[1-^{13}C]pyruvate cannot be used to study TCA cycle metabolism as the 1st carbon position is oxidized to CO_2 in the production of acetyl-CoA from pyruvate catalyzed by pyruvate dehydrogenase. The second carbon enters the TCA cycle and therefore [2-^{13}C]pyruvate is a better probe of TCA cycle metabolism.

The rate of pyruvate-to-lactate flux is affected by multiple factors including perfusion, membrane transport, metabolite pool sizes, enzyme expression and co-factor availability. Where there are conflicting processes, results could be difficult to interpret, for example, following treatment the lactate pool size decreases but perfusion of [1-^{13}C]pyruvate increases.

2. **The answer should discuss the following points:**

   Imaging is essential in clinical oncology and is already used to screen, diagnose, stage and detect response to treatment in most different types of cancer. Hyperpolarized substrates could improve the sensitivity and specificity of imaging for cancer diagnosis, define cancer phenotypes/genotypes (potentially avoiding the need to biopsy), monitor the tumor microenvironment and improve early detection of treatment response in cancers.

   The following hyperpolarized substrates should be discussed:

   [1-^{13}C]pyruvate—a probe of upregulated glycolysis and the Warburg effect, a typical phenotype of solid tumors. A wide range of potential clinical applications have been proposed for hyperpolarized [1-^{13}C]pyruvate particularly in cancer. These include aiding diagnosis, providing prognostic information, stratifying patients for treatment and monitoring response to treatment. Many of these potential uses have been demonstrated in preclinical studies and [1-^{13}C]pyruvate was the first hyperpolarized imaging agent to be used in patients.

   [1,4-$^{13}C_2$]fumarate—increased hydration to malate is seen in areas of cell death where disrupted membrane integrity results in leakage of fumarase from the mitochondria. Its potential use is mainly in the detection of therapeutic response.

   [1-^{13}C]glutamine—determining IDH mutation status in glioma.

   pH sensitive substrates, for example, hyperpolarized [^{13}C]bicarbonate and [1-^{13}C]zymonic acid as markers of the tumor microenvironment.

   [6-^{13}C]arginine—a marker of myeloid derived suppressor cells.

Chapter 8

Solutions

1. Not necessarily. Low production of HP-bicarbonate could reflect either damage to PDH or to mitochondria, or it might simply reflect overwhelming competition by other sources of acetyl-CoA such as fatty acids or ketones. One possible way to differentiate between these two possibilities is to fully activate PDH with a pharmacological agent such as dichloroacetate, insulin plus glucose, or perhaps propionate. If the bicarbonate signal does not increase after activation of PDH, it would suggest that the complex or mitochondria are damaged.
2. Passage of 1 mol of HP-[1-^{13}C]pyruvate through PDH would yield 1 mol of HP-bicarbonate, while passage of 1 mol of HP-[1-^{13}C]pyruvate through PC followed by combined forward flux through the TCA cycle and PEPCK should yield 5 mol of HP-bicarbonate. The problem has been to distinguish between forward flux through the TCA cycle (net oxidation) versus direct passage of HP-[1-^{13}C] pyruvate → HP-[1-^{13}C]OAA → HP-[1-^{13}C]PEP (first step in gluconeogenesis). If there was full scrambling of HP-[1-^{13}C]OAA through fumarate prior to production of PEP, then 50% of the 5 mol HP-bicarbonate would reflect loss of CO_2 in converting OAA to PEP, while the other 50% (2.5 mol) would reflect forward flux in the TCA cycle.
3. One possible probe might be uniformly deuterated-[1,3-^{13}C]-D-glucono-δ-lactone. The advantage of this probe is that upon uptake into cells and passage through the PPP_{ox}, two different HP products would be formed, HP-[^{13}C]bicarbonate and HP-[2-^{13}C]ribulose-5-phosphate. Both ^{13}C-enriched products are nonprotonated so should display relatively long T_1's (10 s or more) and both products are formed in the same enzyme-catalyzed reaction in the PPP_{ox} so the kinetics of appearance of each HP-product should be similar.

T. Jue, D. Mayer (eds.), *Dynamic Hyperpolarized Nuclear Magnetic Resonance*, Handbook of Modern Biophysics, https://doi.org/10.1007/978-3-030-55043-1

Chapter 9

Solutions

1. Using the using the Henderson-Hasselbalch equation, pH = pKa + log([HCO_3^-]/[CO_2]), we can rearrange to give the ratio of [HCO_3^-] and [CO_2] as $10^{(pH - pKa)}$, which at pH 7.2 and with a pKa of 6.17, gives a ratio of approximately 10:1.
2. In addition to the normal challenges of imaging hyperpolarized agents in the beating heart (i.e., needing to account for cardiac and respiratory motion, careful consumption of the hyperpolarized magnetization), there is a need to suppress the large pool of hyperpolarized signal in the chambers of the heart or this will lead to large partial volume artefacts that will overwhelm the perfusion signal that you are trying to measure. Read the following articles for more details—(1) Fuetterer, M., Busch, J., Peereboom, S.M., Von Deuster, C., Wissmann, L., Lipiski, M., Fleischmann, T., Cesarovic, N., Stoeck, C.T., Kozerke, S.: Hyperpolarized ^{13}C urea myocardial first-pass perfusion imaging using velocity-selective excitation. J. Cardiovasc. Magn. Reson. (2017). https://doi.org/10.1186/s12968-017-0364-4 and (2) Lau, A.Z., Miller, J.J., Robson, M.D., Tyler, D.J.: Cardiac perfusion imaging using hyperpolarized 13c urea using flow sensitizing gradients. Magn. Reson. Med. (2016). https://doi.org/10.1002/mrm.25713.
3. Reductions in PDH flux will limit the incorporation of the ^{13}C label into the TCA cycle and so any changes in incorporation of the ^{13}C label into the TCA cycle must take account of this. This has been achieved previously by normalizing the data obtained from [2-^{13}C]pyruvate experiments by the flux through PDH, as assessed by [1-^{13}C]pyruvate experiments in the same animal. Read the following article for further details—Dodd, M.S., Ball, D.R., Schroeder, M.A., Le Page, L.M., Atherton, H.J., Heather, L.C., Seymour, A.M., Ashrafian, H., Watkins, H., Clarke, K., Tyler, D.J.: In vivo alterations in cardiac metabolism and function in the spontaneously hypertensive rat heart. Cardiovasc. Res. 95, 69–76 (2012). https://doi.org/10.1093/cvr/cvs164.

T. Jue, D. Mayer (eds.), *Dynamic Hyperpolarized Nuclear Magnetic Resonance*, Handbook of Modern Biophysics, https://doi.org/10.1007/978-3-030-55043-1

Chapter 10

Solutions

1. Read the following articles: (1) Shulman RG. 2005. Glycogen Turnover Forms Lactate during Exercise. Exerc Sport Sci Rev. 33: 157-62. (2) Bendahan, D., Chatel, B., and Jue, T. 2017 Comparative NMR and NIRS analysis of oxygen-dependent metabolism in exercising finger flexor muscles Am J Physiol 313: R740–R753.
2. Assuming the lactate shuttle model, increased lactate utilization is expected in the exercised muscle, producing increased bicarbonate. Moreover, increased [5-^{13}C]glutamate from [2-^{13}C]pyruvate is expected due to the accelerated oxidative phosphorylation of pyruvate.
3. Small and tightly-regulated pool sizes of the TCA intermediates and their high turnover rates may prevent building up ^{13}C-labeled signal of the metabolites. Glutamate and acetyl-carnitine are capable of having large pool sizes, resulting in relatively large accumulation of ^{13}C-labels.

T. Jue, D. Mayer (eds.), *Dynamic Hyperpolarized Nuclear Magnetic Resonance*, Handbook of Modern Biophysics, https://doi.org/10.1007/978-3-030-55043-1

Index

T. Jue, D. Mayer (eds.), *Dynamic Hyperpolarized Nuclear Magnetic Resonance*, Handbook of Modern Biophysics, https://doi.org/10.1007/978-3-030-55043-1

The manufacturer's authorised representative in the EU is Springer Nature Customer Service Centre GmbH, Europaplatz 3, 69115 Heidelberg, Germany. If you have any concerns regarding our products, please contact ProductSafety@springernature.com

Printed and bound by CPI Group (UK) Ltd, Croydon, CR0 4YY
22/09/2026
02221062-0013